MANMADE

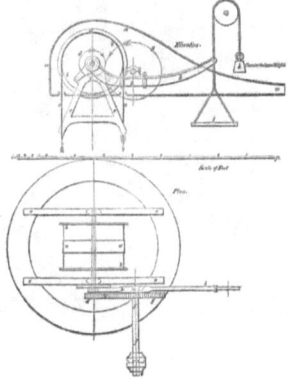

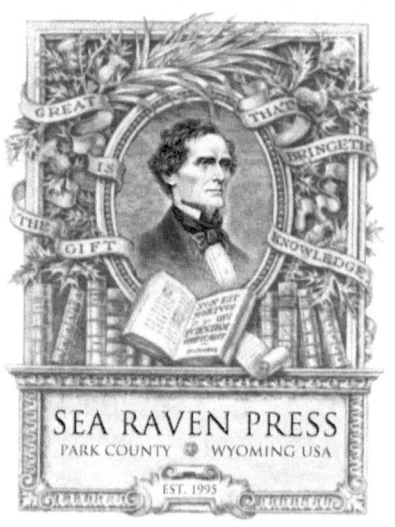

"Books invite all; they constrain none."
Hartley Burr Alexander (1873-1939)

LOCHLAINN SEABROOK WRITES IN THE FOLLOWING GENRES

American Civil War
American History
American Politics
American South
Ancient History
Biblical Exegesis
Biblical Hermeneutics
Biography
Christian Mysticism
Coffee Table
Comparative Religion
Cooking
Diet and Nutrition
Education
Ethnic Studies
Etymology
European History
Family Histories
Film
Genealogy
Ghost Stories
Health and Fitness
Humor
Law of Attraction
Life After Death
Matriarchy
Men
Metaphysics
Military History

Mysteries and Enigmas
Natural Health
Natural History
Onomastics
Paleography
Paranormal
Patriarchy
Philosophy
Photography
Poetry
Politics
Presidential History
Quiz
Reference
Religion
Revolutionary Period
Science
Self-help
Spirituality
Spiritualism
Technology
Thanatology
Thealogy
Theology
UFOlogy
Victorian Period
Wildlife
Women
World History

Mr. Seabrook does not author books for fame and glory, but for the love of writing and sharing his knowledge.

SeaRavenPress.com

MANMADE

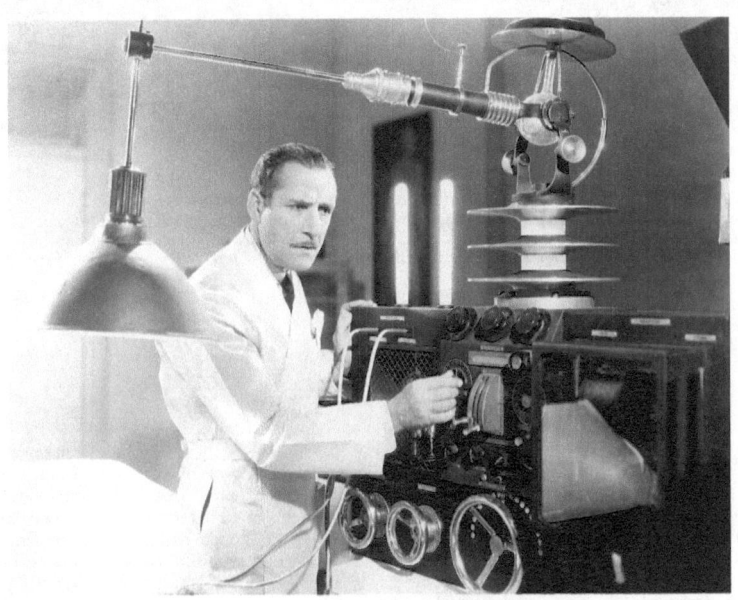

Male Inventors Who Created the Modern World

A One-of-a-Kind Dictionary Celebrating Men's Many Vital Contributions to Everyday Society

LOCHLAINN SEABROOK
Bestselling Author, Historian, Artist

Diligently Researched and Generously Illustrated
by the Author for the Elucidation of the Reader

2025

Sea Raven Press, Park County, Wyoming, USA

MANMADE

Published by
Sea Raven Press, LLC, Cassidy Ravensdale, President
Park County, Wyoming, USA
SeaRavenPress.com

Copyright © all text and illustrations Lochlainn Seabrook 2025
in accordance with U.S. and international copyright laws and regulations, as stated and protected under the Berne Union for the Protection of Literary and Artistic Property (Berne Convention), and the Universal Copyright Convention (the UCC). All rights reserved under the Pan-American and International Copyright Conventions.

PRINTING HISTORY
1st SRP paperback edition, 1st printing, April 2025 • ISBN: 978-1-955351-51-5
1st SRP hardcover edition, 1st printing, April 2025 • ISBN: 978-1-955351-52-2

ISBN: 978-1-955351-51-5 (paperback)
Library of Congress Control Number: 2024945085

This work is the copyrighted intellectual property of Lochlainn Seabrook and has been registered with the Copyright Office at the Library of Congress in Washington, D.C., USA. No part of this work (including text, covers, drawings, photos, illustrations, maps, images, diagrams, etc.), in whole or in part, may be used, reproduced, stored in a retrieval system, or transmitted, in any form or by any means now known or hereafter invented, without written permission from the publisher. The sale, duplication, hire, lending, copying, digitalization, or reproduction of this material, in any manner or form whatsoever, is also prohibited, and is a violation of federal, civil, and digital copyright law, which provides severe civil and criminal penalties for any violations.

Manmade: Male Inventors Who Created the Modern World, by Lochlainn Seabrook. Includes an introduction, illustrations, endnotes, index, and bibliography.

ARTWORK
Front and back cover design and art, book design, layout, font selection, and interior art by Lochlainn Seabrook. All images, image captions, graphic design, and graphic art copyright © Lochlainn Seabrook. All images selected, placed, manipulated, cleaned, colored, tinted, and/or created by Lochlainn Seabrook. Cover photo: "In the Laboratory," Everett Collection. Photo page 37: James Dewar.

All persons who approve of the authority and principles of Colonel Lochlainn Seabrook's literary work, and realize its benefits as a means of reeducating the world about facts left out of mainstream books, are hereby requested to avidly recommend his titles to others and to vigorously cooperate in extending their reach, scope, and influence around the globe.

The views documented in this book concerning men, male inventiveness, and the Masculine Principle are those of the publisher.

WRITTEN, DESIGNED, PUBLISHED, PRINTED, & MANUFACTURED IN THE UNITED STATES OF AMERICA

DEDICATION

To my fellow males,
and my fellow male inventors.

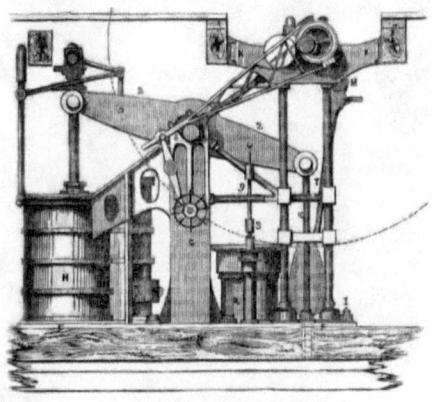

Epigraph

THE MALE INVENTOR

"The inventor looks upon the world and is not contented with things as they are. He wants to improve whatever he sees, he wants to benefit the world; he is haunted by an idea. The spirit of invention possesses him, seeking materialization."

Alexander Graham Bell
1847-1922

CONTENTS

Fair Use & Disclaimer ❧ page 11
Notes to the Reader, by Lochlainn Seabrook ❧ page 13

Introduction, by Lochlainn Seabrook ❧ page 15
    TV, Cell Phones, Computers, & Dating Apps ❧ page 15
    Rebar, Eyeglasses, Batteries, & Microphones ❧ page 16
    Drums, Recording, Classical Music, & Styrofoam ❧ page 16
    Vacuum Cleaners, Matches, Paint, & Bicycles ❧ page 17
    Hard Disk Drives, Plastic, Blue Jeans, & Masking Tape ❧ page 18
    Chocolate, Cellophane, Asphalt, & Fertilizer ❧ page 19
    Periodic Table, Ski Lifts, Vehicles, & Drones ❧ page 19
    Steel, 4-Wheel Drive, Speedometers, & Gasoline ❧ page 20
    Seatbelts, Motorcycles, Rubber Tires, & Dive Gear ❧ page 21
    Assembly Lines, Zippers, Mouth Wash, & Elevators ❧ page 22
    Sports, Smoke Detectors, Cheese Slicers, & Barcodes ❧ page 22
    Christmas Cards, Metal Detectors, Schools, & Guns ❧ page 23
    Honey, GPS, Lawnmowers, & Ballpoint Pens ❧ page 24
    Bridges, Robots, Dishwashers, Violins, & Rodeos ❧ page 25
    Credit Cards, National Parks, & Operating Systems ❧ page 26
    Ear Muffs, Light Bulbs, Garden Hoses, Games, & Books ❧ page 26
    Paper Clips, Cable TV, Movies, & Photography ❧ page 27
    TV Shows, Geodesic Domes, Birth Control, & Stamps ❧ page 28
    Public Services, Ceiling Fans, Aluminum, & Wi-Fi ❧ page 29
    Lasers, Vaccines, Thermometers, & Aspirin ❧ page 30
    Medical Treatments, Instruments, & Procedures ❧ page 30
    The Benevolent Patriarchy ❧ page 32
    Male Inventors: Civilization's Unsung Heroes ❧ page 32
    Male Inventors & the 99.99 Percent Rule ❧ page 33
    The Author as Inventor ❧ page 33
    How Male Inventions are Typically Created ❧ page 33
    Men: The Archetypal Problem-Solvers ❧ page 35

ENTRIES

A ❧ page 39
B ❧ page 43
C ❧ page 59
D ❧ page 71
E ❧ page 79
F ❧ page 85
G ❧ page 97
H ❧ page 111
I ❧ page 127
J ❧ page 129
K ❧ page 137
L ❧ page 151
M ❧ page 165

N ❧ page 185
O ❧ page 193
P ❧ page 199
Q ❧ page 215
R ❧ page 217
S ❧ page 225
T ❧ page 245
U ❧ page 253
V ❧ page 255
W ❧ page 259
X ❧ page 269
Y ❧ page 271
Z ❧ page 275

Notes ❧ page 279
Bibliography ❧ page 281
Index ❧ page 293
Praise for the Author ❧ page 381
Meet the Author ❧ page 385
Learn More ❧ page 389

FAIR USE

☞ PERTAINING TO THE USE OF THE NAMES OF INDIVIDUALS IN THIS BOOK
The men listed in the following pages are public figures whose entire life stories (when known) are publically available in the public domain—as are their patents (for American patent holders visit: www.uspto.gov/patents). In other words, the information provided in this book related to public inventors and the public innovations they described, developed, discovered, created, and/or patented, is part of the public forum, and and public discussion of them is therefore permissible under the Fair Use Act of 1976.

☞ PERTAINING TO THE USE OF BRAND NAMES IN THIS BOOK
Under Section 107 of the Copyright Act of 1976, allowance is made for fair use for purposes such as criticism, comment, news reporting, scholarship, and research. Fair Use is a use permitted by copyright statute that might otherwise be infringing. Education and scholarly research, which are the primary purposes of this book, tip the balance in favor of Fair Use. And, as with the individuals' entries themselves (see above), all models, brands, brand names, and company names have been treated by the author with balance, indifference, and impartiality. No editorializing.

☞ PERTAINING TO THE COPYRIGHTED MATERIAL IN THIS BOOK
Despite the public nature of the material contained herein, the research, style, writing, and wording of the text, as well as all images (this includes the artistic embellishment and manipulation of all images) in this book, are original, and are therefore the copyrighted work of the author.

DISCLAIMER

🖋 There is currently no consensus among historians, educators, or popular culture writers regarding the information in this book. Indeed, there are ongoing questions and disagreements pertaining to the details of both deceased and living inventors. Therefore its contents must be regarded as being for entertainment, educational, and archival purposes only. The contents of this book may not, and should never be, construed as complete, indisputably accurate, or purely historical in any way, shape, manner, or form.

🖋 Pertaining to the entries in this book: No attempt has been by the author to augment, minimize, embroider, dramatize, understate, overstate, or distort in any way, either positively or negatively, the accomplishments of the men listed in this book. The author has nothing but the greatest respect for the creative imaginations of these individuals. Thus, in the grand tradition of chronicling authentic history without bias, the author has written out his biographical inventor entries in the most impartial, unprejudiced, and objective manner possible. No preference or deference is involved, and none should be assumed.

🖋 It is not the intention of either the publisher or the author to supply specific biographical, historical or chronological data, but rather to provide you the reader with generalized information geared toward helping you better understand how men have benefitted the modern world via their talents, ingenuity, and inventiveness.

🖋 Neither the publisher nor the author make any claims as to the biographical, historical, or chronological statements made in this book.

🖋 Do not rely solely on the contents of this book concerning male inventors and their work. This volume is a collection of widely varied opinions, views, beliefs, and assumptions, gathered, consolidated, and organized by the author for the convenience of you the reader—nothing more. Always do your own research.

🖋 The information in this book is not intended for anything but entertainment and educational purposes. Uses that do not include one or both of these two purposes fall outside this book's intended parameters.

❦ Any action on your part in response to the information provided in this book is solely at your own risk and discretion.

❦ Due to the discordant nature of the author's source material, neither the publisher nor the author are responsible for the accuracies or inaccuracies of the material contained herein.

❦ The reader accepts full responsibility for any possible risks related to the information in this book.

❦ Neither the publisher nor the author endorse, recommend, or have any connection to, the individuals, inventions, brands, ideas, designs, devices, instruments, products, companies, organizations, books, or any other material, mentioned in this book.

❦ The reverse is also true. None of the individuals listed or mentioned in this book have any connection to either the publisher or the author. As such, none of the individuals listed or mentioned in this book should be viewed as either endorsing or approving of the author's views, statements, and opinions as recorded in this or any of his other literary works. Again, there is absolutely no association.

❦ Neither the publisher or the author have reviewed or evaluated the authenticity, legitimacy, credibility, validity, pertinence, or relevance of any of the individuals or material cited in this book. It is up to you the reader to validate or invalidate, as well as form your own judgements relating to, this book's contents.

❦ Do not attempt to duplicate or improve on any of the inventions mentioned in this book. If you must, please consult or work with a professional. If you are a professional yourself, proceed with caution. A percentage of the inventions listed here are potentially hazardous, while many others are extremely dangerous if not lethal. And, in fact, a number of the inventors listed in this book were injured or even died while experimenting on or testing their inventions. You always experiment at your own risk.

❦ Because new discoveries are constantly being made about the individuals and inventions discussed in this book, the reader understands and accepts that some of the information supplied herein may be, and probably has become, obsolete or proven incorrect or invalid since its writing and publication. Caution is advised.

❦ This book has not been peer-reviewed by inventors, historians, scientists, or technologists, and was never intended to be; therefore any and all evaluations, assessments, and conclusions related to the material cited in this book are the responsibility of you the reader.

❦ Neither the publisher nor the author assume any obligation or liability—and make no warranties—with respect to the accuracy, usefulness, quality, trustworthiness, or completeness of the information, brands, brand names, products, models, processes, ideas, proposals, designs, patents, apparatuses, businesses, or companies, referred to or described in this book, including what may be obsolete data, inaccuracies, contradictions, typos, errors, omissions, opinions, theories, hypotheses, outdated research, etc. It is your responsibility and yours alone to be aware of the above mentioned possible issues as well as determine the value of this book's content.

❦ Owning, possessing, borrowing, reading, or using this book or its contents in any manner whatsoever infers that you have read, understand, and agree to this Disclaimer and all of the terms and conditions contained therein—without exception.

❦ If you (or your client or associate) has been included in this book and would like your name and entry removed, simply contact us and we will immediately delete it and release a new edition of the book.

THE PUBLISHER

NOTES TO THE READER

"The limits of the possible can only be defined by going beyond them into the impossible." —ARTHUR C. CLARKE

☛ INVENTIONS: Most of the men I have chosen for my dictionary were prolific inventors, some credited with dozens, hundreds, and even thousands of innovations. Listing *all* of these for each individual would extend my book to thousands of pages and would not serve my purpose: to simply and clearly highlight *male* inventors and their countless important contributions to modern society. Thus, I have not attempted to formulate *complete* invention lists for my inventors. Rather I have chosen to focus primarily on the invention or inventions, discovery or discoveries, product or products, companies or organizations, for which an inventor, discoverer, creator, or founder is best known; that is, for which he is most famous.

☛ AWARDS: Though many of my male inventors and founders were given numerous awards and honors throughout their lives, for simplicity's sake I only make note of Nobel Prize winners and nominees.

☛ SOURCES: I used hundreds of old and new sources for this work, a number of which disagree with one another on my inventors' name spellings, birth and death dates, their occupations, and even their inventions themselves. Many of these statistics are still being disputed to this day. I have attempted to correct these issues wherever possible; but, as with most material where large time spans and multiple chroniclers are involved—the majority of the latter which are neither historians, authors, or even writers—accuracy can seldom be guaranteed. The most that can be hoped for is a mere modicum of exactitude.

☛ NATIONALITY: To simplify the biographical aspects of my individual entries, I list inventors' nationalities (and sometimes ethnicities) by their *birthplace*, not who their ancestors were or where they eventually moved later in life, went to school, immigrated, settled, worked, retired, or died. An example: An inventor born in Scotland (a Scotsman), but who later moved to, lived in, worked in, and passed away in Norway (and is therefore usually labeled a "Norwegian"), I identify as Scottish—even if he spent his entire adult life in Norway. By the same token, again for simplicity's sake, an inventor of Danish descent (a Dane) who was born in Italy I will usually list as Italian.

☛ CO-INVENTORS: Inventions, like the founding of organizations and companies, are rarely the work of a single person, despite claims to the contrary. In fact, probably a majority of the world's innovations have resulted from collaborative efforts involving two or more people; indeed, in many cases entire teams and departments have contributed. However, only on rare occasions are the individual personal names of these team members recorded in my sources—some of which are

Sheet rubber manufacturing, circa late 1800s.

hundreds of years old and poorly chronicled. Additionally, since these fellow contributors can run into the many dozens, even hundreds, for each invention, I have been forced to leave them out of my entries due to space and time, not to mention the problem of confusing and often contradictory data—which invariably only makes matters worse. Nevertheless, where possible and where known with any accuracy, I have attempted to list (or at least mention) *primary* co-inventors and co-founders, either alongside one another or in their own separate entries.

☞ DATES: An invention date (shown in parentheses) may indicate the year an inventor first "conceived" it, the year he "realized" it, the year he "proposed" it, the year he "designed" it, the year he actually invented it, the year he issued scientific papers describing it, the year he patented it, the year it was publicized and introduced to the public, or the year it was officially adopted and began being sold commercially to the public for the first time. My original sources are seldom clear on these points (perhaps because the exact date of invention is not always known to begin with), and so wide latitude must be granted pertaining to this particular detail. In many cases, the best that can be hoped for is a roughly estimated date.

☞ INTERPRETATION: Considering the important points made above, rather than viewing my book as a biographical inventors dictionary of ironclad historical facts, it would be safer, as well as more appropriate and logical, to read it as my own personal interpretation of the myriad of conflicting, usually imperfectly documented, and almost always confusing source material that was at my disposal while penning this lexicon.

☞ MY BOOK IS COLOR-BLIND: Unlike many technology writers who editorialize their work in order to promote their personal views and ideologies, in my entries I have adhered to the facts and pertinent details as best as they are currently known. As such, I ignore the race of my male inventors, regarding this trait as not only irrelevant to my topic, but completely unrelated to the inherent creativity, inborn intelligence, and native inventiveness that form the basis of human innovation. Male inventors, from the celebrated to the unknown and nameless, hail from every race. There is absolutely no connection between skin color and ingenuity. For this same reason I also disregard the political persuasions, ethnicities, personal lives, and religions of my subjects. While these topics certainly have biographical merit, my book is not a biographical dictionary per se; rather it is a highly abridged inventors dictionary, and as such these elements are therefore of little value here.

L.S.

INTRODUCTION
MAN: THE HUMAN SWISS ARMY KNIFE

"The history of the world is the biography of great men."
Thomas Carlyle

TV, CELL PHONES, COMPUTERS, & DATING APPS
Have you ever used or do you own one of the following? A television, a cell phone, a computer, a computer mouse, a tablet, a USB drive, a floppy disk, a dot matrix printer, a laser printer, a 3D printer, a flatbed scanner, a color scanner, a fax machine, voicemail, a calculator, an adding machine, a smart card, an electric stove, a gas stove, a microwave oven, a washing machine, a refrigerator, an electrical wall plug, an electrical wall outlet, a digital wrist watch, polarized sunglasses, an air conditioner, or a thermos?

Have you ever gone to see a movie at a movie theater or watched a movie on DVD? Have you ever logged onto the Internet, visited or created a Website, or used email or an Internet chat network to communicate? Have you ever owned or used a device with a liquid crystal display? Have you ever owned, played, or listened to a CD, or installed a program onto your computer from a CD? Have you ever utilized MP3 technology, such as listening to music on an MP3 player? Have you ever used a word processor? Have you ever used any type of device that required a microprocessor?

Do any of your computer devices use a hard drive (internal or external) to store data? Is your computer keyboard equipped with a small rubber-capped "pointing stick"? Do you have a device that connects to the Internet via Ethernet? Have you ever used a computer with an IP address, or one that used Internet protocol (IP) or Transmission Control Protocol (TCP) to communicate with other computers on the Internet? Have you ever used an online computer bulletin board service? Do you own or operate, or have you ever used, a Website with an address that begins with "http" or "https"? Have you ever used or relied on speech recognition computer software or facial recognition software and cameras?

Have you or anyone you know ever used the popular dating applications Hinge, eHarmony, or Tinder? Have you ever owned or used something made from polyvinyl chloride or PVC, such as a garden hose, a vinyl record, a home appliance, a camera body, synthetic leather, camping equipment, packaged food, packaged medications, sports footwear, or shower curtains? Have you ever used a pull chain light socket, or owned a home or worked in a building whose floors were covered with linoleum? Have you ever owned, worn, or admired a synthetic gemstone?

REBAR, EYEGLASSES, BATTERIES, & MICROPHONES
Have you ever lived in a home or building constructed after 1950, that used plastic siding, plastic plumbing, plastic roofing, plastic insulation, or plastic window and door frames? Have you ever lived in, worked in, or owned a concrete structure strengthened with rebar? Have you ever lived or worked in a building with a revolving door at the entrance? Have you ever owned or used an electric extension cord?

   Have you ever used or owned a slide rule or a Vernier Scale, used an "x" in a multiplication calculation, or drawn up a Venn Diagram to illustrate the mathematical relationship between two or more data sets? Have you ever invested in stocks, relied on a Wall Street ticker tape, or monitored the Dow Jones Averages? Do you wear, or know anyone who wears, eyeglasses, bifocals, or soft contact lenses, or who uses Braille or Blissymbols to read?

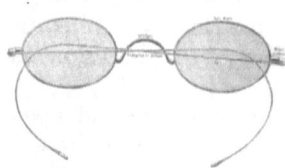
Eyeglasses, circa 1895.

Have you ever used a device that required batteries, either non-chargeable or rechargeable? If you are an outdoor enthusiast have you ever worn a windproof, breathable, waterproof jacket, gloves, shoes, or pants? Have you ever owned or used a gas camping lamp to see or a ferrocerium rod to start a fire? Have you ever owned or used an item made from microfiber, Vinalon, neoprene, Spandex, Lycra, or Elastane, or worn riveted jeans, owned artificial pearls, or sat on a chair or car seat made from artificial leather?

   If you are a musician do you own or have you ever played a piano, a harpsichord, a clavichord, a concertina, an accordion, an electric organ, a pipe organ, or a synthesizer keyboard? Do you own or use a metronome or a microphone, in particular a Shure microphone? Have you ever notated or read a musical score? Have you ever owned or played an electric guitar (solid body or hollow body), in particular, a Fender, a Les Paul, a Gretsch, or a Rickenbacker guitar?

DRUMS, RECORDING, CLASSICAL MUSIC, & STYROFOAM
Have you ever owned, played on, or listened to someone else play on a set of Ludwig drums, Pearl drums, or Gretsch drums? Have you ever owned, played, or enjoyed the sounds of a drum machine, a harmonica, or a Theremin? Have you ever owned any type of musical equipment (such as an electric guitar, microphone, or amplifier) whose electrical signals were powered by a pentode (a vintage vacuum tube)?

Have you ever used magnetic recording tape, a cassette tape, a cassette tape recorder/player, multi-track recording, over-dubbing, an eight-track tape recorder, reverb, or digital delay? Have you ever listened to a either a monaural or a stereo recording? Have you ever owned, used, or listened to music or a human speaker through a loudspeaker? Are you an admirer of classical music, in particular, the Romantic Period, which highlighted the song cycle, the solo sonata, the quartet, the concerto, and the symphony?

Have you ever sung to a karaoke machine, or listened to someone else who has? Have you ever owned or used a personal portable stereo listening device? Have you ever owned or used a device equipped with a foil electret microphone (found in, for example, most computers, laptops, and smartphones)? Have you ever owned or used a Swiss Army Knife? Have you ever worn either a natural perfume or a synthetic perfume? Have you ever taken antibiotics to knock out a bacterial infection, or used an antibiotic ointment to prevent a cut from becoming infected? If you are an American citizen, have you ever been protected by or benefitted from the U.S. Constitution or the Bill of Rights? Have you ever used or benefitted from anything made with Styrofoam?

Have you ever mailed a letter at the Post Office and had your letter sorted and processed by an optical character recognition (OCR) machine? Have you ever used any of the following words: belittle, lengthily, monotonously, ottoman, Anglomania, electioneering, authentication, indecipherable, pedicure, or neologize?

VACUUM CLEANERS, MATCHES, PAINT, & BICYCLES

Have you ever used or benefitted from an electric fence? Have you ever gotten your electricity from a three-phase electric power system (used by nearly all power companies)? Have you ever enjoyed the vintage aesthetic of a flip-disk display on the side of a bus or on a train station departure board? Have you ever owned or used a bagless vacuum cleaner? Do you own or have you ever used a foam fire extinguisher, or have your belongings, home, or life ever been saved by one?

Have you ever used a wooden safety match to light a barbeque, campfire, candle, or stove? Do you own or have you ever used a voltmeter to measure the voltage in a circuit? Have you ever owned or used a galvanometer to measure electrical currents, or an electrometer to measure voltage? Have you ever owned or used a tachymeter, a telemeter, or a range-finder, or benefitted from a Schmitt Trigger?

Have you owned or utilized a product that relies on the refrigerant Freon? Have you ever owned or used a propane-gas powered refrigerator? Have you ever used, depended on, or benefitted from blood or perishable food or medicines that were transported and delivered using mobile refrigeration, such as is found on refrigerated, chiller, and "reefer" trucks and trailers? Have you ever owned, ridden in, or driven a vehicle whose engine used a slotted armature device? Have you ever owned, driven, or benefitted from a forklift (truck)? Have you ever lived in or worked in a skyscraper? Have you ever used a disposable safety razor or a Q-tip?

Have you ever owned or used any type of glassware, or more specifically faceted drinking glasses? Have you ever lived in or owned a house with glass windows, or worked in a building with glass walls or windows? Have you ever used commercial paint to paint an inside or outside wall? Have you ever lived in or owned a house or building constructed with joint-strengthening gang nail plates?

Have you ever owned, driven, or lived or worked on land developed by any type of earthmoving equipment, such as a bulldozer? Have you ever owned or used something powered by a chain drive (e.g., a bicycle, motorcycle, forklift, crane, conveyor belt, or agricultural equipment)? Have you ever owned, used, or benefitted from a chainsaw, in particular a Stihl chainsaw?

HARD DISK DRIVES, PLASTIC, BLUE JEANS, & MASKING TAPE

Have you ever owned, used, or benefitted from something equipped with an electromagnet (e.g., engines, scientific instruments, generators, loudspeakers, hard disk drives, medical equipment. etc.)? Have you ever owned or used a pocket watch? Have you ever been taught about the motion and position of the planets in our solar system with the use of an orrery or planetarium? Have you ever used or been taught physics by a teacher using a cloud chamber?

Have you ever used sunscreen, modern cookware, hair-straightening cream, golf clubs, a gas mask, or anything made from any type of plastic? Have you ever gotten a permanent wave, or used artificial eyelashes or artificial eyebrows? Have you or anyone you know ever had a dental implant (artificial tooth root) or a prosthetic implant (artificial limb)?

Royal typewriter, circa 1923.

Have you ever owned, worked on, or made something that used, a knitting frame, or have you ever owned a knitted article of clothing?

Have you ever vaped (used an electronic cigarette), flown a hang glider or a sailplane, flown in an aircraft steered with a control stick, or owned or used a typewriter? Have you ever had your foot measured in a shoe store with a silver metal Brannock Device? Have you ever owned, driven, or benefitted from a caterpillar tractor? Have you ever roller skated, "Rollerbladed," skated at a roller skating rink, or belonged to a roller skating club? Have you ever owned or used UMBO shelving or UMBO furniture, or a product from IKEA? Have you ever worn a pair of blue jeans, or any item of clothing made from blue denim?

Have you ever made something using masking tape or cellophane tape, or used a sugar packet for your tea or coffee, eaten instant noodles or a breakfast cereal, consumed Coco-Cola, soda or carbonated water, a Nestlé's chocolate product, Nutella, or condensed soup, eaten a Russian salad, or prepared food using either a Japanese RFIQ cooker or a Turbococina? Have you ever used and enjoyed a manmade sweetener, such as saccharin or Sweet'N Low?

CHOCOLATE, CELLOPHANE, ASPHALT, & FERTILIZER
Have you ever eaten cotton candy, or enjoyed a food or sweet treat made with milk chocolate, or eaten a chocolate bar, drank chocolate milk, or used cocoa powder or cocoa butter in a recipe? Have you ever consumed a product made from fermented milk, such as kefir or yogurt? Have you ever eaten or cooked with margarine? Have you ever owned, used, or relied on a cream separator or a dairy milking machine?

Have you ever wrapped and stored a perishable item, like food or flowers, in cellophane? Have you ever served or consumed Vermouth? Have you ever consumed any type of fried food—that is, a consumable made with trans fat, such as French fries, doughnuts, cakes, cookies, crackers, pizza, chips, nondairy creamers, pre-prepared dough, pies, pastries, or peanut butter? Have you ever eaten food prepared with, or yourself prepared food using, any type of Teflon-coated cookware?

Have you ever walked on an asphalt or cement sidewalk, parked on an asphalt or cement driveway or parking lot, or driven on a cement or asphalt roadway? Have you ever driven on an American interstate highway, or benefitted from America's interstate highway system? If you are a runner or athlete, have you ever used a starting block? Have you ever used an ATM machine? Have you or anyone you know ever benefitted from Alcoholics Anonymous?

If you are a gardener, botanist, or farmer, have you ever used chemical fertilizer on your products, eaten food that was grown using chemical fertilizer, or used a farm elevator, a manure spreader, a hay baler, a haymaking machine, a cultivator, a chaffcutter, a seed drill, a reaping machine, a tree-pruner, or a plow? If you are a rancher, cattleman, grazier, or stockgrower, have you ever fed your animals AIV fodder? Have you ever used or benefitted from cloud seeding or from methane gas (e.g., for heating, fuel, electricity production, etc.)?

PERIODIC TABLE, SKI LIFTS, VEHICLES, & DRONES
If you are involved in any of the modern sciences, have you ever used or had need of binomial nomenclature, the polymerase chain reaction method, Periodic Law, or the Periodic Table? Have you ever worked in an industry that cleaned particulate matter from the air using an electrostatic precipitator? If you are a winter sports enthusiast have you ever ridden on a ski lift?

Do you own, have you driven, or have you ever been a passenger in a car, truck, tractor, motorboat, sailboat, icebreaker (ship), snowmobile, all-terrain vehicle (ATV), armored tank, hydroplane, airplane, sea plane, jet, turbojet, supersonic passenger jet, blimp, dirigible, hot air balloon, spacecraft, or rocket? Have you ever flown in an aircraft powered by a valveless pulse engine, or one equipped with winglets, supercritical wings, straight wings, delta wings, or oblique wings? Have you ever owned or ridden in a boat equipped with an inboard or an outboard motor, or one that derived its thrust from a marine screw propeller?

Has your life or property ever been saved by a "super scooper"—a firefighting amphibious aircraft that scoops up water from lakes and seas and drops it on wildfires? Do you own or drive an aerodynamic car (nearly all cars today are built aerodynamically)? If you lived during the 1950s or 1960s, do you remember the world-altering first flight of the hypersonic rocket-powered aircraft known as the North American X-15? Have you ever been part of an aircraft investigation involving a "black box"—that is, either a flight data recorder or a cockpit voice recorder?

Automobile engine, circa 1904.

Have you or anyone you know ever been saved by an aircraft ejection seat? Have you ever used a parachute to sky dive, or have you or anyone you know ever been saved by one? Have you ever owned, used, or benefitted from a small drone, or flown in man-sized drone? Have you ever ridden on or seen someone (like a stuntman) ride a rocket-powered motorcycle? Have you ever used or been saved by a line-throwing apparatus or by a lifeboat? Have you ever benefitted from a hyperboloid structure?

STEEL, 4-WHEEL DRIVE, SPEEDOMETERS, & GASOLINE

Have you ever owned or used a product produced by U.S. Steel or International Harvester? Have you ever used a product made with magnetic steel? Have you ever owned, ridden in, or driven a vehicle with bumpers? Have you ever owned, driven, or ridden in a vehicle powered by a steam engine, a gas engine, an electric engine, a crankless engine, or one that employed a differential gear train, a harmonic drive gear, a magnetic clutch, a dual-clutch transmission, or a catalytic converter, or one that used antilock brakes?

Have you ever owned or driven a rear-wheel drive, four-wheel drive, or all-wheel drive vehicle, or operated machinery that travels on mecanum wheels or omniwheels? Have you ever owned, ridden in, or driven any type of vehicle installed with shock absorbers or that came with air conditioning?

Have you ever used or depended on a water pump or a power plant, or used an industrial machine tool of any kind? Has your home or business ever been saved from flooding by a city drainage pump? Have you ever owned or used a suitcase with wheels, owned or ridden in a glass-bottom boat, or "walked on water" using foot floats or water-walking shoes?

Have you ever used or owned a product that employs semiconductors—a material that is found in nearly every electronic instrument, device, and machine that runs on electricity? Have you ever used, or know anyone who has used or been affected by, a torpedo or a bouncing bomb during wartime? Have you ever used or owned a screw-cutting lathe?

First reversible 4-cylinder 4-cycle diesel engine ever made, circa 1907.

Have you owned or driven a vehicle with a speedometer, or one with an engine that contained a fuel injector, valves, a solenoid, a fuel pump, an electric generator, or relays? Have you ever owned or driven a vehicle that used lead gasoline, ethyl gasoline, or diesel fuel? Have you ever benefitted from any type of fuel produced via the Fischer-Tropsch Process? Have you ever owned or driven a vehicle with a diesel engine, or that was air-cooled, coated with Duco paint, had an electric starter ignition, or a two-cycle diesel engine?

SEATBELTS, MOTORCYCLES, RUBBER TIRES, & DIVE GEAR

Have you ever ridden in or owned a commercial bus or semi-trailer truck? Have you ever owned or ridden in a vehicle using a half-track assembly? Have you ever used a gauge block to calibrate a measurement, used an oscilloscope to analyze voltage variations, or used a dynamometer to measure the torque or power of an engine? Have you ever been subjected to a Breathalyzer test or ridden in a vehicle with seatbelts or airbags?

Have you ever used an iconoscope or a kinescope? Have you ever owned or operated an electric cash register, lived in an area that uses three-way traffic lights, used or owned an arc lamp, enjoyed the Art Deco ambiance of neon lights, or engaged in the martial art known as Sambo? Are you a citizen of, do you live in, or have you visited, the United States of America?

Do you own or have you ever ridden on a motorcycle, bicycle, unicycle, or moped? Have you ever owned, used, or ridden in or on a vehicle equipped with radial tires, or rubber tires—and more specifically air-filled rubber tires? Did your tires have valve stems (for inflating them with air)? Have you ever owned or used a Segway product, like a Segway KickScooter, eBike, or GoKart? Have you or anyone you know ever used the motorized wheelchair known as the iBOT?

22 ∞ MANMADE

Have you ever participated in underwater diving using scuba gear, swim fins, a snorkel, or a dive mask? Have you ever been to Marineland or SeaWorld? Have you ever swum, surfed, dived, or floated at a water park, or enjoyed going down a water slide—particularly a tubular water slide?

ASSEMBLY LINES, ZIPPERS, MOUTH WASH, & ELEVATORS

Have you ever owned or driven a Ford Motor product, a BMW product, a Mercedes Benz product, a Tesla product, a Chevrolet product, an Oldsmobile product, a General Motors product, or any type of John Deere equipment? Have you or anyone you know ever worked on an assembly line, or owned and enjoyed a product made on an assembly line? Have you ever visited or worked at the Smithsonian Institution?

Have you ever owned or used a Michelin product, or enjoyed the many toy models at Disneyland? Have you ever used a lint roller, a dry cleaner to clean a garment, or worn an item that uses a zipper to fasten clothing? Have you ever owned or worn a Mackintosh raincoat?

Have you ever used kerosene fuel for any purpose? Have you ever camped out and used a kerosene lamp or a kerosene stove? Have you ever used or benefitted from paraffin wax (used in lubricants, candles, insulation, fuels, cosmetics, etc.), creosote (used in pesticides, antiseptics, shampoos, wood treatments, etc.), or phenol (used in mouth washes, sore throat lozenges, cleaning products, etc.)?

Have you ever ridden in a gyrocar, or on a cable car, a trolley car, a trolleybus, a rack railway, an electric elevator (with automated doors), or on Switzerland's Pilatus Railway, or on Mount Washington's Marsh Rack Railway in New Hampshire?

SPORTS, SMOKE DETECTORS, CHEESE SLICERS, & BARCODES

Have you ever played or enjoyed watching baseball, (American) football, tennis, basketball, rugby, badminton, pickleball, hockey, golf, or (British) soccer? Have you skated on an ice rink, or been to a hockey game, where the famous ice resurfacer known as the "Zamboni" was used to clean the ice? Have you ever owned, used, depended on, or been saved by a heat detector, fire alarm, or smoke detector?

Have you ever lived in a home or worked in a building that had either cast-iron heating radiators or central heating? If you have ever been part of a demolition crew, mining operation, or military unit, have you ever used Semtex explosive, or detonated an explosive using electric wires? Have you ever used an airsickness bag, or used or benefitted from a communications satellite?

Have you ever used a sink faucet, a cheese slicer, or a rubber ice cube tray? Have you ever used a bottle opener on a bottle with a disposable bottle cap? Have you ever had your child's IQ tested, or purchased a product with a child-proof bottle cap? Have you ever flown a gyrocopter? Have you ever owned or piloted a helicopter, flown in a helicopter, been rescued by a helicopter, or had something delivered by a helicopter—in particular a Bell helicopter or a Sikorsky helicopter? Have you ever owned, piloted, or ridden in a Hovercraft?

Have you ever used or had need of a rain gauge or a water gauge? Have you ever used, created, or owned a barcode, a UPC code, or a QR code? Have you or a cashier ever used a barcode scanner on an item? Have you ever owned or used a welder to fuse metal pieces together, or have you ever owned, driven, or ridden in a vehicle that was constructed with parts made from welded metal? Have you ever owned or used an arc welder, or benefitted from something created with arc welding technology? Have you ever owned or used a blowtorch to heat various materials?

CHRISTMAS CARDS, METAL DETECTORS, SCHOOLS, & GUNS
Have you ever written or received a Christmas card? Have you ever used Western Union to transfer or receive money? Have you or anyone you know ever been the beneficiary of microfinance or microcredit? Have you ever used a metal detector, or been scanned by a handheld metal detector at a sporting event, government building, or airport? Have you ever admired or been saved by the bright sweeping light of a lighthouse? Have you ever used a solar cell, a solar panel, a magnifying glass, a reading aid, a photo-copier, or a movie projector, or watched a film wearing 3D movie glasses?

Harvard University, 1906.

Have you, or has anyone you know, attended Washington and Lee University, Harvard University, Rutgers University, Brown University, Vanderbilt University, the Massachusetts Institute of Technology (MIT), Vassar College, Emerson College, Columbia University, Stanford University, Johns Hopkins University, Carnegie Mellon University, Brandeis University, Dartmouth College, Yale University, Cornell University, University of Pennsylvania, University of Chicago, or Princeton University? Have you ever lived in or visited the city of Lomonosov, Russia, or attended Moscow State University?

Have you ever attended a performance at Carnegie Hall, or had any type of interaction with the Carnegie Institution for Science, the Carnegie Museum of Natural History, the Carnegie Endowment for Peace, the Carnegie Trust for the Universities of Scotland, the Carnegie Museum of Art, or the Carnegie Corporation of New York? Have you ever lived in, worked in, or visited the city of Washington, D.C.?

Have you ever owned or fired a modern pistol, BB gun, rifle, assault rifle, recoilless rifle, machine gun, tranquillizer gun, or mortar? What about an underwater pistol or an underwater rifle? Have you ever fired a weapon that used old-fashioned gunpowder (black powder), or a modern one that uses smokeless gunpowder? If you are a vintage gun enthusiast have you ever owned or fired a muzzle-loading musket, a wheellock, a flintlock, or a percussion lock weapon? Have you ever fired a Pritchett bullet or an Enfield rifle? Have you ever owned or used a firearm made by Remington, Savage, Browning, Glock, Winchester, or Colt? Have you or anyone you know ever been injured by shrapnel?

HONEY, GPS, LAWNMOWERS, & BALLPOINT PENS

Have you ever used or depended on a seismograph to detect seismic activity, or has your life, or the life of anyone you know, ever been saved by an earthquake early warning system? Have you ever used or benefited from the earthquake-measuring instrument known as the Richter Scale?

If you are a beekeeper, have you ever owned or used a frame beehive or a queen bee excluder? Have you ever used or enjoyed honey or beeswax? Have you ever owned or played on a trampoline that had a safety net enclosure around it? As a child were you ever taught by an educator using the operant conditioning method?

Have you ever used GPS to track your time, establish your location, find your way on a map, or navigate a hiking trail or an unfamiliar roadway? Were you ever saved, or do you know of someone whose life was saved, by a search and rescue team using GPS?

Do you own or have you ever used a silent intruder alarm that automatically calls the police? Have you ever depended on the international crime prevention group, the Guardian Angels, for your safety? Have you ever benefitted from radiocarbon dating, which is used in, among other fields, archaeology, biology, anthropology, forensic science, climatology, oceanography, paleontology, and environmental science?

Have you ever used a pre-cell phone landline (a traditional wired telephone), caller-ID, an analog (tape) telephone answering machine, or a digital telephone answering machine? Have you ever made a long distance phone call? Have you ever seen or enjoyed the first photos of earth from the moon, or the first photos of man on the moon? Have you ever owned or used a catadioptric telescope or a Schmidt Camera?

Do you remember being taught the theory of evolution, the three laws of planetary motion, or the Bohr Model in school? Have you ever used or benefitted from a spectroscope or an interferometer? Have you ever owned an ornamental armillary sphere, or been taught astronomy by a teacher who using a real armillary sphere? Have you ever owned or used a lawnmower, a rubber band, a stroboscope, or a strobe light?

Have you ever used a product that relied on quantum theory or quantum mechanics, such as a transistor radio, a bread toaster, or solar cells? Have you ever owned or used a ballpoint pen, a fountain pen, a mechanical pencil, or a Magic Marker? Have you ever purchased a product in a plastic bag that used a small, removable, notched clip (known as a bread tag) to seal it at one end?

BRIDGES, ROBOTS, DISHWASHERS, VIOLINS, & RODEOS

Have you ever lived in town or city that kept the roads clean using a street sweeping truck? Have you ever ridden in a taxi that kept track of your fare using a taximeter? Have you ever driven on a toll road and paid your toll using an automatic coin collector?

A Bartolomeo Giuseppe Guarneri violin, circa early 1700s.

Has or is your home or business ever been powered by electricity from a hydroelectric dam, nuclear power station, or wind turbines? Have you ever owned or ridden in a ship or aircraft powered by a gas turbine? Have you ever driven over a bridge (in particular, a wire cable suspension bridge), ridden in a bathyscaphe, a submarine, a submersible, worked on an oil rig, worked in an oil refinery, or benefitted from an oil, gas, sewage, or water pipeline? Have you ever used, owned, or relied on a robot of any kind in any capacity? Have you ever used or benefitted from swarm bots or swarm technology?

Do you own or have you ever used an electric garage door opener, an automatic dishwasher, an electric water pump, or an electric toothbrush? Have you ever used or owned an alkaline battery, a lithium battery, or a Zamboni Pile? Have you ever owned, played, or enjoyed the sounds of an electric guitar, trumpet, saxophone, flute, French horn, violin, viola, cello, or bass viol? Do you live or work in a structure protected by a lightning rod? Have you ever hung a mirror or picture frame using a plastic expanding wall plug? Have you ever swept out your chimney or hired a professional to clean it?

Have you ever watched or participated in a rodeo? Have you ever benefitted from desalinated water, or consumed water that was purified by osmosis, or water that was purified at a wastewater treatment plant? Have you or anyone you know ever taken or benefitted from synthetic Vitamin C? Have you ever used Morse Code, semaphore, a walkie-talkie, listened to AM radio or FM radio, or finger "pinched" an image on your cell phone to zoom in or out? Have you ever used a touchscreen on a cell phone, tablet, or computer screen, or a touchscreen kiosk at a fast food outlet, a mall, a restaurant, or a theme park? While traveling have you ever worked with time zones in order to plan your trip?

CREDIT CARDS, NATIONAL PARKS, & OPERATING SYSTEMS
Have you ever used a compass, or consumed black or white pepper for their medicinal health benefits? Have you ever owned, purchased, used, or been given a credit card, debit card, gift card, driver's license, hotel room key, store loyalty card, or identification badge? Have you ever used or owned a padlock, a cylinder lock, a keyless combination lock, or a pin tumbler lock? Have you ever owned or used a Murphy Bed?

Have you ever ridden on a passenger train that traveled on railroad tracks equipped with railroad spikes and T-rails? Have you ever enjoyed the sound of a train's air-horn? Have you or anyone you know (who lived before 1968) ever ridden in a Pullman Sleeper railroad car? Have you ever been to Yellowstone National Park, Yosemite National Park, Grand Canyon National Park, Sequoia National Park, or Mount Rainier National Park?

Have you ever used or benefitted from a network router, a neural network simulator, a broadband network, Network Time Protocol, cloud computing, ARPANET, HTML, HTTP, CSS, BASIC, PHP, or JavaScript? Have you ever used or benefitted from the Altair BASIC programming language, the C++ programming language, the Pascal programming language, the Python programing language, the B programming language, the C programming language, or the Java programming language?

Have you ever used or benefitted from the MS-DOS operating system, the Windows CE operating system, the UNIX operating system, the Linux operating system, the GNU operating system, the Mach operating system, the NeXTSTEP operating system, macOS, iOS, the RAR file format, the WinRAR file archiver, or the FAR file manager? Have you ever used Google's Web search engine, the Microsoft Network, or owned a computer equipped with the Unified Extensible Firmware Interface Standard?

EAR MUFFS, LIGHT BULBS, GARDEN HOSES, GAMES, & BOOKS
Have you ever ridden in or driven a vehicle through an underwater tunnel? Have you ever worn ear muffs, or used a microscope, telescope, binoculars, electric light bulb (metal filament or LED), fluorescent light, electric generator, or air pump? Have you ever depended on a

mercury-vapor lamp for safety or visibility, such as in a sports arena, street, store, or factory?

Have you ever owned or used a machine that employs gaskets? Have you ever used a garden hose, rubber footwear, or any type of adhesive? Have you ever owned or relied on a product that was made using sealant, insulated wire, or a conveyor belt?

Have you ever owned or used an IBM computer? Have you ever owned or used a Microsoft product, a Panasonic product, a Samsung product, a Dell product, an HP product, an Nvidia product, a Canon product, a Minox product, a Sigma product, a General Electric (GE) product, an Intel product, a GoPro product, a Toshiba product, a Dolby product, or a Sony product? What about Apple and its specific products, such as the iMac, iPod, MacBook, iPhone, or iPad tablet? Have you ever used Paypal, Starlink, or the social media platform X (formerly known as Twitter)? Have you ever owned or read an ebook?

Have you ever owned or used a Wham-O product, or a Nintendo Entertainment System, or played one of the following popular games: Monopoly, Scrabble, PlayStation, Donkey Kong, Twister, Mastermind, Tetris, the Legend of Zelda, Ouija Board, Rubik's Cube, Uno, Mario, Rummikub, or Star Fox?

Have you ever played a crossword puzzle, or owned or played with such toys as a Frisbee, Lego, Meccano, Tinkertoys, Hornby Model Trains, Erector Set, Dinky figures, Lincoln Logs, American Flyer Trains, a gyroscope, Playmobil, or Kiddicraft Self-Locking Building Bricks? If you are a gamer have you ever used a "Game Boy" console or a cross-shaped control pad, or played the electronic game system known as "Game and Watch"?

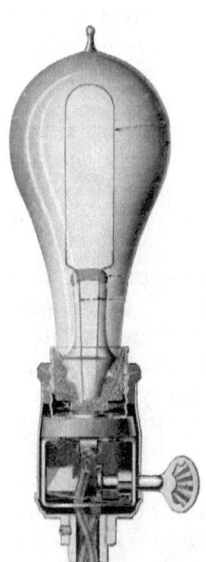

Edison lamp bulb and socket, circa 1881.

Have you ever read a book, magazine, or newspaper, or any other type of ink-printed material, or used an item, such as a cleaning product or refrigerator, that utilized chloroform? Have you ever used or benefitted from an explosive? Have you ever worn clothing imprinted with dazzle camouflage, enjoyed art that employed dazzle camouflage, lived or worked in a building decorated with dazzle camouflage, or served on an ocean-going military vessel painted in dazzle camouflage?

PAPER CLIPS, CABLE TV, MOVIES, & PHOTOGRAPHY

Have you ever used a quadrant, an octant, or a sextant to navigate at sea, enjoyed a malted beer or whiskey, or used either a punch card or punch card reader? Have you ever used or benefitted from a heat lamp, a sun lamp, or light therapy? Have you ever owned or used a paper clip, a paper hole punch, or a ring binder? Have you ever spoken or benefitted from the international manmade language known as Esperanto?

Have you ever owned or watched a modern flat-screen TV (plasma, LED, or OLED), or perhaps an early model television with an old-fashioned "rabbit ears" antenna? Have you ever owned or used a roof-mounted, wire "fishbone" antenna to watch TV or communicate via ham radio? Have you ever paid for and used cable TV or satellite TV?

Voigtlander camera, circa 1928.

Have you ever used a TV remote control, an infrared computer port, a thermal imaging camera, an infrared camera, night vision equipment, or a confocal scanning microscope? Have you ever read or written a sonnet? Have you ever written on, drawn on, painted on, or typed on ordinary everyday writing paper, or used it to wrap or print something?

Have you ever written, filmed, acted in, worked on, or gone to a movie? If you are a film editor, have you ever worked with a Moviola? Have you ever taken a photograph (either color or black and white), filmed a video, used a polarized camera filter, used a wide angle lens, used a light meter (photometer) to shoot a scene, or had your picture taken in a photo booth? If you work in entertainment, broadcast, or online media, have you ever used a Teleprompter?

Have you ever enjoyed a Victorian Daguerreotype photograph or created a photogram? Have you ever used a Geiger counter to check for radiation levels? Have you ever enjoyed a painting created with acrylic paint, or used acrylic paint yourself? If you are an artist, have you ever used collapsible metal paint tubes to store and carry your paints? Have you ever used aerosol spray paint? Have you ever enjoyed or owned a piece of artwork by a painter from the Impressionist period? Have you ever used a sandblaster or benefitted from sandblasting (used mainly to clean and restore surfaces)?

TV SHOWS, GEODESIC DOMES, BIRTH CONTROL, & STAMPS

Have you ever watched or enjoyed any of the following hit TV series? *Star Trek, Game of Thrones, The Wire, Seinfeld, The Sopranos, The Simpsons, One Day at a Time, Arrested Development, Cheers, The Jeffersons, Breaking Bad, The Mary Tyler Moore Show, The Avengers, Doctor Who, Monty Python's Flying Circus, All in the Family, South Park, Saturday Night Live, Captain Kangaroo, The Tonight Show,* or *The Twilight Zone?*

Have you ever constructed, lived in, or owned a geodesic dome, or used music therapy or aromatherapy to relax or heal? Have you ever relied on calculus? Do you own or have you ever used a Newtonian telescope, or enjoyed the pictures sent back by the Hubble Space

Telescope or the James Webb Space Telescope?

Have you ever used birth control pills, a male condom, a female condom, a female diaphragm, a birth control sponge, or an intrauterine device (IUD)? Have you ever mailed a letter in a lickable envelope? Did your letter have a postage stamp on it? Has your Post Office ever authorized your letter or package using a postage meter? Have you ever used, owned, or benefitted from an atlas? Have you ever put a bumper sticker on a vehicle or enjoyed one on someone else's vehicle?

Have you ever seen, used, owned, or enjoyed a hologram? Have you ever played a video game or used a video game console or controller? Have you ever adopted a pet from Britain's Royal Society for the Prevention of Cruelty to Animals (RSPCA) or from the American Society for the Prevention of Cruelty to Animals (ASPCA)?

PUBLIC SERVICES, CEILING FANS, ALUMINUM, & WI-FI

Have you ever trained an animal using the concept of conditioned reflex? Have you ever used cat litter or an "invisible" electric pet fence? Have you ever depended on a public library, a public hospital, or a volunteer fire department? Have you ever been to the Library of Congress, or used or benefitted from its services?

Have you ever cooled yourself under a ceiling fan, had your teeth cleaned by a dentist's dental tool, been anesthetized with nitrous oxide, used a Bunsen Burner, a glass laboratory flask (pycnometer), a Winogradsky Column, or a Petri dish, written with a fountain pen, used carbon paper, or worked on a sewing machine? Do you live in a part of the world that utilizes Daylight Savings Time?

Have you ever used or owned anything made of steel (like a stainless steel pan), or consumed a beverage from an aluminum soda can, used aluminum foil for baking, or cooked food in an aluminum frying pan or pot? Have you ever owned or used a car, aircraft, or boat made with aluminum parts? Have you ever used parchment paper to bake food, or used a Mason jar to preserve food?

A Singer sewing machine, circa 1914.

Have you ever attended a plastination exhibition where the perfectly preserved internal organs and muscles of human bodies were on display? If you are a doctor, a nurse, or any other type of health practitioner, were you ever instructed using plastinated body parts or science films in medical school?

Have you, and or anyone you know, ever benefitted from an organ transplant? Have you or anyone you know ever been involved in a situation that required either regular fingerprinting or DNA fingerprinting? Have you ever used a hand-cranked powered radio or a foot-powered radio?

Have you ever communicated using a 4G or 5G cellular network, wi-fi, fiber-optics, or a geostationary satellite, a low earth orbit satellite (LEO), or a medium earth orbit satellite (MEO)? Do you own, wear, or use any type of clothing or textile, or have you ever used a multiple spindle weaving frame? Have you every used a flush toilet, or owned or ridden on a boat or ship protected by marine lightning rods? Have you ever made use of trigonometry, or used or benefitted from radar technology or satellite communications?

LASERS, VACCINES, THERMOMETERS, & ASPIRIN
Have you ever purchased, owned, or used something that was engraved by a laser? Have you ever undergone any type of surgery that required a laser? Have you or anyone you know ever attended a laser show? Have you ever used an adjustable pipe wrench or an adjustable spanner wrench?

Have you ever gotten a vaccine for bubonic plaque or cholera, or have you or anyone you know been saved by the Heimlich Maneuver? Have you or anyone you know ever been saved by a firefighter or firefighting team? Have you ever owned or used a thermometer, a hygrometer, a barometer, a barograph, or an anemometer, or eaten food that was cooked, warmed, dried, broiled, fried, grilled, or roasted using infrared radiation? Have you ever owned, used, or benefitted from a device containing a centrifugal fan (e.g., hair dryers, furnaces, air purifiers, clothes dryers, air conditioning units, etc.)?

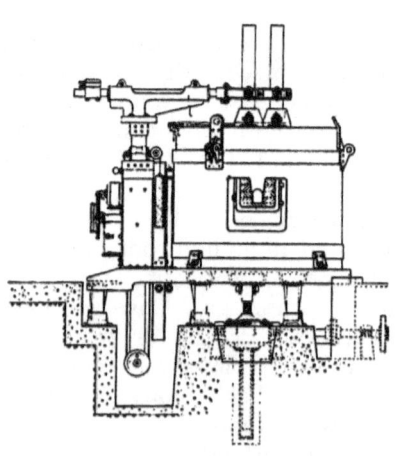

Furnace schematic, circa 1926.

Have you or anyone you know been successfully treated for cholera, typhus, tuberculosis, malaria, or anthrax? Have you ever been vaccinated for rubella, streptococcus, pandemic flu, human cancer, chickenpox, pneumococcus, meningococcus, mumps, hepatitis A, hepatitis B, or measles? Have you ever taken aspirin for a headache? Have you ever used or do you know anyone who has used LSD, methamphetamine, or heroin?

Has a behavioral psychologist, using the concept of classical conditioning, ever treated you or anyone you know for depression, anxiety, phobias, OCD (obsessive-compulsive disorder), or substance abuse? Have you ever undergone psychoanalysis by a Freudian or Jungian psychoanalyst, or benefitted from studying such psychology-related concepts as individuation, the 12 archetypes, or the collective unconscious? Have you ever taken a the word association test or a Rorschach Test, or been categorized as an introvert or an extrovert?

MEDICAL TREATMENTS, INSTRUMENTS, & PROCEDURES
Have you, or anyone you know, ever suffered from a health issue and been diagnosed or treated with one of the following devices, instruments, techniques, medications, chemicals, gases, compounds, or procedures:

artificial heart
artificial kidney
atomic force microscope
auscultatory method
balloon catheter
Body Mass Index (BMI)
brain-machine interface
carbon dioxide laser
carbon monoxide laser
cardiac defibrillator
cataract surgery
chiropractic
chromatography
cryogenics
CT scan
dialysis
disposable hypodermic syringe
distraction osteogenesis
Drinker Respirator
ECG
EEG
electron microscope
electron paramagnetic resonance
exoskeleton
Feulgen Reaction
gene targeting
genetic testing
Golgi Method
Gomori Trichrome Stain method
Gram Staining Method
green fluorescent protein
hearing aid
heart-lung machine
hemodialysis

holography
Holter monitor
HPV vaccine
hypnosis
Ilizarov surgery
Ilizarov apparatus
in situ hybridization
insulin pump
intra-aortic balloon pump
iron lung
IVF
laser
lithotomy
magnetic healing
micro CT scan
microneurosurgery
microplate
moss bioreactor
MRI
nitrous oxide
Nivalin
nuclear medicine scan
ophthalmoscope
oximeter
pacemaker
pap test
patch clamp technique
penicillin
phase-contrast microscope
plastic surgery
polio vaccine (injection or oral)
portable cardiac defibrillator
portable kidney dialysis machine
positron emission tomography (PET)
pseudoscope

radial keratotomy
Reichstein Process
rhinoplasty
robotic arm prosthesis
rotavirus vaccine
Schick Test
Schiff Test
Schlenk Flask
Site-Directed Mutagenesis
Southern Blot
spin-flip Raman laser
stem cells
stereoscope
swivel-frit
synchrotron
syringe
thermometer (either mercury-in-glass, digital, strip, or ear)
tuberculin skin test
ultracentrifugation
urinary catheter
Van Gieson's Stain
Walden Inversion
Warthin-Starry Stain
wireless heart monitor
Wright's Stain
X-ray
Yuzpe Regimen
Ziehl-Neelsen Stain

THE BENEVOLENT PATRIARCHY
If you said yes to even one of the above questions, questions that could be multiplied thousands of times over, you have an individual with an X and a Y chromosome to thank for it: a biological male!

The old adage is indeed true: "It's a man's world." Not a male-dominant, anti-female patriarchy, as some would have you believe. But rather a male *idea*-dominant, *pro-human* patriarchy; that is, a *patriarchy of male ideas*, nearly all innovations which, being products of the "benevolent patriarchy,"[1] have been and continue to be advantageous to humanity.

MALE INVENTORS: CIVILIZATION'S UNSUNG HEROES
While most of the men I have included in this book received countless plaudits, laurels, and awards—not to mention world renown, lasting fame, honorary degrees, statues, and often fantastic wealth—during their lifetimes, millions of others have passed through history unknown, unnamed, and unsung, quietly performing daily problem-solving chores for their female partners, wives, families, communities, and towns. In their own right, these men too were hero-inventors; clever, industrious, selfless males who created, maintained, and repaired the complex machinery that powers society—never expecting riches, and certainly not a plaque, reward, monument, or even a mere "thank-you."

Nearly every single idea, item, and product embraced and utilized by humanity today was proposed, discovered, developed, created, or patented by men, making both their inventions and the modern world itself literally *manmade*. The global historical significance behind these statements is beyond dispute. Let us look more closely at this topic for a moment.

MALE INVENTORS & THE 99.99 PERCENT RULE

According to current mainstream science, modern humans appeared around 300,000 years ago, with some 117 billion people having lived on earth between that time and today. Add to this the bold facts that written records only go back approximately 5,300 years (to the 6th Millennium BC; that is, 3300 BC), and that many hundreds of thousands of inventions were discovered, developed, and created long before then.

Obviously, we will never know the complete or precise details concerning who invented what, when, where, how, and why. This difficulty is highlighted by, for example, the bow and arrow, which currently dates to some 70,000 years ago. And yet it is almost impossible to determine who first came up with an idea that is a mere century old. A myriad of other similar illustrations could be given.

To reemphasize then: Despite the fact that we will never know the names of the progenitors of most inventions, it is obvious from both history and the information I present in this book, that *at least 99.99 percent of all inventions ever made by humans have been made by biological men; that is, humans whose natal sex was male.*

THE AUTHOR AS INVENTOR

Manmade is not merely an exercise in invention technologies on my part. One of the many reasons I wrote this book is that, being a male inventor myself, I have always been fascinated by inventing, inventions, and inventors, particularly those connected to the field of mechanical engineering, my main area of interest when it comes to technology. In my case, beginning in early childhood, I began mentally concocting—long before I had ever heard of them—a wide variety of mechanical innovations, including the windmill, the hydro turbine, the wind turbine, and the ramjet, to name but a few.

Buchanan Windmill, circa 1890.

Like Benjamin Franklin, however, I never patented any of my ideas and designs, and many of them (including dozens of my inventions not mentioned here), since "invented" and claimed by others, have gone on to become great commercial and global successes.

HOW MALE INVENTIONS ARE TYPICALLY CREATED

Because, due to numerous biological evolutionary factors, men are naturally inventive, it would be no exaggeration to say that there have been billions upon billions of male inventors throughout history and prehistory. Of these a mere fraction are known, and from this group only a mere fraction are well-known. It is from this last group that I have personally hand-selected a tiny fraction of male inventors for inclusion in my book, men I number among the more interesting, creative, and forward thinking of the naturally inventive male of our species.

A question I am often asked is: Where do men get their ideas? Where does their inspiration and motivation to invent come from?

Most male inventions (such as eyeglasses, firearms, artificial hearts, etc.) emerge due to *necessity*. For example, we have Lewis Edson Waterman, who invented the modern fountain pen after his dip ink pen leaked on a contract, costing him time and energy—and more importantly a customer. Victorian gentleman William Lawrence Murphy was courting a woman whom he wanted to invite to his small studio apartment. Unfortunately, his bed was in plain view; unfortunate because, at the time, it was against societal rules for an unmarried woman to be present in a man's bedroom. Murphy solved the problem by constructing a bed that could be folded up and hidden in his closet, giving birth to the "Murphy Bed." (The two later married.) Philippe-Guy E. Woog was having gum pain, but could not find a curative treatment, even from dentists. After some research he discovered that the coarse unrefined foods of early man

Early printing press, circa 1904.

were natural tooth cleaners and gum stimulators, qualities both lacking in the overly-processed diet of modern humans. To remedy the situation he invented an artificial gum massager, one that aided in keeping the teeth clean, but also helped prevent gum disease. He called it the "Broxodent." Today we call it the electric toothbrush.

Alternatively, many male innovations are *accidental*. For instance, we have the case of Benjamin Chew Tilghman, who, while stationed at a military base in the desert, noticed how glass windows were being pitted by wind-blown sand—leading to his invention of sandblasting. There was George de Mestral, who, after a day of sport hunting in the field, thought of the idea of Velcro while trying to remove stubborn cockleburs from his clothing. Barnes Wallis came up with the idea of the bouncing bomb while skipping marbles over water. Then there was the case of Roy Joseph Plunkett, who, while researching new ways to produce refrigerant in his laboratory, discovered an unknown white powder in a metal cylinder left out overnight—a waxy residue that turned out to be the heat-resistant material later known as Teflon.

It is intriguing to consider that in thousands upon thousands of cases, the modern world is a result of "happy accidents" just like these.

MEN: THE ARCHETYPAL PROBLEM-SOLVERS

If you are a male, or if you are someone who loves and cherishes men, or if you are a male or female inventor seeking inspiration, or if you are simply someone who enjoys technological history, it is my hope that you will be invigorated, enlightened, and uplifted reading about my male inventors—whom I refer to as "human Swiss Army knives." For they, and their unnamed, everyday, "average Joe" problem-solving counterparts, have, for hundreds of thousands of years, made the world a better, safer, and happier place for, not just men, but for *everyone*: men, women, and children of all ages.

There can be no rational debate over the historicity of this statement. As my book *Manmade* shows, it is irrefutable. To reiterate: Nearly 100 percent of the items, products, designs, and ideas all of us own, use, and depend on everyday were invented by men!

Despite this overt fact, as I write these words various elements of society continue to bombard us with such outworn statements as "men are useless" and "women don't need men." Are such claims true? Open your mind, read on, and decide for yourself. A world of discovery awaits you.

Lochlainn Seabrook
Rocky Mountains, Wyoming, USA
April 2025
Mater Artium Necessitas, ancient proverb

MANMADE

Male Inventors
Who Created the Modern World

ABALAKOV, VITALY MIKHAYLOVICH. 1906-1986. Russian mountaineer and chemist: Invented ironmongery, the ice screw, camming devices, and the Abalakov thread or V-thread (circa 1950s), a temporary gearless anchor, all used in either mountain and/or ice climbing.
ABBE, ERNST KARL. 1840-1905. German physicist and engineer: Invented the apochromatic lens system in 1868 (used in microscopes), the refractometer in 1869 (measures the water content of liquids), and the Abbe condenser in 1870 (also used in microscopes). Nobel Prize winner.
ADAMIAN, HOVANNES. 1879-1932. Armenian engineer: Invented the tricolor principle (circa 1928), which was used in the development of the color television.
ADAMS, JOHN. 1735-1826. American Founding Father, first vice president of the United States (1789), second president of the United States (1796), main author of the Massachusetts Constitution (1780), politician, political theorist, diplomat, attorney, gentleman farmer: Founded the Library of Congress (1800), which was based on James Madison's original idea (1783).

John Adams.

ALBERTI, LEO BATTISTA. 1404-1472. Italian polymath, scientist, priest, cryptographer, architect, mathematician, cartographer, author, poet, artist, linguist, and inventor: Invented the anemometer (1450),

used to measure wind speed, and the "Alberti cipher" (a polyalphabetic code system), and is considered the founder of the science of cryptography in the West.
ALDERSON, SAMUEL W. 1914-2005. American engineer and physicist: Invented the crash test dummy (1949). An anthropomorphic device used in crash simulations, it has saved thousands of lives.
ALEXEIEFF, ALEXANDRE. 1901-1982. Russian artist and animator: Invented the pinscreen animation technique (circa 1932), which was used in early stop motion animation films.
ALEXEYEV, ROSTISLAV. 1916-1980. Russian engineer and inventor: Invented ground effect vehicles (circa 1920s-1930s), which are hybrids of aircrafts and ships using hydrofoils. Alexeyev is known as the "father of the ground effect vehicle."
ALFEROV, ZHORES I. 1930-2019. Russian physicist: Co-invented the room temperature continuous wave operating diode laser (circa 1970). Nobel Prize winner.
AL-HATHAM, IBN. Circa 965-circa 1039. Iraqi astronomer, mathematician, researcher, and inventor: Invented the camera obscura (precursor of the modern camera), the pinhole camera (a camera with a pinhole but no lens), and the magnifying glass. Al-Hatham is known as the "father of modern optics."
ALIKHANOV, ABRAM. 1904-1970. Soviet-Armenian physicist: Invented the nuclear reactor (1942), which uses heat energy released by splitting atoms to produce electricity. Alikhanov is known as one of the "fathers of Soviet particle physics."
ALLEN, PAUL G. 1953-2018. American businessman, computer scientist, filmmaker, investor, entrepreneur, author, and philanthropist: Co-founder of the Microsoft Corporation (1975), more popularly known as Microsoft.
ALLEN, STEVE. 1921-2000. American TV and radio entertainer, comedian, writer, musician, and actor: Co-created the hit TV talk show *The Tonight Show* (1954).
AMATI, ANDREA. Circa 1505-1577. Italian stringed instrument maker, luthier, and inventor: Sadly for history, the inventor of the violin remains anonymous, the first known reference to the modern version being depicted in a painting by Gaudenzio Ferrari (circa 1535). Nonetheless, Amati comes closer than anyone else to earning the title, as one of his violins, the earliest known violin still in existence, dates from around 1565. Amati was also the probable inventor of the first viola, the first cello, and the first bass viol.
AMES, BRUCE NATHAN. Born 1928. American biochemist: Invented the Ames Test (1973), which is used to test for cancer-causing chemicals.
AMICI, GIOVANNI BATTISTA. 1786-1863. Italian astronomer, botanist, and optics designer: Invented the Amici roof prism (circa 1843), a dispersive optical prism used in spectrometers; he also invented

immersion microscopy (1847).

ANDO, MOMOFUKU. 1910-2007. Taiwanese businessman and inventor: Invented chicken ramen (1958), also known as instant noodles.

ANDREW, SAMUEL. 1656-1738. American educator and clergyman: Co-founder of the Collegiate School (1701), later renamed Yale University (1718) after its main benefactor, American colonial leader Elihu Yale.

ANGER, HAL. 1920-2005. American biophysicist and electrical engineer: Invented diagnostic cameras, such as the well counter (radioactivity measurements) and the gamma camera.

ÅNGSTRÖM, KNUT JOHAN. 1857-1910. Swedish physicist: Invented the pyrheliometer (1893). Also known as the Ångström pyrheliometer, it is a device for measuring solar radiation. Still in use.

ANIM-MENSAH, ALEXANDER. Dates unknown. Currently living. Ghanaian (West African) chemical engineer, author, and inventor: Responsible for over 30 inventions related to chemical and water use, product testing, energy management, refrigeration and heat pumps, membrane science, performance monitoring, and waste minimization and environmental impact reduction.

ANSCHÜTZ-KAEMPFE, HERMANN. 1872-1931. German scientist, art historian, and inventor: Invented the gyrocompass (circa 1906), a revolutionary navigation instrument.

ANSCHÜTZ, OTTOMAR. 1846-1907. German photographer, engineer, and inventor: Invented the projecting electrotachyscope (circa late 1880s), an early motion picture projection system.

APPERT, NICOLAS. 1749-1841. French chef, confectioner, and inventor: Invented food preservation (using hermetically sealed glass bottles). Appert is known as the "father of food science."

Nicolas Appert.

ARCHIMEDES. Circa 287-212 BC. Greek mathematician, astronomer, and inventor: Invented the Archimedes' screw, a displacement pump that raises water level. Archimedes is known as the "father of mathematics" and the "father of hydrostatics."

AREZZO, GUIDO OF. Circa 991-1033. Italian monk, composer, medieval music theorist, writer, and inventor: Invented Guidonian musical notation, which revolutionized music writing, leading to the modern form of musical notation.

ARGAND, AMI. 1750-1803. Swiss chemist and physicist: Invented the Argand burner or Argand lamp (1784), the first scientifically manufactured oil lamp.

ARMSTRONG, EDWIN HOWARD. 1890-1954. American electrical engineer and inventor: Developed the principle of frequency modulation, which we know today as FM radio.

ARMSTRONG, WILLIAM GEORGE. 1810-1900. English industrialist and mechanical engineer: Invented the hydraulic accumulator (1842), a gas pressured vessel that stores, maintains, and recaptures energy; used in hydraulic systems to increase efficiency and reliability.

ARNOTT, NEIL. 1788-1874. Scottish physician and inventor: Invented the Arnott stove (circa 1838), the Arnott ventilator (a ventilating chimney valve), the smokeless grate, and the hydrostatic bed (1838)—also known as a waterbed.

ARTIN, EMIL. 1889-1962. Austrian (of Armenian descent) mathematician (algebraist) and educator: Invented Artin rings (1944) and the braid theory (circa 1920s), used in the field of algebraic theory and considered an important contribution to modern abstract algebra.

ASPDIN, JOSEPH. 1788-1855. English businessman, stonemason, and inventor: Credited with inventing the first artificial cement (1824), which he named Portland cement. Today it is the most common form of cement in use globally.

ATANASOFF, JOHN VINCENT. 1903-1995. American engineer, physicist, mathematician, teacher, and inventor: Invented the Atanasoff-Berry Computer, or the ABC (developed circa 1939-1942), the first special-purpose electronic computer.

ATHERTON, S. W. Dates unknown. American electrical engineer and inventor: Invented the electric extension cord (1904), revolutionizing the home appliance, construction, and business industries.

AUDIFFREN, MARCEL. Lived Victorian Period. French monk, physicist, and inventor: Invented the first hermetically sealed refrigeration unit. Patented by Audiffren in 1894, it became the first commercially viable household or residential refrigerator.

New York Central Empire State Express, circa 1897.

BABAYAN, BORIS. Born 1933. Azerbaijani computer scientist: Invented the (Russian) super computer.

BABBAGE, CHARLES. 1791-1871. English polymath, engineer, economist, mathematician, philosopher, writer, and inventor: Conceived the first digital computer (1856), known as the Analytical Machine, and also the first computer program; he also invented the cowcatcher (circa 1838) for train locomotives and an early form of the speedometer, as well as aided in the establishment of the modern English postal system. Babbage is known as the "father of the computer."

Charles Babbage.

BABE, VICTOR. 1854-1926. Romanian physician and pathologist: Discovered Babesia (1888), a genus of disease-carrying protozoans carried into the blood by ticks, and which can cause anemia. Babe is known as the "father of serotherapy."

BAEKELAND, LEO. 1863-1944. Belgian Engineer, chemist, and inventor: Invented Velox (1893), the first commercially successful photographic paper, and also Bakelite (1907), the first synthetic plastic. Baekeland is known as the "father of the plastics industry."

BAER, RALPH H. 1922-2014. German Computer programmer, video game developer, and inventor: Baer invented the video game console (1972), and is known as the "father of home video games."

BAEYER, ADOLF VON. 1835-1917. German chemist and teacher: Influenced the dye industry by inventing barbituric acid (circa 1864), synthetic indigo dye, and phenolphthalein; developed fluorescein (1871), a pigment produced by organisms. Nobel Prize winner.

BAGLEY, RODNEY D. 1934-2023. American geological engineer, ceramic engineer, and inventor: Co-developed the cellular ceramic substrate filter used inside catalytic converters, a device that has been estimated to have eliminated billions of tons of toxic emissions produced by the internal combustion engine.

BAIRD, JOHN LOGIE. 1888-1946. Scottish physicist, entrepreneur, and inventor: Was the first to televise electronic motion pictures (1926), creating the modern TV; also invented color television (1928), and researched stereoscopic TV (1946).

BAKER, KENNETH. Dates unknown. Nationality unknown. Lead engineer for the General Motors electric car project known as the GM EV-1 (1990). As demand for EVs was growing, mass production was soon implemented (1996). Though popular with the public, strangely GM quickly decided to pull the plug on the EV-1, recalling all leased vehicles and even destroying the remaining stock with special machinery. Production soon ceased (1999) and the entire program was eventually scrapped (2001), the EV-1 going the way of the dinosaurs. The question remains, why? While GM declared that it had killed the project because, at the time, the car was not commercially viable, devotees of the groundbreaking electric vehicle believe it was possibly something more sinister: pressure from the oil industry. Another suppressed invention?

BALLAS, GEORGE. 1925-2011. American entrepreneur and inventor: Invented the first foliage string trimmer (1971), better known as the "Weed Eater."

BANTING, FREDERICK. 1891-1941. Canadian scientist, orthopedist, pharmacologist, and field surgeon. Co-discovered insulin (1921), which was used as a life-saving treatment for diabetes mellitus. Nobel Prize winner.

BARANOV-ROSSINE, VLADIMIR. 1888-1944. Russian avant-garde artist, futurist painter, sculptor, art critic, manufacturer, and lecturer: Invented the Optophonic Piano (1916), a keyboard controlled instrument that created sounds while projecting moving patterns onto the walls and ceilings of a room. To achieve these effects a bright light was pointed through revolving painted glass disks, after which the images were further altered using mirrors, lenses, and filters.

BARBER, JOHN. 1734-1801. English engineer and inventor: Invented the first gas turbine (1791), a thermal machine that is today used in numerous industries, from aviation and marine propulsion, to transportation and power generation.

BARBIER, CHARLES. 1767-1841. French military officer and surveyor: Invented the point-writing system (circa 1815), a precursor to the Braille writing system, both intended for use by the visually impaired.

BARDEEN, JOHN. 1908-1991. American electrical engineer, physicist, teacher, and inventor: Co-inventor of the transistor (1947), which did away with the more cumbersome vacuum tubes. The transistor, a tiny semiconducting device used for controlling electrical signals, is an integral component of the microchip, and today serves as the very foundation of the computer-information age. Multiple Nobel Prize winner.

BARMIN, VLADIMIR. 1909-1993. Russian scientist, mechanical engineer, and rocket engineer: Invented the first spaceport or cosmodrome, a rocket complex or aeronautical site used for launching and receiving spacecraft.

BARRINGER, ANTHONY R. 1925-2009. Canadian geophysicist: Developed the Induced Pulse Transient system (known as INPUT), an electromagnetic technology used on fixed-wing aircraft to conduct airborne surveys of ore deposits. He also developed or contributed to the development of many other technologies, including E-phase, GASPEC, radiophase, COTRAN, FLUOROSCAN, SURTRACE, AIRTRAC, and LASERTRACE.

BARSANTI, EUGENIO. 1821-1864. Italian scientist, engineer, mathematician, clergyman, businessman, and inventor: Credited with co-inventing the first internal combustion engine (1853). Known as the Barsanti-Matteucci engine, it was constructed on a two-cycle, free piston design as an alternative to the steam engine.

Eugenio Barsanti.

BASCOM, EARL W. 1906-1995. American cowboy, artist, writer, teacher, rodeo performer, actor, printmaker, painter, sculptor, and inventor: Invented the first side-delivery rodeo chute (1916), the first reverse opening rodeo chute (1919), the first hornless bronc saddle (1922), the first one-hand bareback rigging (1924), and the first high-cut riding chaps (1926), most all of which is still in use today. Bascom is known as the "father of modern rodeo."

BASOV, NIKOLAY. 1922-2001. Russian optical physicist, educator, and inventor, specializing in quantum electrodynamics: Co-inventor of the laser and maser (1950s-1960s), and a cofounder of quantum electronics. Nobel Prize winner.

BATTELLI, FREDERIC. Dates unknown. Swiss(?) medical researcher and inventor: Credited with co-discovering the process of heart defibrillation (1899), a concept that paved the way for the development of modern defibrillators.

BAUDOT, JEAN-MAURICE-ÉMILE. 1845-1903. French telegraph engineer: Invented the Baudot code (1870), a system of current-on or current-off signals (of equal duration) that eventually replaced the Morse Code as the most popular telegraphic alphabet.

BAUMANN, EUGEN. 1846-1896. German chemist, pharmacist, researcher, educator, and inventor: Improved upon (some say co-invented) polyvinyl chloride (1872), or PVC, a synthetic thermoplastic polymer that is now one of the most commonly used plastics in the polymer industry.

BAYLIS, TREVOR G. 1937-2018. English businessman, stunt performer, and inventor: Invented the first hand-powered, wind-up radio (1991), replacing electric (AC) and battery operated radios.

BEAUCHAMP, GEORGE D. 1899-1941. American musical instrument designer, musician, and inventor: Invented the world's first electric guitars (1931)—the most famous of which was his lap steel guitar, the A-22, fondly nicknamed the "Frying Pan." The influence and popularity of Rickenbacker guitars was greatly enhanced throughout the 1960s and 1970s by bands such as the Beatles, the Who, Jefferson Airplane, and Creedence Clearwater Revival, who performed and recorded with Rickenbacker's unique stringed instruments. Beauchamp was the founder of the National String Instrument Corporation (1927), and a co-founder of what would become the Rickenbacher Guitar Company (1931).

BEAUFORT, FRANCIS. 1774-1857. Irish astronomer, geographer, hydrographer, and naval officer: Invented the Beaufort scale (1805), a table that estimates wind strength, and also the Beaufort cipher (circa mid 1800s), a multi-alphabet cryptographic code used to encrypt and decrypt messages.

BECK, CLAUDE. 1894-1971. American surgeon, cardiologist, and inventor: An innovator of heart surgery and heart resuscitation methodologies, he is primarily remembered for inventing an external heart defibrillator (1947). Nobel Prize nominee.

BECK, HANS. 1929-2009. German engineer, furniture maker, toy model enthusiast, and inventor: Invented Playmobil toys (1974), popular, small, smiling, fantasy toy figures with bendable arms, legs, and hips, snap on clothes, and interchangeable parts. Beck is known as the "father of playmobil."

BECKMAN, ARNOLD O. 1900-2004. American chemist, electrical engineer, investor, and inventor: Invented two scientific instruments, the electric pH meter (1934) and the quartz spectrophotometer (1940), both used in chemical analysis.

BECQUEREL, ALEXANDRE-EDMOND. 1820-1891. French scientist, physicist, electrical and optical researcher, and inventor: Credited with inventing the world's first photovoltaic cell (1839)—also known as a solar cell. His discovery of the "photovoltaic effect" laid the groundwork, many years later, for the invention of the modern solar panel.

BEETHOVEN, LUDWIG VAN. 1770-1828. Austrian composer,

musician, and pianist: Although known for his stunning compositions (among them: *Eroica, Symphony No. 9 in D Minor, Piano Sonata No. 14 in C-sharp Minor, Fidelio, Symphony No. 5 in C Minor, Violin Concerto in D Major,* and *Piano Concerto No. 5 in E-flat Major*), he was also a gifted musical innovator, credited with inventing the song cycle and the solo sonata, as well as making structural improvements to the quartet, the concerto, and the symphony. Perhaps above all, he was responsible for initiating the musical era known as the Romantic Period (1820-1900), in which he combined earlier established classical forms with his own more modern, groundbreaking brand of musicianship and composition.

Ludwig van Beethoven.

BEKHTEREV, VLADIMIR M. 1857-1927. Russian neurologist, psychiatrist, physiologist, psychologist, and reflexologist: Made numerous contributions to all five of his occupations, proof being the many medical terms that derive from his name, including: Bekhterev's Band, Bekhterev's Nucleus, Bekhterev's Tract, Bekhterev's Disease, Bekhterev's Symptom, Bekhterev's Reflexes, Bekhterev's Test, and Bekhterev's Sign. He invented the term reflexology and is credited with founding the fields of reflexology and objective psychology. He also proposed Bekhterev's Mixture, a medical sedative that affects the central nervous system, meant to be used on patients with epilepsy. Bekhterev is known as the "father of objective psychology."

BELL, ALEXANDER GRAHAM. 1847-1922. Scottish scientist, electrical engineer, physicist, teacher, entrepreneur, and inventor: Invented the telephone (1870s), the audiometer (1879), the photophone (1880), the metal detector (1881), and the graphophone (1886). He is also credited with inventing a number of techniques that aided in educating the hearing impaired. Bell is known as the "father of the telephone." (Note: Bell's invention of the telephone is still highly disputed. He and his firm, the Bell Company, spent decades in and out of court fighting legal claims against him. In 1887, after 550 court challenges, Bell won and his telephone patent was permanently granted.)

BELL, MICHAEL. Born 1938. American film actor, voice actor, stage actor, teacher, and inventor: Co-invented the "Grayway Rotating Drain" (1991), a device that recycles a home's sink and shower water (known as gray water) for reuse.

BELUŠIĆ, JOSIP. 1847-1905. Croatian physics and mathematics professor, entrepreneur, and inventor: Invented the world's first electric speedometer (1888), known at the time as the Velocimeter.

BENARDOS, NIKOLAY. 1842-1905. Russian Engineer and inventor: Invented carbon arc welding (1881), a fusion technique that uses an electric arc to join metals together by melting them. He is credited with at least 200 other inventions in the fields of electrical engineering and agriculture.

BENDALL, MALCOLM. Birth date unknown, currently living. Australian geochemist and green physics inventor: Invented the "Thunderstorm Generator," a free energy generator that is based on plasmoid fusion technology. Similar to Paul Pantone's Geet Fuel Processor, Bendall's innovation can be retrofitted to any type of combustion engine, and uses a combination of water and gas (or diesel) as fuel to create energy from plasmoids. He is also the inventor of Molton Sea Ark Atomic Reconstruction Technology (MSAART), a set of scientific principals that describe the uses of plasmoids as a way of reducing carbon emissions produced by internal combustion engines. While he has many supporters from around the world, like all "free energy technology" inventors, Bendall has been labeled a "crackpot" and his work has been largely disregarded and suppressed by mainstream science.

BENNETT, JR., WILLIAM R. 1930-2008. American physicist, educator, audiophile, and writer: Co-inventor of the first helium-neon gas laser (1960). Invented a dozen other types of lasers, including aiding in the development of the argon ion laser, used to treat patients with diabetic retinopathy. His work opened the door to many modern technologies, such as the compact disk player and grocery store checkout scanners.

BENZ, KARL. 1844-1929. German automotive engineer, mechanical engineer, engine designer, and inventor: Invented the first practical gas-powered (internal combustion) automobile (1886), as well as numerous devices related to car and engine design. Benz co-founded the company Mercedes-Benz (1926)—today a division of Daimler AG—which incorporates his surname. "Mercedes," Spanish for "mercies," was derived from the name of one of the other co-founder's daughters, Mercedes Jellinek.

1954 Mercedes-Benz 300 SL.

BENZINGER, THEODOR H. 1905-1999. German physician and pilot: Invented the ear thermometer (1964).

BERGER, HANS. 1873-1941. German psychiatrist, psychologist, neuroscientist, neurologist, educator, and inventor: Invented the human electroencephalogram (1924). Called EEG for short, it measures the voltage fluctuations of the human brain. Berger is known as the "father of the electroencephalogram." (Note: Berger's invention was a result of his interest in telepathy.)

BERGIUS, FRIEDRICH. 1884-1949. German chemist and educator: Invented the Bergius Process (1913), a method for converting a solid

form of coal, known as lignite, into synthetic liquid hydrocarbons, such as gasoline, diesel fuel, and jet fuel. Nobel Prize winner.
BERLINER, EMILE. 1851-1929. German electrical engineer, industrialist, entrepreneur, author, and inventor: Invented the phonograph record disk (1887), the disk record gramophone (1887), the phonograph (all three which led to the development of the modern record player), the microphone (used in the first telephone), and acoustical tile for auditoriums, work that has greatly impacted the development of the recording industry. He is also credited with inventing the helicopter, lightweight internal combustion engines, and parquet carpet.
BERNERS-LEE, TIM. Born 1955. English computer scientist, computer programmer, physicist, educator, researcher, speaker, and inventor: Co-inventor of the World Wide Web (1989). He is also credited with inventing HTTP or hypertext transfer protocol (1989), the first Web browser (1990), and HTML or hypertext markup language (1991).
BERTHELOT, MARCELLIN. 1827-1907. French scientist, organic chemist, physical chemist, educator, historian, and politician: Invented Berthelot's reagent (1859), an alkaline solution of hypochlorite and phenol that can measure ammonia and urea, important in detecting disease; made many other contributions to the field of analytical chemistry, such as synthesizing benzene (1851), formic acid (1856), methane (1858), and acetylene (1862). He was also instrumental in the development of aromatic compounds, and was the coiner of the words diglyceride, monoglyceride, and triglyceride.
BERTSCH, HEINRICH. 1897-1981. German chemist: Invented the world's first fully synthetic laundry detergent (1960), which was given the brand name "Fewa."
BEST, CHARLES. 1899-1978. Canadian physiologist, medical researcher, and educator: Co-discoverer of insulin (1921), which is still used as a life-saving treatment for diabetes mellitus.
BIELSCHOWSKY, MAX. 1869-1940. German neurohistologist and neuropathologist: Invented the Bielschowsky Silver Stain Technique (circa 1902), which allows nerve fibers and other tissues in the central nervous system to be visually observed; used in the detection of Alzheimer's disease.
BINET, ALFRED. 1857-1911. French neurologist, psychologist, sociologist, and teacher: Co-inventor of the first IQ test (1904), used in experimental psychology to measure intelligence.
BINI, LUCIO. 1908-1964. Italian psychiatrist and educator: Co-inventor of electroconvulsive therapy or ECT (1937), a form of shock therapy used to treat patients with mental disorders.
BINNIG, GERD. Born 1947. German physicist and inventor: Co-inventor of the STM or scanning tunneling microscope (1981)—which allows atoms to be observed in 3D; also the atomic force

microscope or AFM—which allows high-resolution imaging. Nobel Prize winner.

BIRDSEYE, CLARENCE. 1886-1956. American businessman, entrepreneur, engineer, biologist, taxidermist, naturalist, and inventor: Invented quick or flash freezing (1920s), a method that helps preserve the freshness and taste of food. Also invented infrared heat lamps, a technique for removing water from food, and a non-recoiling harpoon gun for whale hunters, and aided in the further development of grocery store and train refrigeration. Founded Birdseye Seafoods (1924). Birdseye is known as the "father of the frozen foods industry."

BÍRÓ, LÁSZLÓ. 1899-1985. Hungarian surrealist painter and inventor: Based on designs by earlier inventors, he developed the first commercially viable ballpoint pen (1931), a writing implement known as the "Biro" in Britain and the "Birome" in Argentina. Bíró co-founded the company Bíró Pens of Argentina (1943). The company's pen patent was eventually purchased by French businessman Marcel Bich, who founded the Bic company (1945). Its first product, the Bic Cristal pen, went on sale to wide acclaim (1950).

BJØRKLUND, THOR. 1889-1975. Norwegian businessman, carpenter, and inventor: Invented the cheese slicer (1925).

BLACKTON, J. STUART. 1875-1941. English cinematographer, film director, screenwriter, film producer, film editor, businessman, actor, animator, screenwriter, artist, and inventor: Introduced animation to film, aiding in the development of the cinematic art form. Founded the Vitagraph Company, which released its first film in 1898. Blackton is known as the "father of American movie animation."

BLATHY, OTTO. 1860-1939. Hungarian mechanical engineer and inventor: Co-inventor of the modern transformer (1885), the tension regulator, the AC watt-hour meter, a motor capacitor for an electric motor, and the turbogenerator. Coined the word transformer.

BLENKINSOP, JOHN. 1783-1831. English mine inspector, mechanical engineer, and inventor: Invented the first practical and commercially successful railway (steam) locomotive (1811), as well as the rack railway (1811)—which is known as the Blenkinsop Rack Railway System.

BLISS, CHARLES K. 1897-1985. Ukrainian chemical engineer, semiotician, and inventor: Invented Blissymbolics (1949), an ideographic writing system that he called Semantography. Though originally intended to be a "universal language," Blissymbols are today used mainly by those with learning disabilities and communication issues.

BLUMLEIN, ALAN D. 1903-1942. English electronics engineer and inventor: Invented stereophonic sound (circa 1930s), the stereo microphone, the panning technique, the transversal filter, and acetate disks. Contributed to the development of numerous others inventions related to electronics, sound, and recording, such as sync-pulse separators, stabilized HT circuits, linear ramp generators, the Emitron (the first electronic camera tube), the 405-line television waveform, and audio, television, and radar electronics. Blumlein is known as the "father of stereophony."

BOCOUR, LEONARD. 1910-1993. American painter, lecturer, art collector, and paintmaker: Co-invented acrylic paint (circa 1946). Made of pigment suspended in acrylic polymer emulsion, it forever altered the art world by offering an inexpensive, quick-drying, water-resistant paint that can be used on almost any surface.

BOEHM, THEOBALD. 1794-1881. German composer, flautist, musician, and inventor: Credited with inventing the modern flute (1848), as well as the Boem System (1847)—an advancement in flute fingering technique.

BOGGS, DAVID R. 1950-2022. American electrical engineer, radio engineer, and computer scientist: Co-invented the Ethernet (1973), a network technology that paved the way for email and Internet "hot spots." His contributions to computer science also led to developments related to the computer, the mouse, word processing, laser printing, and the graphic user interface (GUI).

BOHLIN, NILS. 1920-2002. Swedish mechanical engineer, safety engineer, and inventor: Invented the three-point lap-shoulder automobile seatbelt (1959). The device, which has both delighted and frustrated drivers for 65 years, is still in use today and has saved countless millions of lives.

BOHNENBERGER, JOHANN. 1765-1831. German physicist, mathematician, astronomer, and educator: Invented the gyroscope (1812), a device, using a spinning wheel or disk attached to a gimbal, that measures and/or maintains rotational motion, that is, angular velocity and orientation. The gyroscope has numerous applications and is today used in various industries, including oil, electronics, surveying, navigation, photography, robotics, weaponry, medicine, and science, to name but a few.

BOHR, NIELS. 1885-1962. Danish physicist, researcher, educator, and inventor: He is perhaps best known for his invention of the Bohr Model (1913), a theory proposing that electrons are similar to miniature planets orbiting the sun; in this case, orbiting at fixed levels around an atom's nucleus. The Bohr Model opened the door to the further evolution of quantum physics while expanding our comprehension of atomic physics. Nobel Prize winner.

Niels Bohr.

BOIES, STEPHEN. Dates unknown. Nationality unknown. IBM computer scientist and inventor: Invented voicemail (1973); or, some say, led the team that invented voicemail. Though his innovation, which he called the "Audio Distribution System" or ADS, was originally used mainly inside IBM, it quickly caught on with the outside world, forever altering global communications—particularly corporate communications.

BOMBARDIER, JOSEPH-ARMAND. 1907-1964. Canadian entrepreneur and inventor: The designer and inventor of numerous all-terrain vehicles, his most successful among these were the snowmobile (1935) and the Ski-doo (1959).

BOND, ELIJAH. 1847-1921. American attorney and inventor: Credited with creating the planchette that is used with the Ouija Board (1891), the popular Victorian Spiritualist parlor game.

BONONIENSIS, HIERONYMUS. Lived 1500s. Italian keyboard maker: Invented the harpsichord (circa 1521), an early keyboard instrument and precursor to the pianoforte (piano) that uses a bird feather quill (plectrum) attached to a non-responsive key, whose action, when pressed, plucks a metal string, producing a metallic yet sonorous sound.

BOOTH, CHARLIE. 1903-2008. Australian athlete and inventor: Invented starting blocks (1929), used today in sprinting races.

BORKENSTEIN, ROBERT FRANK. 1912-2002. American photography scientist, researcher, forensic scientist, police officer, educator, and inventor: While he did not invent the alcohol meter, he greatly improved upon it, eventually developing the brand that is today most commonly used by police around the world: the Breathalyzer (1954). His innovation has prevented countless accidents and saved untold thousands of lives.

BORN, BOB. 1924-2023. American businessman, candy maker, investor, confectionary scientist, and inventor: Invented a machine that mechanized the production of marshmallows (brand name "Peeps"), which led to the development of today's marshmallow depositing systems. He also invented "Hot Tamales" (1950), a hot cinnamon-flavored confectionary candy. Though he did not invent "Peeps" (created by an earlier company in 1925), Born is known as the "father of Peeps" for his role in refining and modernizing their production. Son of Sam Born (below).

BORN, SAM. 1891-1959. Russian (Ukrainian) businessman, entrepreneur, candy maker, chocolatier, and inventor: Invented a machine that mechanically inserts sticks in one end of a lollipop (1916), thus creating the first true automated lollipop-making device. Founded the Just Born company (1923), a candy manufacturing business still in production.

BOSE, JAGDISH CHANDRA. 1858-1937. Indian scientist, physicist, plant physiologist, and writer: Invented the crescograph (circa 1928), an instrument that measures plant growth and detects the responses of plants to various stimuli, such as light, toxins, and plant food. He also contributed greatly to the technological development of radio science (wireless telegraphy), including inventing a refined coherer (detects radio waves) and a machine that could send and receive microwave signals. He was also the author of several science fiction stories, and is

known as the "father of Bengali science fiction."
BOULTON, MATTHEW PIERS WATT. 1820-1894. English scientist, inventor, and member of the British Metaphysical Society: Invented the aileron (1868), a hinged flap located on the back edge of an airplane wing that is used to control lateral (sideways) balance. He also worked on inventions related to aerial propulsion and locomotion, propeller blades, generators, projectiles, and boilers.
BOURGEOIS, MARIN LE. Circa 1550-1634. French gunsmith and inventor: Credited with inventing the flintlock gun mechanism (circa 1610), which replaced the more primitive wheellock gun mechanism. The flintlock, in turn, was obsolesed by the percussion lock gun mechanism during the Victorian Period.
BOYDEN, SETH. 1788-1870. American metallurgist, engraver, leather worker, horticulturist, and inventor: Invented malleable cast iron, a special cut-off valve (for steam engines), a leather hide-splitting machine, a nail-making machine, and the first American daguerreotype camera; he also made numerous improvements on the steam locomotive. His work influenced the development of the modern steel industry. Like many other male inventors he was self-taught.
BOYER, HERBERT W. Born 1936. American biotechnologist, researcher, and entrepreneur: Co-discover of recombinant DNA (1973), a gene-splicing technology that aided in the development of the genetic engineering and biotechnology industries, and that led to the emergence of the process known as genetic modification; this in turn resulted in the bioengineered product known as genetically modified organisms (GMO), which includes plants and animals.
BOYLE, WILLARD. 1924-2011. Canadian physicist, photographer, and inventor: Co-inventor of the charge-coupled device (1969), or CCD, a light-sensitive integrated circuit—arranged in a two-dimensional array—that carries electrically charged signals, allowing images to be captured by converting photons to electrons. An important development in the field of digital imaging. Commonly used in, for example, digital cameras, microscopes, bar code readers, scanners, and medical equipment. Nobel Prize winner.
BOZZINI, PHILIPP. 1773-1809. German-Italian physician: Invented the first endoscope (1806), a medical device he used to diagnose his patients by peering inside various bodily openings. His invention, which he called the Lichtleiter ("Lightguide"), led to the creation of the field of endoscopy.
BRADNER, HUGH. 1915-2008. American physicist, educator, and inventor: Invented the neoprene wetsuit (circa 1951), used by scuba divers, surfers, swimmers, kayakers, snorklers, and triathletes. Its insulating ability to regulate body temperature by retaining body heat—thereby avoiding hypothermia—was a major development in the world of water sports.
BRAID, JAMES. 1795-1860. Scottish physician, surgeon, hypnotist,

researcher, author, and philosopher: Although he did not discover or invent hypnotism, he is credited with being the developer of modern hypnotism (1841), hence his title, the "father of modern hypnotism."

BRAILLE, LOUIS. 1809-1852. French educator, organist, and inventor: Inspired by the earlier work of Charles Barbier, this blind pioneer invented a system of writing and printing known as Braille (1821 or 1824). Created for the visually impaired, it consists of raised dots, placed in various configurations, that can be read with the fingers. Braille is also used in musical notation. Braille's innovation opened up a new world of possibilities for the sight-challenged.

Louis Braille.

BRAIN, ARCHIE. Born 1942. English anaesthetist: Invented the laryngeal mask (1981), a supraglottic airway device used to maintain a temporary open pathway during the administration of anesthesia in patients with a blocked airway.

BRANDENBERGER, JACQUES E. 1872-1954. Swiss chemist and textile engineer: Invented cellophane by creating a machine that produced a continuous production of a strong, clear, flexible, protective, waterproof, packaging film (1908). He named it cellophane—a combination of the words cellulose and diaphane (the French word for "transparent").

BRANDENBURG, KARLHEINZ. Born 1954. German computer scientist, electrical engineer, mathematician, and inventor: Credited with inventing (some say co-inventing) MPEG-1 Audio Layer 3 (1991). Better known as MP3, it is a technology that compresses digital audio, which, by decreasing the size of music files, makes them easier to play, store, transmit, and retrieve.

BRANLY, ÉDOUARD. 1844-1940. French physicist, educator, and inventor: Invented the coherer (circa 1890), a device used to detect radio waves, a milestone in early radio communication, which began during Victorian times in what is known as the wireless telegraphy age.

BRANNOCK, CHARLES F. 1903-1992. American businessman and inventor: Invented the Brannock Device (1925). The metal instrument, used to measure feet and determine shoe size, revolutionized the shoe industry.

BRATTAIN, WALTER HOUSER. 1902-1987. American research, experimental, and solid state physicist: Co-inventor of the transistor (1947), which did away with the more cumbersome vacuum tubes. The transistor, a tiny semiconducting device used for controlling and generating electrical signals, is an integral component of the microchip, and to this day serves as the foundation of the computer-information age. Nobel Prize winner.

BRAUN, KARL FERDINAND. 1850-1918. German physicist, electrical engineer, and inventor: Invented the cathode-ray tube or CRT (1897), better known as the cathode-ray oscilloscope—and referred to as the "Braun Tube" in Europe. He also invented the first semiconductor (1874), the first crystal rectifier (1874), and the first phased array antenna (1905). These devices paved the way for the development of radio, television, computer, and communication technologies. Braun is known as the "father of television." Nobel Prize winner.

BRAUN, WERNHER VON. 1912-1977. German physicist, aerospace engineer, architect, and inventor: Invented the V-2 rocket (1936) and the Saturn V rocket (1960s), both which led to advancements in warfare and space exploration.

BREBERA, STANISLAV. 1925-2012. Czechoslovakian chemist: Invented Semtex explosive (circa 1950s), a general purpose malleable, odorless, plastic explosive used chiefly for commercial demolition, mining, and military applications.

BREWSTER, DAVID. 1781-1868. Scottish physicist, mathematician, writer, and inventor: Invented the kaleidoscope (1816) and the stereoscope (1838). Was one of the founders of the British Association for the Advancement of Science (BAAS) in 1831.

BRIN, SERGEY. Born 1973. Russian computer scientist, entrepreneur, and inventor: Co-invented the Google Web Search Engine (1998), a globally popular Internet browser that is accessed 100,000 times a second, 8.5 billion times a day, 2 trillion times a year.

BROOKS, CHARLES B. 1865-1908. American inventor: Invented the first street sweeping truck (1896).

BROOKS, JAMES LAWRENCE. Born 1940. American producer, director, and screenwriter: Co-created the hit TV series *The Mary Tyler Moore Show* (1970).

BROWN II, NICHOLAS. 1769-1841. American businessman, entrepreneur, and philanthropist: His family donated the land for Rhode Island College (1770). After Nicholas, at the time the college treasurer, donated $5,000 to the school (the equivalent of about $133,000 today), it was renamed Brown University in his honor (1804).

BROWN, THOMAS TOWNSEND. 1905-1985. American physicist, electrogravitics researcher, naval officer, UFO researcher, author, and inventor: Invented the Townsend Anti-gravity Machine (1928). Also known as the "Gravitor," Brown claimed that his device could produce thrust, and thus defy gravity, via the influence of electrical fields—a phenomenon later termed the Biefeld–Brown Effect. Despite his assertions and public demonstrations, inevitably Brown was placed in the same box as many other independent thinkers. Pronounced a trickster by mainstream science, his work was ignored, suppressed, and finally dismissed as nothing more than a result of electrohydrodynamics. His many advocates, however, remain steadfast in their support of Brown. Founder of Rand International Limited (1958) and a co-founder of the

National Investigations Committee On Aerial Phenomena or NICAP (1956).

BROWN, WILLIAM C. 1916-1999. American electrical engineer, microwave technologist, and inventor: Invented the Amplitron (circa 1953), also called the cross field amplifier. The tube, which amplified microwave frequencies, found uses in the military and in space exploration. Brown is known as the "father of microwave power transmission."

BROWNING, JOHN MOSES. 1855-1926. American businessman, entrepreneur, firearms manufacturer, gunsmith, and inventor: Co-founder of the Browning Arms Company (1878). A designer and manufacturer of a wide variety of firearms and ammunition, the company has greatly impacted firearms technology (including pistols, rifles, shotguns, and machine guns) into the present day. One of its most popular models is the 9mm Browning Hi-Power Pistol, Model 1935, P-35.

John Moses Browning.

BRUDZIŃSKI, JÓZEF. 1874-1917. Polish scientist, doctor, pediatrician, neurologist, bacteriologist, and politician: Due to his description of and studies on the bacterium Bacillus lactis aërogenes, he is considered by many (disputed by some) as the original discoverer of the science of probiotics (circa 1895). As a result Brudziński is known as the "father of probiotics."

BRUHN, FRIEDRICH WILHELM GUSTAV. 1853-1927. German inventor: Invented the modern taximeter (1891), a device used in taxicabs that automatically computes the distance covered and the fare owed.

BRUNEL, MARC ISAMBARD. 1769-1849. French civil engineer and inventor: Invented the tunneling shield (circa 1818), a protective covering used with boring machines that allows safe excavation in unstable soil, such as water-bearing strata.

BRUSENTSOV, NIKOLAY. 1925-2014. Russian computer scientist: Invented the first modern electronic ternary computer (1958), known as the "Setun."

BUCK, DUDLEY ALLEN. 1927-1959. American scientist, electrical engineer, educator, and inventor: Invented ferroelectric memory (circa 1952), the Cryotron (1954), and content addressed memory (1955), as well as contributing greatly to developments in electron beam lithography.

BUDDING, EDWIN BEARD. 1796-1846. English engineer and inventor: Invented the world's first lawnmower (1830).

BUDKER, GERSH. 1918-1977. Russian physicist, nuclear physicist, educator, and inventor: Invented electron and electron-positron accelerators based on a new means of colliding beams (1965-1967); also invented the VEP-1 electron-electron collider (1950s-1960s), as well as a method that uses electrons to cool heavier particles (1966).

BULL, EDWARD. 1759-1798. English engineer and inventor: Invented the "Bull Engine" (circa 1791), an improved steam engine used principally in mines during the late 1700s.

BUNSEN, ROBERT. 1811-1899. German physicist, chemist, educator, author, and inventor: Invented (or at least contributed to the development and refinement of) the eponymous "Bunsen Burner" (1855). He also invented the Bunsen Battery (1841), the grease-spot photometer (1844), the filter pump (1868), the ice calorimeter (1870), and the vapor calorimeter (1887), and was the discoverer of two important alkali groups: cesium and rubidium.

BURDEN, HENRY. 1791-1871. Scottish engineer, businessman, and inventor: Invented a spike-making machine (1825), the first usable hook-shaped iron railroad spike (1843), and a horseshoe making-machine (1843).

BURDEN, WILLIAM DOUGLAS. 1898-1978. American naturalist, filmmaker, explorer, and author: Co-founder of Marineland (1937), the first location which opened in what is now the town of Marineland, Florida, just south of St. Augustine.

BUREAU, ROBERT. Dates unknown. Nationality unknown. Meteorologist and inventor: Invented the first known radiosonde (1929), a short-wave telemetric radio attached to a weather balloon that transmits atmospheric data (such as temperature, pressure, and humidity) back to a ground receiver.

BURNS, ALLAN P. 1935-2021. American producer and screenwriter: Co-created the hit TV series *The Mary Tyler Moore Show* (1970).

BURR, SR., AARON. 1716-1757. American clergyman and educator: Co-founder of the College of New Jersey (1746), later renamed Princeton University (1896). Burr served as its second president (1748-1757), setting up its study courses and prerequisites for entrance. His son, Aaron Burr, Jr., served as the third vice president of the United States.

BUSA, ROBERTO. 1946–1970. Italian educator, philosopher, clergyman, and innovator: The inventor of the first ebook ("electronic book") is contentiously debated. Arguably the most logical choice, however, would be Busa, for it was he who created the first electronic textual work: the *Index Thomisticus* (started in 1946), a digitized research tool for studying the extensive literary works of Thomas Aquinas, the noted Italian theologian, philosopher, and author. Aided by IBM, the project was completed in 1970 and could be stored on and read from a CD-ROM.

BUSCHMANN, CHRISTIAN FRIEDRICH LUDWIG. 1805-1864. German businessman, entrepreneur, musical instrument designer and producer, musician, and inventor: Credited by many (debated by others) with inventing both the world's first accordion (1822), and the world's

first harmonica (1821), a "mouth organ" now popular with blues, jazz, folk, rock, and country music groups and artists.
BUTTS, ALFRED MOSHER. 1899-1993. American architect and artist: Invented the popular board game Scrabble (1938).
BYUNG-CHULL, LEE. 1910-1987. South Korean businessman and entrepreneur: Founded the Samsung Group (1938), more popularly known as Samsung.

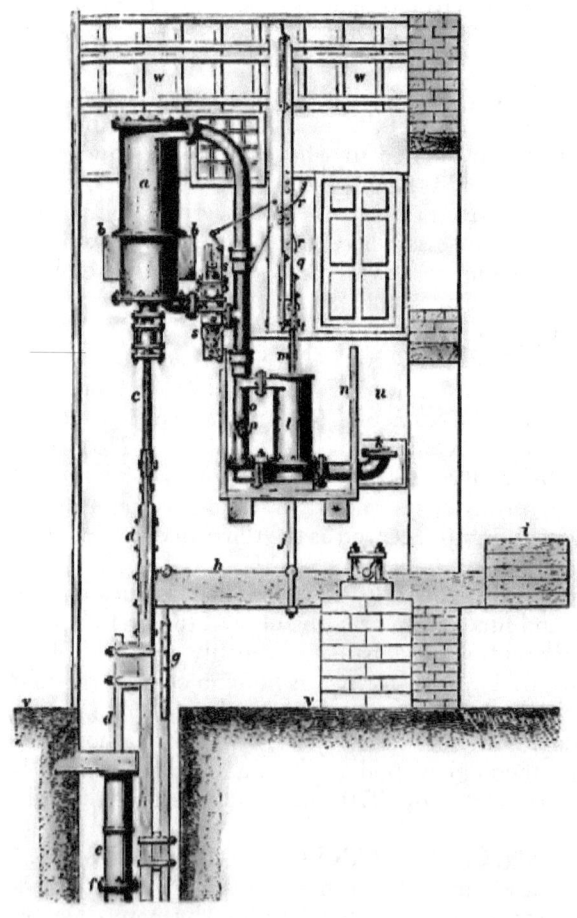

Schematic of Edward Bull's Cornish Engine, 1798.

CAHANA, AVIAD. Dates unknown. American scuba instructor and engineer: Invented the Avelo Dive System, revolutionary neutrally buoyant underwater diving gear that allows for easy-to-maintain buoyancy, while enhancing air consumption, increasing air capacity, and reducing cumbersome equipment and overall weight. Founder of the company xF Technologies, Inc. (2007).

CAILLIAU, ROBERT. Born 1947. Belgian computer scientist, computer programmer, and author: Co-invented the World Wide Web (1989).

CALAHAN, EDWARD A. 1838-1912. American inventor: Invented the stock price ticker system (1863) and the telegraph stock ticker tape (1867).

CALLAN, NICHOLAS J. 1799-1864. Irish physicist, electrical scientist, and Catholic cleric: Invented the induction coil, a device that produces high voltage currents.

CAMP, WALTER CHAUNCEY. 1859-1925. American coach, athlete, sports writer, and football player: Invented the American game of football (1869)—not to be confused with the non-American game of soccer, also known as football. Intended to be an alternative to the much more rigorous game of rugby, Camp not only set up and defined the sport of football, he was also largely responsible for structuring its rules and parameters; he was the first to write a book on the

Walter C. Camp.

topic as well. Camp is known as the "father of American football."

CAMPAGNOLO, TULLIO. 1901-1983. Italian bicycle parts manufacturer, racing cyclist, entrepreneur, and inventor: Invented the quick release rear hub (1930), the bicycle derailleur (1930s), and over 100 other designs and products that were instrumental in improving bicycle performance.

CAMPBELL-SWINTON, ALAN ARCHIBALD. 1863-1930. Scottish electrical engineer, photographer, and inventor: Created the first photograph produced by X-rays in England (1890s), and assisted in the development of the steam turbine, as well as the construction of the steam turbine torpedo boat, the *Turbinia* (1890s). Also recommended using cathode ray tubes for broadcast television (1908), but did not patent the idea.

CANTOR, CHARLES R. Born 1942. American molecular geneticist: Co-invented pulsed-field gel electrophoresis (1984), or PFGE, a genotype technique that separates and sizes complex mixtures of large DNA molecules by applying an electric field; used for molecular typing of bacteria, and also for studying various microorganisms in soil and water.

CAPECCHI, MARIO R. Born 1937. Italian molecular geneticist, biologist, researcher, educator, and inventor: Co-developed gene targeting (early 1980s), which is used to prompt the DNA of a cell to repair or correct itself. Nobel Prize winner.

CAPROTTI, ARTURO. 1881-1938. Italian automobile engineer, architect, and inventor: Invented the Caprotti Valve Gear (circa 1915). By using poppet valves rather than piston valves, this device marked an improvement in steam locomotive engine design and performance.

CARDANO, GEROLAMO. 1501-1576. Italian polymath, physician, physicist, biologist, mathematician, astrologer, philosopher, chemist, and inventor: Invented the Cardan Grille (1550), a method of secret writing employed in steganography. An advancement in the field of cryptography, the Cardan Grille consisted of a piece of paper with rectangle cutouts in which a secret message is written. After removing the grille, the rest of the message is penned in with additional words, which when complete, looks like an ordinary communication. Only someone with the same cutout could read the embedded "secret" message—hidden in plain sight.

CARDEW, PHILIP H. 1851-1910. English electrical engineer and military officer: Invented the hot-wire galvanometer (late 1800s), and telegraphic vibratory transmitter, and helped standardize electrical units across the UK during the early days of the electrical revolution.

CARNEGIE, ANDREW. 1835-1919. Scottish businessman, entrepreneur, industrialist, and philanthropist: Founded the Carnegie Technical Schools (1900), which later merged with the Mellon Institute of Industrial Research, becoming Carnegie Mellon University (1967). Carnegie was also the founder of the Carnegie Institution for Science

(1902), the Carnegie Endowment for Peace (1910), and the Carnegie Corporation of New York (1911), among numerous other philanthropical endeavors. Carnegie was a co-founder of U.S. Steel (1901), and funded and constructed New York City's Carnegie Hall, which was also named after him.

CARLSON, CHESTER F. 1906-1968. American physicist, patent attorney, businessman, and inventor: Invented xerography, or more specifically, the xerographic copier (1938). The xerographic process, or what the inventor called "electrophotography," produces text and images on paper, a well-known dry photocopying technology still in use to this day.

CAROTHERS, WALLACE. 1896-1937. American chemist, engineer, and inventor: Co-invented neoprene (1930), a synthetic rubber, and nylon (1935), the first fully synthetic fiber. The former product revolutionized the sports, medical, and gardening industries, the latter the fashion and electronic industries.

CARPANO, ANTONIO BENEDETTO. 1764-1815. Italian vintner and herbalist: Invented Vermouth (1786), an aromatized wine fortified with various spices, herbs, roots, flowers, and seeds.

CARTWRIGHT, ALEXANDER JOY. 1820-1892. American sportsman: Founder of the New York Knickerbockers Base Ball Club (1845). Though often disputed, Cartwright is also credited with being the founder of baseball itself, or at least the chief developer of the game, for he was the lead architect in setting down what are known as the "Knickerbocker Rules" (1845). These regulations formalized the first baseball rule book, which in turn led to the establishment of the modern game as we know it. Despite these facts, the actual original inventor of baseball is destined to forever remain a mystery, as early forms of the game were already being played in 18th-Century England, with British literary references appearing as early as 1744. Still others maintain that the modern version of the sport originated in Canada, where the first known legitimate game took place in 1838.

Alexander J. Cartwright.

CARVER, GEORGE WASHINGTON. 1864-1943. American agricultural scientist, botanist, educator, environmentalist, painter, and inventor: Mainly known for his work in the fields of farming, agricultural research, crop-cultivation techniques, and soil science (pedology and edaphology), he is also credited with developing educational programs and bulletins intended to aid farmers. Contrary to popular opinion, Carver did not invent peanut butter. However, he was responsible for developing new recipes, as well as new applications, for various crops, in particular, peanuts, potatoes, and soybeans. His push for both wholistic land management and environmental practices, such as organic farming, crop rotation, and the protection of forests and top soil (now known as permaculture), earned the devout Christian numerous honors, awards, and world renown.

CASELLI, GIOVANNI. 1815-1891. Italian physicist, cleric, and inventor: Invented the pantelegraph (circa 1859). A forerunner of the fax machine, this facsimile device was able to send copies of images over telegraph lines.

CAYLEY, GEORGE. 1773-1857. English polymath, aeronautics engineer, aerial navigator, and inventor: Invented the first flyable glider model (1804) and the first full-scale glider or sailplane (1853). He also invented the hot-air or expansion-air engine (1805), the caterpillar tractor (1825), and the light-tension wheel (circa early 1850s)—a precursor to the modern bicycle wheel. A co-founder of the Regent Street Polytechnic Institution (1838), now the Royal Polytechnic Institution, he also made research advances in the fields of electricity, optics, lifeboats, railway equipment, ballistics, acoustics, education, and land reclamation.

CELSIUS, ANDERS. 1701-1744. Swedish meteorologist, physicist, mathematician, chemist, educator, astronomer, and inventor: Invented the Celsius temperature scale (1742), which measures temperature using the freezing point ($0°$) and the boiling point ($100°$) of water—a 100 degree interval for which it is also sometimes called the centigrade scale. The Celsius system is used by most countries (except the U.S.A.) and a majority of scientists around the world—usually in conjunction with the metric measuring system.

CERF, VINT. Born 1943. American computer scientist, engineer, and researcher: Co-invented Internet protocol, or IP (1974), and Transmission Control Protocol, or TCP (1974)—also known as TCP/IP. Cerf has been called one of the "fathers of the Internet."

CERLETTI, UGO. 1877-1963. Italian neuroscientist, psychiatrist, neurologist, educator, and inventor: Co-inventor of electroconvulsive therapy or ECT (1937), a form of shock therapy used to treat patients with mental disorders.

CHADWICK, JAMES. 1891-1974. English scientist, physicist, educator, and researcher: Discovered the neutron (1932), a subatomic particle which, alongside the electron and the proton, makes up one of the three foundational components of all matter. Its discovery led to world-altering developments in the field of nuclear physics, including the creation of nuclear power as well as the atomic bomb. Nobel Prize winner.

CHAMBERLAND, CHARLES. 1851-1908. French microbiologist: Invented the Chamberland Filter (1884). Also known as the Pasteur-Chamberland Filter, it functioned as a bacterial filter, purifying water for medical and industrial purposes.

CHANDLER, JAMES. Dates unknown. Nationality unknown.

Occupation(s) unknown: Credited by some with having co-invented the electronic or digital thermometer (1970).
CHANG, MIN CHUEH. 1908-1991. Chinese scientist and reproduction biologist: A pioneer in the field of in vitro fertilization (IF), he co-invented the oral contraceptive pill (circa 1950s). Better known as "the pill," it revolutionized the birth control industry and greatly contributed to the explosion of the "free love" movement of the 1960s.
CHANG, THOMAS. Born 1933. Chinese physiologist, physician, educator, and inventor: Invented the artificial cell (1957), engineered synthetic particles that mimic biological functions, and which are used in the fields of biology, biotechnology, and various medical therapies.
CHAPMAN, EMMETT. 1936-2021. American jazz musician, songwriter, guitarist, luthier, and inventor: Invented the Chapman Stick (circa late 1960s-early 1970s), an electric, polychordal, multi-string, hybrid guitar-like instrument that allows a musician to play guitar, bass, chords, and melodies, all on the same device.
CHAPMAN, GRAHAM. 1941-1989. English actor, writer, and comedian: Co-created and acted in the hit TV series and comedy troupe *Monty Python's Flying Circus* (1969).
CHAPPE, CLAUDE. 1763-1805. French engineer, cleric, and inventor: Invented the semaphore visual telegraph (1792), a communications technique using relay towers (constructed 5-20 miles apart), telescopes, a code book, and semaphore rigs—posts with moveable or pivoting wooden arms (balanced with counterweights) that spelled out letters according to their positions.
CHARLES, GLEN GERALD. Born 1943. American producer and screenwriter: Co-created the hit TV series *Cheers* (1982).
CHARLES, LES. Born 1948. American producer and screenwriter: Co-created the hit TV series *Cheers* (1982).
CHASE, DAVID H. Born 1945. American producer, director, and writer: Created the hit TV series *The Sopranos* (1999).
CHAUM, DAVID. Born 1955. American mathematician, computer scientist, cryptographer, and inventor: Invented the idea of blind signatures (1982), a type of early digital signature, and more importantly digital cash or ecash (1983). He has also led innovations in mix nets, multiparty computation, and secure voting systems. Chaum is known as the "godfather of cryptocurrency."
CHELOMEY, VLADIMIR. 1914-1984. Russian scientist, rocket engineer, and aerospace designer: Invented the first cruise missiles (1944), the first ICBMs, or intercontinental ballistic missiles (circa 1958), and the Transport-Supply Ship, or TKS (circa 1960s), as well as the first military space station, called Almaz (circa 1964)—though launched under the name Salyut. He also led research and development of military satellites, launch vehicles, and antiballistic missiles.
CHERENKOV, PAVEL. 1904-1990. Russian physicist and inventor: Discover and interpreter of the Cherenkov Effect (1934), which has

aided advancements in medicine, biology, astrophysics, particle physics, and nuclear reactors. The Cherenkov Effect is described by the U.S. government as a phenomenon in which "electrically charged particles, such as protons or electrons, travel faster than light in a clear medium like water. When this happens, the water molecules and particles interact to give off light. Light slows down to 75 percent of its normal speed when it travels through water. This allows the particles emitted from nuclear fuel to move faster than light in water." Nobel Prize winner.

CHERTOVSKY, EVGENIY. 1902-1961. Russian aviation engineer, design engineer, and inventor: Invented the first full pressure suit (1931), which he nicknamed the "Skafandr" (an artificial term, from the Greek words for "ship" and "man").

CHEVROLET, ARTHUR. 1884-1946. Swiss businessman, entrepreneur, race car driver, and inventor: Co-founder, with his brother Louis-Joseph (below), of the Chevrolet Motor Car Company (1911). A Chevrolet vehicle is best known today as a "Chevy."

Chevrolet automobile, circa 1920.

CHEVROLET, LOUIS-JOSEPH. 1878-1941. Swiss mechanic, businessman, entrepreneur, race car driver, and inventor: Co-founder, with his brother Arthur Chevrolet (above), of the Chevrolet Motor Car Company (1911). A Chevrolet vehicle is more familiarly known today as a "Chevy."

CHRISTENSEN, WARD. Born 1945. American computer hobbyist: Co-invented the first online computer bulletin board service (1978).

CHRISTIANSEN, OLE KIRK. 1891-1958. Danish carpenter, entrepreneur, designer, and engineer: Invented the popular building block toy known as Lego, a name deriving from the Danish words *leg godt*, meaning "play well." The idea began to take shape in 1934, with the Lego brick, so well-known today, getting its start in 1949.

CHRISTIE, SAMUEL HUNTER. 1784-1865. English physicist, mathematician, astronomer, and educator: Created the Wheatstone Bridge (1833), an electrical circuit made up of four resistors and a voltmeter. The device is used to procure precise measurements of resistance, and is often employed in such fields as wind tunnel instrumentation.

CIERVA, JUAN DE LA. 1895-1936. Spanish aerospace engineer, civil engineer, aircraft pilot, and inventor: Invented the first autogyro (1923), a type of rotorcraft that was created as an alternative to the helicopter. Also known as the autogiro, gyrocopter, or gyroplane, it possesses an unpowered, horizontally rotating propeller on top (which rotates freely in the slipstream) that provides lift, with forward movement being supplied by a conventional mounted engine and propeller. Widely used during World War II.

CLAGGET, CHARLES. 1740-1795. Irish composer, musician, and inventor: Invented the teliochordon (a type of pianforte), the chromatic trumpet, the chromatic French horn, and the aiuton, a perpetually tuned metallic organ (all circa 1788). He was also responsible for numerous improvements on a variety of musically related items (mainly concerned with tuning), such as an adjustable fingerboard for violins, the Royal Teleochordan Stop for the harpischord, refinements on the tuning fork, glass keys (for keyboard instruments), and the valve system for the early version of the trumpet.

CLARK, EUGENE BRADLEY. 1873-1942. American mechanical engineer, businessman, and inventor: Invented the "Tructractor," the precursor to the modern forklift truck (1917). Founder and managing director of the Clark Equipment Company (1916).

CLARK, H. FRED. 1937-2012. American medical scientist, veterinarian, and educator: Co-invented the rotavirus vaccine (1998), a pentavalent vaccine for the highly contagious pathogen that causes gastrointestinal infections and inflammation of the stomach and intestines.

CLARK, LELAND. 1918-2005. American biochemist and inventor: Invented the Clark Electrode (circa 1953-1956), a device used to measure a liquid or gas for its oxygen concentration.

CLARKE, ARTHUR C. 1917-2008. English scientist, physicist, science fiction writer, screenwriter, actor, presenter, explorer, and inventor: First to conceive of the communications satellite, which would use the corresponding idea of the geostationary satellite (1945), a manmade object synchronized with the earth's rotational speed, thereby remaining in fixed orbit—from where it could relay radio signals. Clarke's idea, which has greatly advanced both satellite technology and communications technology, is still in use, with hundreds of geostationary satellites currently in orbit around our planet.

CLAUDE, GEORGES. 1870-1960. French physicist, chemist, electrical engineer, and inventor: Invented the neon light (1910), a small glass gas discharge lamp containing neon gas at low pressure, which produces luminescence when electric light is passed through the gas.

CLEESE, JOHN. Born 1939. English actor, producer, writer, and comedian: Co-created and acted in the hit TV series and comedy troupe *Monty Python's Flying Circus* (1969).

CLÚA, JAIME FERRÁN Y. 1851-1929. Spanish doctor and bacteriologist: Developed the first vaccine for cholera (1885).

COANDĂ, HENRI-MARIE. 1886-1972. Romanian physicist, aerospace engineer, and inventor: Chiefly remembered for discovering the Coandă Effect (1932), in which a vacuum is created when a jet stream follows a curve. Used in air flow amplification technology pertaining to, for example, the field of aerodynamics, as well as numerous other compressed air applications and industries.

COCKERELL, CHRISTOPHER. 1910-1999. English engineer and inventor: Invented the Hovercraft (1954), an air-cushion vehicle or ACV that utilizes large fans or blowers to create a "pillow" of air beneath it, allowing it to move swiftly and efficiently over land, water, ice, and mud. Commonly used in such fields as sports, ferrying, water navigation, and military.

COFFEY, AENEAS. 1780-1852. Irish engineer, surveyor, exciseman, distiller, and inventor: Invented the Coffey Still (1830), a continuous still design that greatly improved upon the traditional batch distiller while aiding in the evolution of distilling science.

COIGNET, FRANÇOIS. 1814-1888. French builder, businessman, entrepreneur, and inventor: Credited by some with inventing the idea of reinforced concrete (circa 1853). Better known as rebar (short for "reinforcing bar"), it was a mesh of iron wire (today, of steel rods) that is embedded in wet concrete to fortify its tensile strength, thereby greatly enhancing its load-bearing capacity. The rebar system, which is used primarily to strengthen and stabilize concrete foundations, columns, arches, pipes, walls, girders, floors, and beams, greatly advanced the concrete construction industry, permitting the erection of heavier, wider, taller, and more durable structures than ever before (especially beneficial in earthquake-prone and high wind areas). Today's massive modern metropolises, with their huge skyscrapers and expansive bridges, owe their very existence largely to the invention of rebar.

COLE, HENRY. 1808-1882. English civil servant, author, designer, and inventor: Invented the Christmas card (1843); also founded London's Museum of Ornamental Art and the South Kensington Museum—from which evolved today's Victoria and Albert Museum (1852).

COLMAR, CHARLES XAVIER THOMAS DE. 1785-1870. French mathematician, engineer, businessman, entrepreneur, and inventor: Invented the "Arithmometer" (1820). An early form of the calculator that could perform basic math (adding, subtracting, multiplying, dividing), it was the most widely used calculator up until the early 1900s.

COLT, SAMUEL. 1814-1862. American businessman, industrialist, entrepreneur, design engineer, gun manufacturer, and inventor: Invented the Colt revolver (1831), a revolutionary handgun design that employs a revolving cylinder that holds multiple cartridges. The repeating pistol allowed the shooter to fire up to six rounds (hence its nickname, the "six-shooter") without reloading, a boon to, for example, soldiers, police, and hunters. Colt was also the first American manufacturer to open a

Samuel Colt.

facility in England (1851), and was the founder of the Colt Patent Fire-Arms Manufacturing Company (1855)—today known as the Colt Manufacturing Company. Among the firm's most popular firearms is the Colt 1911, the famed military revolver that aided America in both World Wars.

CONGREVE, WILLIAM. 1772-1828. English artillery officer and inventor: Proposed the idea of plating warships with protective armor (1805), though he is best known for his invention of the Congreve rocket (1808), a portable military weapon with a warhead attached to a stick. This incendiary stick-guided artillery device could be fired from a variety of platforms, such as vehicles and racks or stands, making it an important advancement in both rocketry and explosives.

CONSTANTINESCU, GEORGE. 1881-1965. Romanian scientist, civil engineer, and inventor: Invented the sonic oil drilling technique (1913), as well as the hydraulic interrupter gear (circa 1917), a device that allowed an aircraft pilot to fire a mounted machine gun through rotating propellers. He is best known, however, for inventing the theory of sonics (1918), a branch of continuum mechanics that involves transmitting mechanical energy through vibrations—in this case, through solids, liquids, and gases. A prodigious creative, he also aided in the refinement of the internal combustion engine and, like many of the inventors in this book, was the holder of over 100 patents. Constantinescu is known as the "father of the theory of sonics."

COONS, ALBERT. 1912-1978. American physician, pathologist, bacteriologist, immunologist, educator, and author: Co-invented a method for marking antibodies with fluorescent dyes (1941)—a technique known as immunofluorescence—leading to important advancements in the field of biomedicine.

COOPER, MARTIN. Born 1928. American engineer, entrepreneur, and inventor: Invented the first handheld mobile phone (1973). Cooper is known as the "father of the handheld cellular phone."

COOVER, HARRY. 1917-2011. American chemist and inventor: Invented superglue (1942), also referred to as cyanoacrylate adhesive and Eastman 910. The instantly bonding compound, well-known to consumers and manufacturers, is also used (in nontoxic FDA approved form) by combat medics to treat wounds, by surgeons to rejoin arteries and veins, and by dentists to temporarily fix broken crowns and dentures.

COPEMAN, LLOYD GROFF. 1865-1956. American engineer, industrial designer, entrepreneur, and inventor: Invented an electric thermostat (1909) that led to improvements in the electric stove, shortly after which he founded the Copeman Electric Stove Company (1912). Also invented a toaster with an automatic bread turner (1912), the rubber ice cube tray (1928), and the Flexor-Line clothesline (1943)—a portable, elastic, lightweight washing line (made from braided rubber surgical tubing) that is still popular with travelers, campers, and hikers.

Copeman received over 600 patents during his lifetime.

CORNELISZOON, CORNELIS. 1550-1607. Dutch inventor: Invented the wind powered sawmill (1594), a new, faster, more efficient mill design that was able to cut timber using wind energy. Also helped develop and improve grain milling, water mills with pumps, and animal powered mills.

CORNELL, EZRA. 1807-1874. American businessman, entrepreneur, mechanic, farmer, salesman, philanthropist, and inventor: Founded the company Western Union Telegraph Company (1856), now Western Union, and was a co-founder of Cornell University (1865)—which was named after him.

COTTER, JOSEPH. Dates unknown. English scientist, physicist, electrical engineer, and quantum engineer: Credited with leading the team that developed quantum navigation (2018). Cotter's highly accurate "quantum compass," which uses quantum sensors to calculate location without the need for satellites, is revolutionizing navigation technology, and has become a valuable tool in regions where GPS is undependable or temporarily unavailable.

COUCOULAS, ALEXANDER. Born 1933. American research engineer, writer, and inventor: Invented thermosonic bonding (circa mid 1960s), a technique that uses heat, force, and ultrasonic energy to connect two materials together. Coucoulas is known as the "father of thermosonic bonding."

COULTER, WALLACE H. 1913-1998. American engineer, entrepreneur, and inventor: Invented the Coulter Principle (1948), a technique for automating the numbering and classification of microscopic particles that are suspended in fluid. This led to advancements in laboratory hematology (the science, study, and testing of blood), a field for which he is known as one of the founding fathers.

COUSTEAU, JACQUES. 1910-1997. French explorer, ocean exploration pioneer, naval officer, filmmaker, photographer, and inventor: Co-inventor of the aqualung (1943), a self-contained underwater breathing device, and the Nikonos underwater camera (circa early 1960s). Cousteau helped raise awareness of issues related to ocean ecology and marine life preservation, not to mention world population, and was instrumental in popularizing marine sports.

COVER, JR., JOHN HIGSON. 1920-2009. American physicist and inventor: Invented the Taser (1974), also known as a conducted energy device (CED), a non-lethal handheld weapon that shoots electric probes or darts that are connected to the gun by thin, insulated, high voltage wires. Used chiefly by law enforcement agencies to temporarily stun or disable violent individuals, but also popular with consumers as a means of protection and self defense.

CREIGHTON, JAMES. 1850-1930. Canadian engineer, athlete, journalist, sportsman, and attorney: While the origins of the stick game now called "hockey" date into the distant past, and are therefore unknowable, Creighton is remembered as the man who invented the modern sport of ice hockey (circa 1875), establishing its rules, conditions, and regulations—making him the true "father of ice hockey."

CRISTOFORI, BARTOLOMEO. 1655-1731. Italian harpsichord maker, keyboard instrument builder, musician, and inventor: Invented the pianoforte (circa 1709)—now known in its shortened form as the piano, and in his day called, in Italian, the *gravicembalo col piano e forte*: the "harpsichord that plays soft and loud." Unlike a harpsichord (which creates one-volume tones by pressing a non-responsive key, causing a plectrum to pluck a string), his pianos created tones by pressing a highly responsive, pressure-variable key, causing a hammer to strike a string. This allowed for the creation of a wide dynamic range (hence the original name, pianoforte, meaning "soft-loud"), completely revolutionizing music, and more specifically, keyboard design, keyboard construction, keyboard music, keyboard composition, and keyboard performance.

Bartolomeo Cristofori.

CROOKES, WILLIAM. 1832-1919. English physicist, chemist, and inventor: Discoverer of the element thallium (1861). Due to his study of electrical discharges (through rarefied gas) he was able to observe the dark space around negative electrodes or cathodes (a phenomenon today referred to as Crookes Dark Space), leading to the invention of the Crookes Tube (1870). Better known today as the cathode ray tube, it seems to have been instrumental in the monumental discovery of X-rays. He also invented the Crookes Radiometer (1873), an apparatus that transmutes light radiation into rotary motion, which became useful in measuring instruments and applications. Finally, he was the discoverer of plasma, which he termed "radiant matter" (1879). Also once known as the fourth state of matter, in 1929 scientists gave this superheated ionized gas the name plasma.

CRUMP, S. SCOTT. Born circa mid to late 1950s. American businessman, mechanical engineer, and inventor: Invented fused deposition modeling (1988). Also known as fused filament fabrication (FFF), it is an extrusion-based 3D printing technique that deposits melted plastic around a printing surface, generating a 3D model. Commonly employed in manufacturing, education, and prototyping.

CTESIBIUS OF ALEXANDRIA. Lived circa 270 BC. Greek physician research engineer, and inventor: Invented the first pipe organ (3rd Century BC). His "water organ," known as the hydraulis, used water weight to force air through the instrument's pipes, generating sounds. He also made improvements on the water clock and invented a compressed air-powered catapult.

CUGNOT, NICOLAS-JOSEPH. 1725-1804. French military engineer and inventor: Designed and invented the first steam-powered road vehicle (1769), a massive tractor with a two-piston engine, constructed to transport artillery. The large tricycle style body was arranged with a single front wheel that carried out the functions of both driving and steering.

CULLEN, WILLIAM. 1710-1790. Scottish physician, chemist, and educator: First to propose the idea of artificial refrigeration (1748), which opened up avenues to the eventual invention of the refrigerator.

CZOCHRALSKI, JAN. 1885-1953. Polish chemist and metallurgist: Invented the Czochralski Process (1916), a method for measuring the crystallization rate of metals. More importantly the Czochralski Process allows the creation of single crystals of various materials, useful in the production of semiconductors, optics, synthetic gemstones, salts, and metals.

Two electricity generating engines, 120 horsepower each, Hamburg, Germany, 1913.

DAGUERRE, LOUIS. 1787-1851. French physicist, painter, theater scene designer, and inventor: Invented daguerreotype photography (1839), which involves a complex "film" developing process using an iodized silver plate, mercury fumes, and salt. He was also the inventor of the diorama (1822) and its natural byproduct, the diorama theater (also 1822). Daguerre, whose invention was the most common and popular type of photography throughout most of the Victorian Era (1837-1901), is known as one of the "fathers of photography."

Louis Daguerre.

DALÉN, NILS GUSTAF. 1869-1937. Swedish physicist, chemist, industrialist, engineer, businessman, entrepreneur, and inventor: Known for four inventions: 1) the Dalén Light (circa early 1900s)—designed for lighthouses, it is a device that uses acetylene, a gas that creates a bright white light. 2) Agamassan (1909), a substance that soaks up acetylene, rendering concentrated gas nonexplosive. 3) The automatic sun valve or Solventil (circa 1912), used in conjunction with the Dalén Light. 4) The AGA cooker (1922), an improvement on the stove. Nobel Prize winner.

DAMADIAN, RAYMOND. 1936-2022. Armenian physicist, physician, biophysicist, entrepreneur, and inventor: Invented the first magnetic resonance imaging machine (1974). Better known as an MRI, it is a noninvasive scanning technique that uses powerful magnetic fields and radio waves to produce highly detailed pictures of the inside of the body; useful in detecting disease such as tissue, vascular, muscular, organ, and osteological abnormalities without exposure to harmful radiation.

DANIELL, JOHN FREDERIC. 1790-1845. English chemist, meteorologist, and inventor: Invented the Daniell cell (1836), a battery used to generate and store electricity. It led to numerous advancements in battery technology and electrical telegraphy.

DARBY, GEORGE ANDREW. English inventor: Invented what could be called either the first heat detector or the first smoke alarm (1902), an electrically-powered device called the "Heat-Indicator and Fire Alarm." When it detected a potentially hazardous rise in temperature near its location, it sounded an alert, allowing the residents to escape safely.

D'ARMATI, SALVINO. 1258-1312. Italian inventor: Though disputed by some, he is widely credited with having invented the first pair of eyeglasses (circa 1284). The frame of his stemless eyewear, which balanced on the bridge of one's nose, was probably made of bone or metal, into which were encased two nonprescription quartz lenses.

DARROW, CHARLES B. 1889-1967. American heating engineer, board game designer, and inventor: Credited by some with inventing "Monopoly" (1934), held to be the world's most popular board game, with hundreds of millions of sets sold, and played by over 500 million people.

DARWIN, CHARLES. 1809-1882. English scientist, naturalist, botanist, geologist, author, and explorer: Not an inventor of dazzling products, but a proposer of dazzling ideas, the most significant example being his theory of natural selection (circa 1837). This view he put forward after returning from his famous scientific ocean voyage aboard the HMS *Beagle* (1831-1836), during which time he discovered what he held to be evidence of the biological evolution (of all species) based upon his central idea, what has come to be referred to, in modern shorthand, as the "survival of the

Charles Darwin.

fittest." This notion, the foundational hypothesis upon which the idea of evolution is based, is still taught in schools worldwide—though it is frequently and vigorously contested by those who embrace the religious doctrine of creationism. While Darwin is traditionally credited with being the sole developer of natural selection, he must share the honor with Welsh scientist Alfred Russel Wallace, who independently conceived of the idea around the same time.

D'ASCANIO, CORRADINO. 1891-1981. Italian aerospace engineer, educator, and inventor: Invented the Vespa motor scooter (1946), an inexpensive recreational and commuting vehicle.

DAVENPORT, THOMAS. 1802-1851. American blacksmith, entrepreneur, publisher, and inventor: Invented the first electric motor (1834), the first electric railway (1835), the first electric printing press (1837), and the first trolley car (1881). Also known as a street car, the latter was a public transport vehicle that ran on a track and was powered by electrified overhead cables.

DAVID, LARRY. Born 1947. American actor, comedian, producer, and writer: Creator of the hit TV series *Curb Your Enthusiasm* (2000); co-creator of the hit TV series *Seinfeld* (1989).

DAVIDSON, ROBERT. 1804-1894. Scottish chemist and inventor: Invented the first electric railway locomotive (1837), powered by galvanic battery cells.

DA VINCI, LEONARDO. 1452-1519. Italian polymath, scientist, physicist, architect, painter, sculptor, philosopher, engineer, botanist, chemist, geologist, mathematician, anatomist, artist, cartographer, astronomer, composer, poet, and inventor: Invented items and proposed ideas that led to developments in a wide variety of fields. Examples: the aerial screw (helicopter), the anemometer (for measuring wind speed), the hygrometer (for measuring humidity), the ornithopter (a flying machine), the parachute (military, sports, safety), the triple barrel canon (a light, powerful, 3-shot weapon), the giant crossbow (a fearsome 81-foot wide bow that launched large stones and firebombs), the armored car (predecessor of the modern tank), an accurate clock (for the medieval period), a horse colossus (a 24-foot bronze statue that was never made), a robotic soldier made with gears, wheels, pulleys, and cables (precursor to the modern robot), a leather diving suit and mask with a snorkel, helmet, tubes, and a diving bell (precursor to scuba gear), the fountain pen, and a portable revolving bridge (used by the military), among many other remarkable innovations.

Leonardo da Vinci.

DAVIS, JACOB. 1868-1908. Latvian tailor and inventor: Invented riveted jeans (1871). By attaching copper rivets to the corners of pockets, the pocket openings (and fly base) were strengthened, adding to the jeans' longevity and quality.

DAVY, HUMPHRY. 1778-1829. Cornish chemist, educator, and inventor: Invented the Davy Miners Lamp (1815), a device with a wire gauze chimney that encloses the flame, making it safe for use in mines. The lamp contributed to both the mining industry and to the industrialization of the West. Davy also discovered the physiological effects of nitrous oxide (laughing gas), and proposed a simple theory that explained electro-chemical action.

DAY, JOSEPH. 1855-1946. English engineer and inventor: Invented the two-stroke internal combustion engine (1889), a lightweight, simple machine today commonly used in small applications like handheld tools (e.g., chainsaws), sports vehicles (e.g., motorcycles), and lawn equipment (e.g., mowers).

DE FOREST, LEE. 1873-1961. American electrical engineer, entrepreneur, and inventor: Invented the Audion (1906), a thermionic grid-triode vacuum tube that greatly increased the reception of wireless signals. He founded the De Forest Wireless Telegraph Company (1902), and developed an electrolytic device that could detect Hertzian waves (early 1900s). He also invented the triode vacuum tube (circa 1912) and an optical recording system he called Phonofilm (circa 1920), as well as made contributions to the fields of sound recording, motion pictures, medicine, and the military. The Audion vacuum tube, however, is arguably his greatest achievement: Years before the invention of the transistor (1947), the Audion not only made live radio broadcasting a reality, it went on to become the foundational element of all electronic communication, processing, and sensing devices—including the computer, the telephone, radar, television, and radio. De Forest is known as the "father of the radio" and the "grandfather of television."

DELL, MICHAEL S. Born 1965. American businessman, investor, and entrepreneur: Founded Dell Computer Corporation (1984), more popularly known as Dell.

DEMIAN, CYRILL. 1772-1849. Romanian musical instrument maker and inventor: Credited by some, contested by others, with inventing the world's first accordion (1829), a wind instrument that is hand-powered, and employs reeds, bellows and a keyboard.

DEMIKHOV, VLADIMIR. 1916-1998. Russian scientist, medical pioneer, researcher, and inventor: Created the first artificial heart as well as performed the first surgical implantation of an artificial heart in a non-human animal (1938)—a dog.

DENISYUK, YURI NIKOLAEVICH. 1927-2006. Russian scientist, physicist, educator, and inventor: Invented the Denisyuk Hologram (1958), that is, 3D volume holography, made popular in films such as *Star Wars* and *Star Trek*. Besides holography he also made contributions in the fields of optical logic, non-linear optics, coherent optics, and high density data storage. Known as one of the "founders of modern holography."

DENNARD, ROBERT H. 1932-2024. American electrical engineer, computer scientist, and inventor: Invented dynamic random-access memory or DRAM (circa 1967), an arrangement of semiconductor memory cells densely consolidated on a silicon chip. The memory cell's transistor allowed the storage and identification of binary data to be read as an electrical charge, revolutionizing computer technology.

DÉRI, MIKSA. 1854-1938. Hungarian electrical engineer, textilist, power plant builder, entrepreneur, and inventor: Inventor of the constant voltage AC electrical generator (1883) and the single phase type

of repulsion motor (circa late 1800s); also co-inventor of an improved closed iron core transformer (1885).
DEWAR, JAMES. 1842-1923. Scottish chemist, physicist, and inventor: Co-developer of cordite (1889), a smokeless explosive comprised of nitrocellulose, nitroglycerine, and petroleum jelly, that is used in ammunition. More importantly, he invented the Dewar Flask (1892), a vessel more popularly known today as a carboy, a vacuum bottle, or by the brand name "Thermos." Dewar was also the first person to both liquefy hydrogen gas (1898) and solidify it (1899), shortly after making the important discovery that charcoal is useful in creating vacuums (1905).
DIANIN, ALEKSANDR P. 1851-1918. Russian chemist: First to synthesize bisphenol A or BPA (1891), which is produced by combining phenol with acetone in the presence of an acid catalyst. Some 60 years later, it was found that a hard transparent resin, known as polycarbonate, is created when BPA reacts with phosgene (carbonyl chloride), a discovery that revolutionized modern day society. While some believe that high levels of BPA are a health hazard, this highly versatile microwaveable plastic continues to be widely used in everything from bottles, food containers, eyeglass lenses, and windows, to computers, DVDs, water supply pipes, and thermal paper.
DICKINSON, JONATHAN. 1688-1747. American Founding Father and clergyman: Co-founder of the College of New Jersey (1746), later renamed Princeton University (1896). Dickinson served as its first president.
DICKSON, WILLIAM KENNEDY LAURIE. 1860-1935. French-born Scottish cinematographer, engineer, film producer, film director, photographer, actor, and inventor: Co-inventor of the Kinetograph (1890), a camera that took photos of objects in motion. The resulting celluloid film was then viewed (through peepholes) in another one of his co-inventions, the Kinetoscope (1891), in which quickly flashing still images produced the appearance of a motion picture. These two inventions formed the foundation for the upcoming explosive development of the motion picture camera—greatly impacting the entire film industry to this day. His third co-invention was the Kinetophone (1895), which combined the Kinetoscope with a phonograph that played sounds recorded on cylinders. This particular device paved the way for "talkies," motion pictures with synchronized sound (1920s), out of which the modern film era was born.

William K. L. Dickson.

DIEHL, PHILIP. 1847-1913. German mechanical engineer and inventor: Among many other innovations, he invented the Diehl

Incandescent Lamp (1882), the ceiling fan (1882), the first dental motor (1885), and the first electrical sewing machine motor (1889). Founded the Diehl Manufacturing Company (circa 1887).

DIESEL, RUDOLF. 1858-1913. German thermal engineer and inventor: Invented the Diesel engine (1893), a simple, powerful, fuel efficient internal combustion engine that revolutionized the transportation, truck, automobile, farming, military, locomotive, construction, sea vessel, and electrical utility industries.

DISNEY, WALT. 1901-1966. American animation pioneer, film director, film producer, TV producer, TV show host, family entertainer, artist, film actor, film editor, voice actor, caricaturist, screenwriter, businessman, cartoonist, illustrator, entrepreneur, and inventor: Co-founded the Walt Disney Company (1923), co-invented (with William Garity) the multiplane camera (circa 1937) used in animation, created and supervised the construction and opening of California's Disneyland Theme Park (1955), and, before he passed away, drew up plans for Florida's Walt Disney World (1971). Most notably, Disney was the co-creator of the globally popular cartoon character, Micky Mouse (1928).

Mickey Mouse, the loveable rodent that made Walt Disney famous.

DOBELLE, WILLIAM H. 1943-2004. American scientist, biological engineer, and biomedical inventor: Invented the Dobelle Eye (1978), a prosthetic device that aids the visually impaired by imitating the electrical stimulation between the eye (retina and optic nerve) and brain. The still largely experimental instrument is comprised of three parts, a camera (mounted on eyeglasses), a portable computer processor, and brain implants (surgically embedded in the visual cortex).

DÖBEREINER, JOHANN W. 1780-1849. German chemist: A largely self-taught scientist, he invented the Döbereiner's Lamp (1823), one of the first (cigarette and cigar) lighters. Inside a small glass jar he placed several chemicals (such as zinc and sulfuric acid) that reacted, forming hydrogen gas, which was then ignited by a flint mechanism on top. Though it aided in subsequent lighter design, Döbereiner's Lamp, a rather bulky and heavy contraption, was soon replaced by wooden matches.

DOI, TOSHITADA. Born 1943. Japanese scientist and electrical engineer: Co-invented the compact disk (1979), a flat, circular, 4.75 inch plastic disk that stores and plays back digital information as either audio (CD) or video (DVD); it is also used for computer data (CD-ROM). This digital, portable, optical disk data storage medium is considered one of the most significant developments in the realm of sound, visual, and computer technologies.

DOLBY, RAY M. 1933-2013. American physicist, electrical engineer, and inventor: Invented the Dolby noise-reduction system (1965), a device that reduces background noise (commonly known as "hiss") that is naturally created by the analog recording process.

DOLIVO-DOBROVOLSKY, MIKHAIL. 1862-1919. Russian-Polish electrical engineer and inventor: Credited with independently inventing (along with several other individuals) the three-phase electric power system (1887), which uses three alternating-current (AC) conductors to transmit electricity. Unlike residential homes, which typically use a single-phase power system, the three-phase system is more efficient at generating and transmitting greater power, and is thus used primarily by commercial centers and industrial buildings.

DOORNE, HUB VAN. 1900-1979. Dutch metal engineer, entrepreneur, and inventor: Invented the Variomatic Transmission (1958), a continuously variable transmission or CVT that, while providing smooth acceleration, tends to be lightweight, fuel efficient, simple, and cost-effective compared to other types of transmissions.

DORNBERGER, WALTER. 1895-1980. German engineer, rocket scientist, military officer, and inventor: His work and ideas inspired the building and design of both the North American X-15 (1959), a hypersonic rocket-powered aircraft, and the U.S. Space Shuttle (1981-2011).

DORRANCE, JOHN THOMPSON. 1873-1930. American scientist, chemist, and business executive: Invented condensed soup (1897), a thick stock from which most of the water has been removed, but which is added back by the consumer during preparation. The condensing process allows this type of soup to be sold in smaller cans, lowering the cost and making them easier to store.

DOW, CHARLES. 1851-1902. American reporter and finance journalist: Invented the Dow Jones Industrial Average (1884), which resulted from his records of average U.S. stock prices. He was also a co-founder of *The Wall Street Journal* (1889).

DOYOYO, MULALO. Born 1970. South African engineer, educator, and inventor: Co-inventor of Cenocell, a cementless concrete that is produced from the waste ash that is leftover after the coal-burning process. Cenocell is considered a "green material," one with a number of advantages: It is lightweight, strong, and fire-resistant, as well as an excellent insulator.

DRAGOMIR, ANASTASE. 1896-1966. Romanian inventor: Based on earlier renditions, he co-invented the modern ejection seat (circa 1929), an emergency device that allows a pilot to quickly eject from a plane or jet and safely land on the ground using a parachute.

DRAIS, KARL FREIHERR VON. 1785-1851. German forest official and inventor: Invented the Laufmaschine, or "running machine" (1817), the first bicycle. Also known variously as the Velocipede, the Dandy Horse, and the Draisienne, it lacked the pedals and gears of a modern bike, and was powered solely by the legs of the rider, who sat on the seat of the wooden two-wheeled machine, then "walked" forward, steering with a large single handle mounted over the front wheel.

DREW, RICHARD G. 1899-1980. American scientist, lab technician, and sandpaper manufacturer: Invented masking tape (1925), known by the brand name Scotch Masking Tape, as well as transparent cellophane tape (1930), known by the brand name Scotch Tape. These two products completely altered the commercial world, affecting everything from the automobile, medical, and art industries, to the construction, office, and electrical industries.

DRINKER, PHILIP. 1894-1972. American engineer, educator, and inventor: Co-invented the first iron lung (1928), known as the Drinker Respirator, a biomedical respiratory device that is used to treat polio patients.

DUNLOP, JOHN BOYD. 1840-1921. Scottish veterinary surgeon and inventor: Invented the first practical pneumatic tire (1888), an air-filled, reinforced rubber tube that made bicycling easier and more comfortable. His idea went on to be used in the development of pneumatic tires for cars, trucks, motorcycles, airplanes, etc.

DUQUET, CYRIL. 1841-1922. Canadian scientist, jeweler, goldsmith, musician, clockmaker, and inventor: Though disputed by some, he is credited with inventing the telephone handset (1878), and for later creating a unit that incorporated both the receiver and the transmitter in one hand piece.

DURANT, WILLIAM C. 1861-1947. American businessman, financial tycoon, and entrepreneur: Founder of General Motors (1908)—formerly known as the General Motors Corporation or GMC. Durant was also a co-founder of Chevrolet (1911). Among his many commercial acquisitions were Buick (1904), Oldsmobile (1908), Cadillac (1909), and Frigidaire (1918).

William C. Durant.

DUSHKIN, ALEXEY. 1904-1977. Russian architect and city planner: Created the first deep column station (circa 1938), a form of subway station possessing an innermost central hall, two connecting side halls, and linked circular passages set amidst columnar rows.

DYSON, JAMES. Born 1947. English businessman, designer, and inventor: Invented the bagless vacuum cleaner, a type of vacuum design known as a cyclone separator. Cyclonic action allows gases and solids, for instance, to be distinguished and compartmentalized, extremely useful not just in the vacuum industry, but in the dairy industry, food-drying technology, and incinerator technology, as well. He is also credited with inventing (some say improving) the bladeless fan (2009), which he called the "Dyson Air Multiplier."

EASTMAN, GEORGE. 1854-1932. American businessman, entrepreneur, and inventor: Invented roll film (circa 1885), a small flexible spool of film that replaced the far more cumbersome photo process that entailed using heavy, fragile glass plates. He also invented the Kodak Camera (1888) and the Brownie Camera (1900), all innovations that revolutionized photography by lowering costs and simplifying the art, thereby opening up the field to amateurs and hobbyists.

EASTON, SR., ROGER L. 1921-2014. American physicist, politician, and inventor: Lead developer (some say the sole inventor) of the Global Positioning System (1973), better known as GPS. Owned by the U.S. government and operated by the United States Air Force, GPS is a space-based radio-navigation system that supplies precise data concerning one's location, speed, and time. GPS has been key in further developing technologies surrounding national security, the military, electronics, agriculture, emergency services, commercial products, meteorology, mapping, environmental protection, surveying, biology, package and food delivery, transportation, mining, seismology, sports, marine concerns, home appliances, public safety, railways, banking, space, recreation, travel, power grids, automobiles, construction, and aviation, to name only a few.

EBERHARD, MARTIN. Born 1960. American engineer, businessman, and entrepreneur: Co-founded Tesla, Inc. (2003), an electric vehicle company. Served as its first CEO.

ECKERT, JOHN ADAM PRESPER. 1919-1995. American scientist, computer scientist, physicist, electrical engineer, and inventor:

Co-invented the first programmable digital computer (1948), which was named ENIAC—an acronym for Electronic Numerical Integrator and Computer. This precursor to the modern computer laid the groundwork for the evolution of today's computers.

EDISON, THOMAS ALVA. 1847-1931. American businessman, scientist, physicist, mathematician, entrepreneur, film producer, and inventor: Invented the electrographic vote recorder (1868), the automatic telegraph (circa 1870), the first electric copy machine (1876), the phonograph (1877)—a device that led to the modern record player, the carbon transmitter or carbon microphone (circa 1878), and the magnetic iron ore separator (1880). He was also the co-inventor of the Kinetoscope (1891), which aided in the evolution of motion picture cameras and in turn the rise of the film industry. Contrary to popular opinion, he did not invent the electric light bulb. Rather he improved on designs that were created by earlier inventors, eventually developing the first truly useful incandescent light bulb (1879). Edison was a co-founder of General Electric (1892), more popularly known as GE.

Thomas A. Edison.

EDMAN, PEHR VICTOR. 1916-1977. Swedish medical scientist and biochemist: Invented Edman Degradation (1950), a chemical technique for determining the amino acid sequence of proteins.

EDSON, MARCELLUS GILMORE. 1849-1940. Canadian chemist, pharmacist, and inventor: Invented "peanut paste" (1884), an early forerunner of what would later become known as peanut butter.

EDWARDS, ROBERT GEOFFREY. 1925-2013. English physiologist and biology researcher: Co-invented in-vitro fertilization, or IVF (circa 1960s), a method by which a man's sperm is combined with a woman's egg outside the woman's body—with insemination taking place in an environmentally-controlled laboratory chamber. IVF has greatly aided the plight of thousands of couples experiencing infertility issues. Nobel Prize winner.

EICH, BRENDAN. Born 1961. American computer scientist and computer programmer: Invented JavaScript (1995), a programming language used on the World Wide Web, and which allows the creation of dynamic content, such as multimedia and animated images. Works in conjunction with two other important computer languages: HTML and CSS.

EINSTEIN, ALBERT. 1879-1955. German scientist, physicist, mathematician, philosopher, educator, author, and inventor: Though primarily a theoretical inventor, he had over 50 patents, with original ideas that included the general theory of relativity, the special theory of relativity, the quantum theory of light, the Bose-Einstein condensate, Brownian movement (the fluctuation-dissipation theorem), the photoelectric effect, wave-particle duality, Avogadro's number, and the mass/energy relationship. Arguably among the top ten most influential scientists of the modern era, Einstein's views and theories opened the door to a myriad of innovations in fields as diverse as atomic energy, atomic clocks, cathode ray tubes (CRT TVs), electromagnet technology, the stock market, solar power, lasers, space exploration, GPS navigation systems, night vision technology, paper towels, photoelectric technology, and cosmology. Einstein is known as the "father of relativity." Nobel Prize winner.

Albert Einstein.

EINTHOVEN, WILLEM. 1860-1927. Dutch physiologist, professor, and inventor: Invented the string galvanometer (1901), and the first practical electrocardiograph (1903) or ECG, both important diagnostic tools used to measure the electrical activity of the heart. Einthoven is known as the "father of modern electrocardiography." Nobel Prize winner.

EISENHOWER, DWIGHT D. 1890-1969. President of the United States of America: Creator (or head of) of DARPA (1958), or, "Defense Advanced Research Projects Agency," a primitive forerunner of cloud computing technology. He is also credited with inventing (or at least proposing) the American interstate highway system (1954), often called "the greatest public works project in history." Eisenhower is thus known as the "father of the interstate highway system."

EISENSTADT, BENJAMIN. 1906-1996. American entrepreneur and philanthropist: Invented the sugar packet (circa mid 1940s) as well as the popular artificial sweetener called "Sweet'N Low" (1958).

EISLER, PAUL. 1907-1992. Austrian electrical engineer and inventor: Invented the printed circuit board (1936), or PCB, a thin, usually fiberglass sheet that connects various components in a single circuit via electronic printed or etched pathways over which signals travel. The device, which saves space and simplifies the computing process, greatly enhanced computer technology and today is found in a majority of electronic products, as well as in nearly every field of human endeavor, from computers, cell phones, and energy, to communications, industrial, and military.

ELIAVA, GIORGI. 1892-1937. Georgian microbiologist and educator: Co-discovered Phage Therapy (circa 1920s), a treatment that employs viruses to treat bacterial infections, such as those related to the lungs, eyes, ears, and urinary tract. Though revolutionary in its time, bacteriophage therapy, as it is also known, has largely been replaced today by antibiotics.

ELLIOTT, JAMES BEDFORD. 1846-1906. English bicycle racer, bicycle designer, entertainer, and inventor: One of the possible inventors of the unicycle (circa late 1800s), a vehicle with one wheel that is propelled forward and backward using foot pedals.

ELLIS, WILLIAM WEBB. 1806-1872. English clergyman: Although various forms of football and handball have been around for at least 1,000 years, if not much longer, Ellis is credited with inventing the modern game of rugby (1823). The sport, which was codified 22 years later (1845), was named after Rugby School (located in the town of Rugby, England), which Ellis was attending as a student at the time. To this day a statue of Ellis adorns the Rugby School campus (erected 1997), and the Webb Ellis Cup, the trophy handed out to the winners of the Rugby World Union Tournament, memorializes his name.

ELMQVIST, RUNE. 1906-1996. Swedish physician and inventor: Invented the first fully implantable pacemaker (1958), a battery-powered device that helps control the heart's rhythm and rate. The small machine is surgically implanted in the chest or abdomen as a treatment for those with heart issues, such as heart blockages or heart failure.

ELSENER, KARL. 1860-1918. Swiss businessman, entrepreneur, cutler, knife maker, parliamentarian, and inventor: Invented the famous Swiss Army Knife (1891), a device that not only revolutionized the knife industry, but as an all-in-one miniature toolkit, it has altered the way we travel, work, and play. Its compact size and utilitarian aspects have made it a necessity, for example, on NASA space missions, on hiking and camping expeditions, and on military operations.

EMERSON, CHARLES WESLEY. 1837-1908. American physician, clergyman, and author: Founder of the Boston Conservatory of Elocution, Oratory, and Dramatic Art (1861), which was later renamed Emerson College (1939).

EMERSON, JOHN HAVEN. 1906-1997. American inventor: This self-taught inventor created the Emerson Respirator (1931), an improvement on the original iron lung (1928), a biomedical respiratory device that is used to treat polio patients.

ENGELBART, DOUGLAS. 1925-2013. American scientist, engineer, and inventor: Invented the computer mouse (1963), a small palm-sized, handheld input device that controls an onscreen cursor. It is used mainly for moving and selecting text, clicking on icons, moving computer files and folders, and dragging-and-dropping objects, among other functions.

ERCOLINO, MICHAEL D. 1906-1982. American businessman and inventor: Invented the conical V beam TV antenna (1940s). An improvement on earlier antenna designs, it helped capture broadcast signals, allowing TV viewing across greater distances.

ERICSSON, JOHN. 1803-1889. Swedish mechanical engineer, naval architect, and inventor: In England he invented the screw propeller (circa 1836), a revolving, bladed, rotary device that propels airplanes through the air and ships through the water. Later, after moving to the

U.S., he invented the first armored turret warship (circa 1862). Named the *Monitor*, the steam-powered vessel relied on his earlier invention, the screw propeller, for propulsion. Employed by the U.S. government during America's War Between the States (1861-1865), the *Monitor* played a role in the inconclusive naval battle with the C.S. (Confederate) government's ironclad warship the *Virginia*, which took place at Hampton Roads, Virginia, March 8-9, 1862.[2]

ERLENMEYER, EMIL. 1825-1909. German organic chemist, pharmacist, and educator: Invented the Erlenmeyer Flask (1901), a small, stoppered, glass laboratory bottle with a cylindrical neck, flat bottom, and a conical body. Used mainly for stirring, cooling, and heating solutions, and also for cultivating microorganisms, the Erlenmeyer Flask is designed to prevent liquids from spilling out during mixing, and to sit securely on flat heated surfaces. He was also the first to synthesize glycolic acid (derived from grapes), as well as the first to propose the formula for naphthalene. Lastly, he developed the important biological convention known as the Erlenmeyer Rule: "Alcohols in which the hydroxyl group is attached directly to a double-bonded carbon atom become aldehydes or ketones."

EVANS, MARTIN JOHN. Born 1941. English scientist, geneticist, and educator: Co-developed gene targeting (early 1980s), a method that recombines genes in an effort to make changes to the genome. The process uses laboratory mice, also known as "knockout mice," to create animal models of human diseases. The technique greatly aided in research that led to treatments for the human diseases that were modeled. Nobel Prize winner.

EVINRUDE, OLE. 1877-1934. Norwegian businessman, entrepreneur and inventor: Developed what would become the first commercially viable outboard motor (1906), a self-contained propulsion unit installed on the outside of a boat's hull, and which is used to power small watercraft. The outboard motor continues to be extremely popular in the boating, fishing, and military industries.

Ole Evinrude.

An early steam-powered ship.

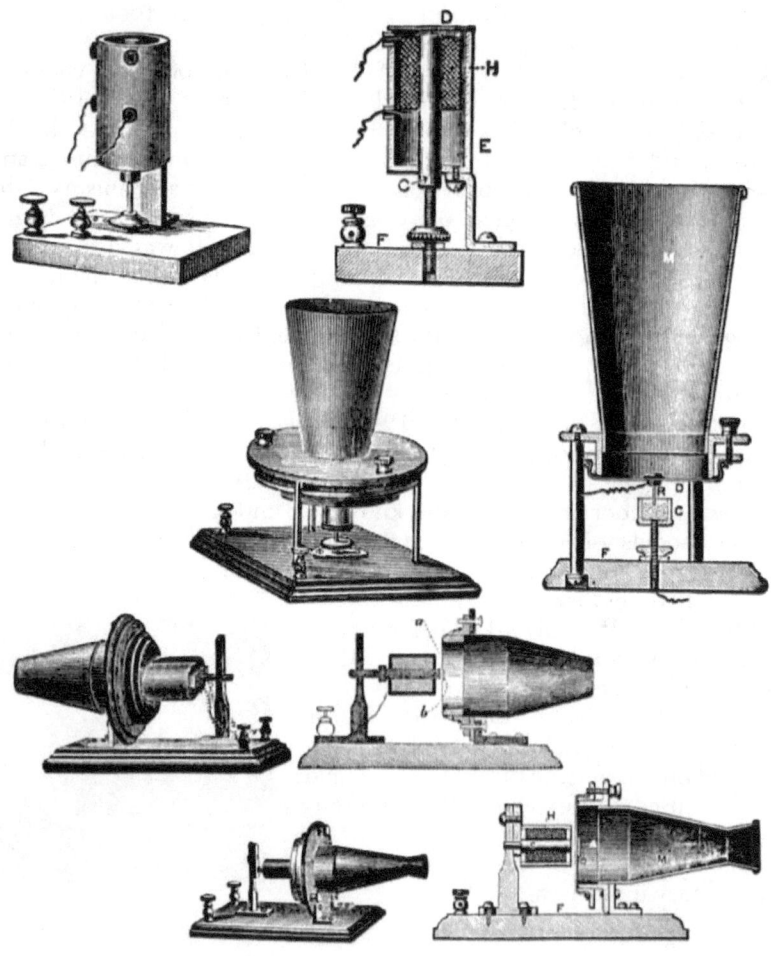

An assemblage of early telephones that were exhibited at the Philadelphia Centennial Exposition, 1876.

FABRY, CHARLES. 1867-1945. French physicist: Co-invented the Fabry-Pérot Interferometer (1896), an instrument that employs multi-beam light interference, making it useful in the fields of high resolution spectroscopy, telecommunications, and astronomy. Fabry also co-discovered the ozone layer (1913), establishing that it filters out ultraviolet radiation from the sun in the upper atmosphere.

FACE, JR., SAMUEL A. 1923-2001. American painter and inventor: Invented the Face Floor Profile Numbering System (1979), or F-number system, the current North American standard measurement guide for specifying the levelness and flatness of concrete floors. He also invented the Lightning Switch (circa 2006), a wireless light switch that uses the physical energy from a person's finger—rather than electricity—to power itself.

FAGGIN, FEDERICO. Born 1941. Italian scientist, physicist, electrical engineer, computer scientist, businessman, and inventor: Invented the microprocessor (1971), or CPU: central processing unit. A single semiconductor chip (in essence, a miniature computer), it serves as an electronic device's brain, merging and instructing input, output, arithmetic, control, and storage functions. Its invention transformed the electronics industry (in relation to, for example, appliances, communications, banking, toys, vehicles, audio, video, batteries, tools, etc.), and in particular the computer industry.

FAHLBERG, CONSTANTIN. 1850-1910. Russian chemist and researcher: Credited by many with being the sole discoverer of the manmade sweetener saccharin (1879). (Some, however, maintain that Fahlberg shares this honor with his research partner at the time, Ira

Remsen.)

FAHRENHEIT, DANIEL GABRIEL. 1686-1736. Polish (later Dutch) physicist, scientific instrument maker, and inventor: Invented the Mercury-in-glass thermometer (1714), a sealed glass tube with a mercury-filled reservoir and calibration along the side in Fahrenheit degrees. (Today it is being replaced by electronic and digital thermometers due to the dangers of mercury). He also invented the eponymous Fahrenheit Scale (1724), a temperature gradient that uses 32° for the freezing point of water and 212° for the boiling point of water. (This is gradually being replaced now by the Celsius scale.)

FARADAY, MICHAEL. 1791-1867. English chemist, physicist, author, and inventor: Known primarily as a researcher in electromagnetism and electrochemistry, and for being the discoverer of the liquefaction of gases and refrigeration (1823), benzene (1825)—used to create light for homes, electric current induction (1831), electrolysis (1833)—which uses electricity to separate matter, diamagnetism (1845), and the eponymous Faraday Effect (1845)—the interconnection between light and electromagnetism. He went on to invent the electric motor (circa 1821); the first homopolar generator or electric dynamo (1831), also called the Faraday Disk—an electromagnetic induction machine that generates current by converting rotational mechanical energy into electrical energy; the electric transformer (1831)—which turns current into voltage; and the Faraday Cage (1836), a device that blocks electric waves. A unit of capacitance, called the "farad," is named in his honor. While Faraday may not be a household name like da Vinci, Edison, or Bell, in my opinion he was certainly one of the most influential inventors of all time. For the innovations of this humble but extraordinary, largely self-taught scientist (who turned down a knighthood) have arguably impacted today's society more than any other inventor in modern history, eventually (either directly or indirectly) making possible such things as power stations, nuclear power plants, hydroelectric dams, automobile engines (both gasoline- and electric-powered), truck engines, train engines, boat engines, plane engines, motorcycle engines, the cell phone, the refrigerator, the air conditioner, the electric garage door opener, wind turbines, the electric water pump, the automatic (cow) milking machine, fans, the electric toothbrush, the electric guitar, the sewing machine—nearly, in fact, all modern electric devices that depend on mechanical energy; affecting everything from farming to aerospace engineering.

Michael Faraday.

FARINA, JOHANN MARIA. 1685-1766. Italian chemist, businessman, perfumer, and inventor: Invented the first cologne (1709). His popular citrus-scented blend of essential oils, known in the 1700s as

Eau de Cologne ("Water from Cologne"), is still being produced and sold to this day.

FARNSWORTH, PHILO. 1906-1971. American scientist, telecommunications pioneer, and inventor: Invented the first all-electronic television system (1927), permanently reshaping the modern world.

FAY, SAMUEL B. 1856-1953. English inventor: The self-taught innovator is credited with inventing the first paper clip (1867), helping declutter offices around the world ever since.

FELDKAMP, LEE. 1942-2021. American scientist, physicist, nuclear engineer, and inventor: Invented micro CT technology (circa early 1980s), a new type of computer-assisted tomography, another important advancement in medical science.

FENDER, LEO. 1909-1991. American businessman, entrepreneur, musical instrument designer and maker, amplification systems designer and maker, musician and inventor: Invented the Fender Telecaster (1948), the Fender Precision Bass (1951), and the Fender Stratocaster (1953), three of the world's bestselling electric guitars. Co-founder of the K and F Manufacturing Corporation (1944) and founder of Fender Manufacturing (1946)—today known as the Fender Musical Instruments Corporation.

FENG, MILTON. Born 1950. Taiwanese electrical engineer, educator, and inventor: Invented the world's fastest light-emitting transistor (2004), or LET, and is the co-inventor of the transistor laser (2004).

FENN, JOHN BENNETT. 1917-2010. American chemist, researcher, and educator: Invented electrospray ionization (1984), an ionization method that is used in mass spectrometry to ascertain the molecular weights and sequence of peptides and proteins, as well as other biological macromolecules. Nobel Prize Winner.

FENTON, HENRY JOHN HORSTMAN. 1854-1929. English chemist: Invented Fenton's Reagent (1894). A chemical reaction created by combining a soluble iron catalyst with hydrogen peroxide, he found that a variety of organic molecules could be oxidized and degraded without the use of high pressures, high temperatures, and complex equipment. This makes it ideal for such applications as wastewater treatment, soil decontamination, and in general, removing pollutants such as chemicals, pesticides, and solvents from the environment.

FERGASON, JAMES. 1934-2008. American scientist, and inventor: Refined and improved the liquid crystal display (1970) via his invention of the twisted nematic liquid crystal display, or TN-LCD. Today the LCD screen is an omnipresent fixture of modern society, and is used on everything from flat-screen TVs, desktop computers, laptop computers, and tablets, to digital watches, cell phones, calculators, and medical equipment, as well as a myriad of other devices. A highly productive inventor with some 150 U.S. patents and 500 foreign patents, Fergason is known as the "father of the modern liquid crystal industry."

FERMI, ENRICO. 1901-1954. Italian physicist, nuclear scientist, educator, and inventor: Invented the nuclear reactor (1942), which greatly advanced nuclear science while ushering in the modern nuclear age—with all of it benefits and dangers. The element fermium (element number 100) is named after him. Nobel Prize winner.

FERNÁNDEZ-MORÁN, HUMBERTO. 1924-1999. Venezuelan scientist, physician, biophysicist, and researcher: Invented the diamond scalpel (1955), used in various scientific applications, including medicine and geology. He also invented the ultramicrotome (1957), a machine that cuts exceptionally thin slices from specimens that have first been embedded in hard material, such as epoxy. These specimen slices are employed by ultramicrotomists, who can then view and analyze them at a microscopic level using an electron microscope.

FERREN, BRAN. Born 1953. American scientist, technologist, engineer, cinematographer, designer, artist, entrepreneur, educator, and inventor: Co-invented pinch-to-zoom technology (2005), a multitouch finger gesture that allows the user to zoom in and out of text, a video, a photo, etc., on the screen of their electronic device.

FERRERO, MICHELE. 1925-2015. Italian businessman, chocolatier, and entrepreneur: Invented Nutella (1964), a hazelnut-cocoa breakfast and dessert spread, typically used on pancakes, crackers, toast, and waffles. He also invented the Kinder Surprise (1974), a large chocolate egg surrounding a small plastic egg that contains a toy inside.

FESSENDEN, REGINALD A. 1866-1932. Canadian electrical engineer, chemist, researcher, radio pioneer, educator, and inventor: He developed the principle of amplitude modulation (1900), today known as AM radio, allowing him to transmit the first radio broadcast over a long distance (1906). Fessenden worked with Thomas Edison and other notable scientists and is known as the "father of voice radio."

FEULGEN, ROBERT J. 1884-1955. German physician: Co-invented the Feulgen Stain (1924), today better known as the Feulgen Reaction. A method for staining nucleic acid (instead of extracting DNA from cells), it is used to measure and identify DNA within cells, making it one of the most trustworthy medical techniques for uncovering cellular abnormalities.

FICK, ADOLF GASTON EUGEN. 1852-1937. German physician, ophthalmologist, physiologist, and inventor: Invented the contact lens (1888). He first experimented with glass scleral contacts (which rested on the sclera and covered the entire surface of the eye); these gradually evolved into corneal contacts (which covered only the cornea). His innovative eyewear revolutionized ophthalmology: At least 125 million people worldwide now wear contact lenses.

FIGUIER, LOUIS. Dates unknown. (He may or may not be confused with a Frenchman named Guillaume Louis Figuier, 1819-1894.) French scientist and inventor: Though the origins of parchment paper, also known today as baking paper, date back to ancient Egypt (and possibly earlier), Figuier is credited with being the co-inventor of modern parchment paper (1847), which he created by immersing ordinary paper in a mixture of sulfuric acid and water. The result of this chemical experiment was a versatile, grease-resistant material that can now be found in most kitchens around the world.

FIRNAS, ABBAS IBN. 810-887. Arabic-Iberian polymath, scientist, physician, engineer, astrologer, author, poet, musician, and inventor: Invented and flew the first known flying machine (875). He also invented transparent glass (using quartz and silica glass), which he made into corrective lens for individuals with impaired vision, as well as a glass-covered underwater dive mask. Additionally he invented a precursor to the metronome, as well as a water clock and a mechanical planetarium for studying the movements of the planets. A moon crater is named after him. Said to have inspired Leonardo da Vinci, Firnas is known as the "father of aviation."

FISCHER, ARTUR. 1919-2016. German inventor: While he has countless creations to his name, he is best known for inventing the Fischer Wall Plug (1958), an expanding plastic screw anchor into which a screw is driven, greatly enhancing its ability to support and hang heavy objects, such as picture frames, flat-screen TVs, mirrors, shelving, cabinets, curtain rods, cupboards, etc. The expanding Fischer Wall Plug led to massive advancements in the construction industry and is today used all over the world.

FISCHER, FRANZ JOSEPH EMIL. 1877-1947. German scientist and organic chemist: Co-invented the Fischer-Tropsch Synthesis (1925), a decarbonization process in which a mixture of carbon monoxide and hydrogen (from any source) are changed into liquid and solid hydrocarbons, creating a $CO/H2$ gas mixture called syngas. By employing this method the energy industry can use carbon-free applications to generate synthetic gas from biomass (e.g., from farms) or from $CO2$ and water. He also co-invented the Fischer Assay (circa 1940s), a scientific testing method that measures the potential oil yield of oil shale.

FISCHER, GERHARD. 1899-1988. German electrical engineer, businessman, and inventor: Invented the hand-held metal detector (1931), a technology now found globally in the metal detectors used at airports, government buildings, and other high security-risk environments.

FISHER, PAUL C. 1913-2006. American politician, entrepreneur, and inventor: Invented the Fisher Space Pen (1965), a pen that uses pressurized ink cartridges filled with thixotropic visco-elastic ink, enabling it to write when held at any angle, while underwater, at zero gravity, on almost any surface, and in the extreme temperature range of between -50° to 160°.

FITCH, JOHN. 1743-1798. American engineer, businessman, entrepreneur, metalworker, silversmith, gunsmith, surveyor, clockmaker, and inventor: Credited with inventing the first steamboat (1787), the first model steam rail locomotive (circa 1794), and the first marine screw propeller (circa 1785). Fitch is said to have inspired Robert Fulton.

FLEISCHMANN, MARTIN. 1927-2012. Czechoslovakian (later British) scientist, chemist, writer, and inventor: Co-developed (with Stanley Pons) cold fusion (1989), a "theoretical" nuclear reaction that would permit fusion at low (i.e., room) temperatures. As the inventors claimed their experiments were successful, the world became excited at the possibility of an infinite source of low cost clean energy. Despite the fanfare, Fleischmann and Pons' innovation was quickly rejected by the scientific community. They were called "frauds," their work "unreproducible," and their claims "unethical." In the aftermath the idea of cold fusion—or low-energy nuclear reaction (LENR), as it is now termed—was discredited and the authors disgraced. Questions remain however, with conspiracy theorists maintaining that the project was shut down by mainstream science due to the threat cold fusion poses to conventional energy producing technologies. Whatever the truth turns out to be, many still consider the idea of cold fusion both intriguing and potentially promising. As of this writing, studies, research, testing, and experimentation are ongoing.

FLEMING, ALEXANDER. 1881-1955. Scottish scientist and bacteriologist: Discovered penicillin (1928), one of the first antibiotics ever identified. This powerful germ-killer transformed medical science and is today used around the globe to treat bacteria-based infections like urinary tract infections, throat infections, skin infections, and nose and ear infections—in the process improving and even saving millions of lives. Nobel Prize winner.

FLEMING, JOHN AMBROSE. 1849-1945. English engineer, physicist, electronics pioneer, author, educator, and inventor: Invented the two-electrode radio rectifier (1904), which he named the "thermionic valve." This electronic innovation—also known as the Fleming Valve, kenotron, or thermionic tube, and today more commonly referred to as a vacuum diode—was the first of its kind. The precursor of the triode, it transformed AC radio signals into direct currents (DC) that were used in the further development of the telephone.

FLEMING, SANDFORD. 1827-1915. Scottish scientist, civil engineer, and writer: Invented the international time standard known as Universal Standard Time (1879), a time-measuring system that is based on the average speed of the earth's rotation. This worldwide time scale reference, once known as Greenwich Mean Time (GMT), and today called Coordinated Universal

Sandford Fleming.

Time (UTC), is the global standard for regulating clocks and for synchronizing timekeeping differences that arise between atomic time and solar time.

FLORINE, NICOLAS. 1891-1972. Russian scientist, aeronautical engineer, mathematician, and inventor: Invented (disputed by some) the first tandem rotor helicopter (1927). This powerful double-rotored aerial workhorse is widely used today by countless industries, including the military, construction, disaster relief, heavy lifting, search and rescue, logging, firefighting, ranching, oil, cargo, evacuation, and transport.

FLOWERS, TOMMY. 1905-1998. English computer scientist, engineer, and inventor: Invented the Colossus (1943), a World War II code-breaking computer that, utilizing a clock pulse, could process up to 25,000 characters per second—an astonishing speed at the time. It was the world's first programmable electronic computer and greatly aided not only Britain, but also the Allied Powers, in the war effort.

FOGARTY, THOMAS J. Born 1934. American cardiovascular surgeon, entrepreneur, inventor, and vintner: Invented the embolectomy catheter (1961). Also known as the Fogarty Balloon Catheter, it is today the global medical industry standard for removing blood clots and is estimated to have saved the lives of some 20 million people.

FOLEY, CHARLES. 1930-2013. American toy and game inventor: Co-invented (with Neil W. Rabens) the popular game Twister (1966).

FÖLSCH, FREDERICK. Circa 1762-1836. English wood turner, ivory turner, fan-maker, pen maker, quill maker, and inventor: Invented (disputed) the first carbon paper (circa 1805), which he called "manifold writer," as well as the first fountain pen (1809), which he called a "reservoir pen" (also disputed).

FORD, HENRY. 1863-1947. American industrialist, businessman, business magnate, entrepreneur, engineer, philanthropist, and inventor: By any definition a titan in the American business world, he is best remembered for founding the still thriving Ford Motor Company (1903), and also for inventing the Quadricycle (1896), the Model T (1908), and the moving assembly line (1913)—the latter which is still in use. One invention that is often left out of our history books, however, was Ford's biotechnological wonder, the "Hemp Car" (1941), which he proudly promoted as "grown from the soil," and "ten times stronger than steel." An automobile made from hemp fiber, wheat, sisal, and resin (plastic) composites, it ran on hemp- and agricultural waste-based ethanol, a renewable fuel that also has the benefit of producing low carbon emissions. Ford's

Henry Ford.

environmentally friendly hemp car was certainly far ahead of its time—which could explain why his prototype was eventually demolished, the program was discarded, and the entire project was suppressed. Conspiracy theories abound and questions continue.

FORLANINI, ENRICO. 1848-1930. Italian aerospace engineer, aeronautics researcher, aeronaut, aircraft pilot, businessman, and inventor: Invented the first helicopter (1875)—the first heavier-than-air vehicle, the first hydroplane or manned hydrofoil ship (circa 1900), and numerous advanced dirigibles known as Forlanini Airships (beginning circa 1901). He is also credited with making the first modern dive mask (1900), which used a flat window comprised of tempered glass. Lastly, he was involved in making improvements in lighting systems, generators, mechanical systems, and threshing machines.

FORSYTH, ALEXANDER JOHN. 1769-1843. Scottish clergyman, gunsmith, chemist, and inventor: Invented the percussion lock firearm mechanism (circa 1806), an ignition technique that works by hammer-striking a packet of potassium chlorate lodged in the breech. The resulting miniature explosion ignites the primary charge, forcing a projectile from the barrel at high velocity. The percussion lock replaced the flintlock firearm mechanism, opening the way for the development of metal cartridge-styled ammunition still used today.

FOSSUM, ERIC R. Born 1957. American physicist, engineer, and inventor: Invented the modern CMOS sensor (1993), a small, fast, inexpensive imaging device that is used by hundreds of millions of people everyday in such items and industries as cameras (DSLR, digital, and video), astronomy, telescopes, inspection systems, vehicle cameras, optical character recognition (OCR), Webcams, scanners, barcode readers, satellites, radar, medicine, military, weapons systems, industrial plants, and robots, among many others.

FOUCAULT, JEAN BERNARD LÉON. 1819-1868. French physicist, astronomer, and experimentalist: Invented the Foucault Pendulum (1851), which proved that the earth rotates on its axis. He also discovered eddy currents (1851), electrical current loops that circulate in conductors like swirling water currents. Lastly, he coined the name gyroscope (1852) for the 1812 invention.

FOURNEYRON, BENOÎT. 1802-1867. French engineer: Invented the water turbine (1827), a rotating device that converts the kinetic energy of moving water into electrical or mechanical energy. Today, best known for its use in hydroelectric dams to supply clean energy to nearby cities, the development of the water turbine was a massive stepping stone in the ongoing development of the electric power generating industry.

Benoît Fourneyron.

FOWLER, JOHN. 1826-1864. English agricultural and civil engineer, grain trader, and inventor: Invented the steam-driven plow (1854), a ground-fixed steam engine that pulled plows over the soil. His innovation reduced costs, eliminated the need for horses, sped up the plowing process, and allowed cultivation of difficult land and soils.

FOWLER, THOMAS. 1777-1843. English mathematician: This self-taught innovator invented the first ternary computer (1840). The wooden machine mechanically calculated using the balanced ternary numeral system, leading to further developments in the computer science of today.

FRANKLIN, BENJAMIN. 1706-1790. American Founding Father, polymath, scientist, statesman, ambassador, author, printer, editor, postmaster, musician, and inventor: Invented swim fins (1717), the Franklin Stove (1742), the lightning rod (1749), a flexible urinary catheter (1752), a 24-hour, three-wheel clock (1757), the glass harmonica (1761), bifocal glasses (1784), and an extension arm for reaching books (1786). He also founded America's first circulating library (1731), America's first volunteer fire department (1736), the University of Pennsylvania (1740), America's first public hospital (1751), and America's first mutual insurance company (1752). Lastly, Franklin devised the concept of Daylight Savings Time (1784), and was the first to recommend that the original 13 American colonies form a confederacy (1754). Note that a version of Franklin's "Albany Plan" was adopted in 1781 when the USA's first constitution, the Articles of Confederation, was ratified, forming what our first president, George Washington, described as a "Confederate Republic." Our third president, Thomas Jefferson, called the new American confederacy a friendly compact between 13 independent "nation-states." Others, such as French diplomat Alexis de Tocqueville, more accurately referred to the USA as "the Confederate States of America," the name that was later adopted by the seceding Southern states, the CSA, in 1861 in their attempt to preserve the original constitutional principles of the Founding Fathers.[4]

Benjamin Franklin.

FRANKLIN, WILLIAM. 1730-1813. American lawyer, military man, New Jersey governor, and son of Benjamin Franklin (above), he was the founder of Rutgers University (1766), which was in turn named after Revolutionary War icon Henry Rutgers (1825).

FRASCH, HERMAN. 1851-1914. German chemist, chemical engineer, mining engineer, pharmacist, and inventor: Invented the Frasch Process (1891), a method for mining deep-lying sulfur from underground deposits. The process involves using superheated water to first melt the sulfur, then applying compressed air to bring it to the surface. Contributed to the growth and development of the U.S. sulfur industry.

FRAUNHOFER, JOSEPH VON. 1787-1826. German physicist, optical glass producer, and inventor: Invented the world's first spectroscope (1814), a scientific instrument that analyzes and measures light spectra, and which has found numerous uses in such fields as astronomy and chemistry.

FRAZER, IAN H. Born 1953. Scottish scientist, engineer, physician, and immunologist: Co-developed the HPV vaccine (1991), or human papillomavirus vaccine, a non-infectious vaccine for cervical cancer.

FRESNEL, AUGUSTIN-JEAN. 1788-1827. French physicist, engineer, mathematician, optics and light researcher, and inventor: Invented the Fresnel Lens (1822), an improved and modified lens that was used in lighthouses to magnify and intensify their light beams. The physically beautiful Fresnel Lens has wide applications and is still used today in, besides lighthouses, such industries as photography, solar panels, space exploration, magnifying glasses, movie projectors, reading aids, and automobiles (e.g., headlights, side mirrors, and taillights).

FREUD, SIGMUND. 1856-1939. Austrian scientist, psychoanalyst, neurologist, intellectual, psychologist, psychiatrist, author, philosopher, researcher, educator, essayist, and critic: Founder of psychoanalysis (1896), a type of psychotherapy that probes and studies the subconscious mind in the treatment of mental illness.

Sigmund Freud.

FRIESE-GREENE, WILLIAM. 1855-1921. English cinematographer, photographer, film director, and inventor: Invented cinematography (1889) by devising a camera that snapped a series of photographs on perforated film rolls. These images quickly passed behind a shutter, forming the precursor of the modern motion-picture camera. He was also the inventor of the first non-artificial color cinematography process (1898).

FROMM, JULIUS. 1883-1945. Polish chemist, businessman, entrepreneur, and inventor: Invented the first seamless condom (1914) by vulcanizing a glass tube dipped in a liquid rubber compound. The result was a product that was not only seamless, but also flexible, tough, and clear. Since various primitive forms of condoms have been around since ancient times, I have nicknamed Fromm the "modern father of safe sex."

FROST, JOHN CARVER MEADOWS. 1915-1979. English scientist, engineer, aircraft designer, and inventor: Among his many aircraft designs he is best remembered by many for a unique innovation named the VZ-9-AV Avrocar (1958). A type of vertical takeoff and landing aircraft (VTOL) that employed the Coanda Effect, the disk shaped UFO-like vehicle garnered patents in three countries and spawned worldwide interest; not only from the public, but from several space agencies and the military—in particular the Royal Canadian Air Force, the United States Air Force, and the United States Army. "Project 1794," as it was classified, was an assault vehicle developed with the aid of NASA, eventually making several test flights (early 1961). Shortly thereafter, however, the VZ-9-AV Avrocar program was abruptly cancelled (late 1961); this despite both Frost's suggested adjustments as well as ongoing enthusiasm from the design team. According to mainstream sources the termination was due to the fact that the vehicle had insurmountable stability problems, which in turn caused funding to dry up. But was this the real reason? Or, as conspiracy theorists speculate, was the sci-fi like project suppressed because its advanced technologies would have obsolesed conventional aircraft, jeopardizing their multibillion-dollar-a-year manufacturers?

FRY, ARTHUR. Born 1931. American scientist and inventor: Invented the Post-it note (1974). Known generically as a "sticky note," it is a piece of paper (of various sizes, shapes, and colors) with a sticky, non-residual, pressure-sensitive, temporarily reusable adhesive on the upper back, permitting it to be attached to paper, metal, plastic, glass, etc.

FULD, WILLIAM. 1870-1927. American businessman, toy and game designer, and inventor: Though disputed by some, he is credited with inventing the name "Ouija" for the popular Victorian parlor game the Ouija Board. Fuld asserted that ouija is an Egyptian word meaning "good luck," or is possibly a derivation of the French word *oui* and the German word *ja*, both meaning "yes."

FULLER, CHARLES W. Lived during the 20th Century. American dentist and inventor: Invented the Gilhoolie (1953), a handy kitchen device that easily opens bottles and jars.

FULLER, RICHARD BUCKMINSTER. 1895-1983. American engineer, futurist, architect, ecologist, philosopher, utopian, cartographer, designer, mathematician, poet, educator, author, visionary, and inventor: Invented the Dymaxion House (1927), a hexagonal structure designed around a mast, Dymaxion Vehicles (1937), the Dymaxion Bathroom (1940), and most notably the geodesic dome (1954), one of the more familiar examples being the Epcot Center, Walt Disney World, Bay Lake, Florida. Fuller is fondly known by his admirers as "Bucky."

Buckminster Fuller.

FULTON, ROBERT. 1765-1815. American engineer, businessman, artist, and inventor: Invented the first successful submarine (1797) and the first commercially viable steamboat (1807), the *Clermont*, simultaneously becoming one of the earliest to use the inboard motor engine design. Note: Though Fulton is often credited with inventing the first steamboat, that honor must go to John Fitch.

FYODOROV, SVYATOSLAV. 1927-2000. Russian eye surgeon, ophthalmologist, politician, and educator: Invented radial keratotomy (1974), a surgical procedure used to correct nearsightedness (myopia).
FYODOROV, VLADIMIR. 1874-1966. Russian scientist, weapons engineer, educator, and inventor: Invented the first assault rifle (1915), a fully operational automatic weapon known as the Fedorov Avtomat.

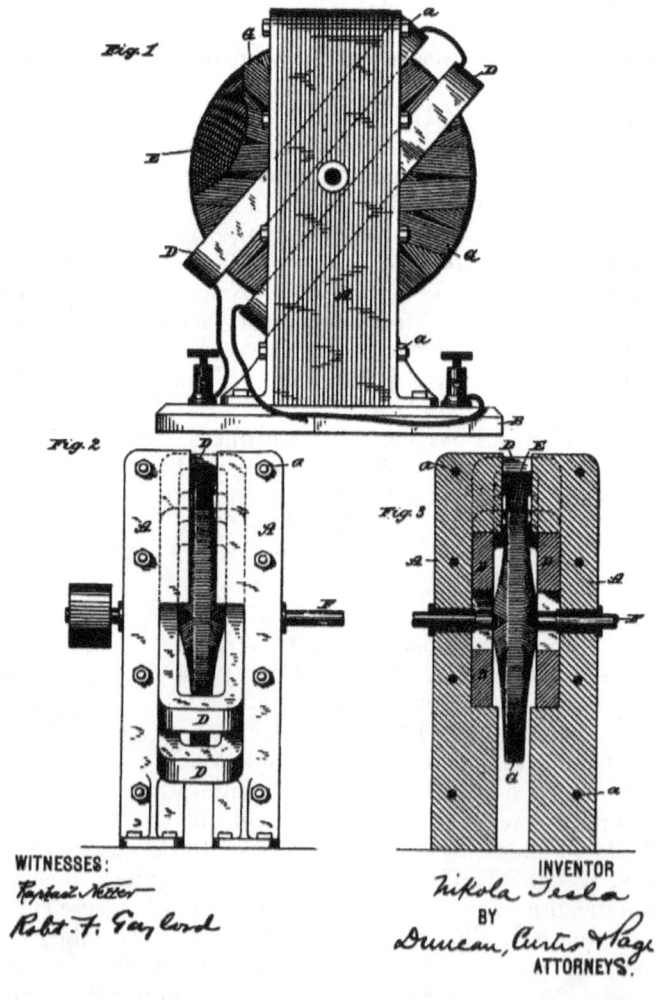

Patent drawing of Nikola Tesla's Electro Magnetic Motor, signed by Tesla and his attorneys, June 25, 1889.

GABOR, DENNIS. 1900-1979. Hungarian physicist, electrical engineer, educator, and inventor: Invented holography (1947). A form of lensless photography, unlike an everyday camera, holographic devices "film" both a light's phase and its intensity. The resulting unidentifiable image of bands and spirals, known as a "hologram," is then illuminated with a laser beam, which arranges the image into a three-dimensional likeness. Holography has altered the modern world with its many important applications in fields such as science, medicine, military, telecommunications, entertainment, the manufacturing industry, advertising and marketing, education, navigation, public safety, and the sales and retail industry. Like many other men listed in this book, Gabor had over 100 patents to his name. Nobel Prize winner.

GALILEI, GALILEO. 1564-1642. Italian polymath, scientist, astronomer, engineer, philosopher, mathematician, and inventor: He is credited with inventing the first thermometer, as well as making improvements on the telescope, water pump, compass, artillery, and clocks, as well as forging advancements in Copernican science, weight balancing, surveying, and meteorology. He was also the discoverer of Io, Ganymede, Europa, and Callisto, Jupiter's major moons. Galileo's innovations were a vital step in advancing the scientific revolution.

Galilei Galileo.

GALITZINE, BORIS BORISOVICH. 1862-1916. Russian physicist, meteorologist, seismologist, educator, and inventor: Invented the first practical electromagnetic seismograph (1906), an instrument used to this day to detect earthquake activity and other types of shifts, motions, and movements in the earth's crust. Scientific data collected from seismographs, along with earthquake early warning systems connected to them, have saved untold numbers of lives.

GALL, JOSEPH G. 1928-2024. American biologist, zoologist, experimentalist, and educator: Invented In Situ Hybridization (1968) also known as ISH, a technique that detects the precise location, as well as allowing the mapping, of genes in chromosomal DNA—along with exposing the cellular distribution of nucleic acid sequences. ISH has advanced science in numerous ways and is today used in such disciplines as scientific research, biology, medicine, pathology, infectious disease management, and gene, phylogenetic, and chromosomal analysis. Gall is known as the "father of modern cell biology."

GALLAGHER, ALFRED WILLIAM. 1911-1990. New Zealand engineer, commercial fisherman, boat-builder, businessman, entrepreneur, and inventor: Invented the electric fence (1937), which is now used globally by farmers, ranchers, gardeners, governments, prisons, the military, pet owners, various businesses, campers, etc.

GALVANI, LUIGI. 1737-1798. Italian physician, surgeon, physicist, obstetrician, comparative anatomist, physiologist, educator, and inventor: A lead researcher in the field of electrophysiology, his experiments with electricity and animal tissues (circa 1770s) led to what must be considered the first electrical battery cell. This occurred when fellow Italian physicist Alessandro Volta borrowed some of Galvani's ideas, leading to the detection of one of the sources of constant current electricity. This discovery in turn led to the invention of the voltaic pile (1800), a type of battery that operates on both chemical and physical principles. In this way Galvani helped pave the way for the upcoming, world-altering electricity revolution. Known as the "father of electrophysiology," the word galvanism (electricity produced by chemical action) derives from his name. Inspired by Galvani's work (in this case, making the appendages of dead animals move by applying electrical currents to their muscles), English authoress Mary Shelley wrote and published her popular Gothic horror novel, *Frankenstein; or, the Modern Prometheus* (1818).

GARBUZOV, DMITRI. 1940-2006. Russian physicist: Co-invented the room temperature continuous wave operating diode laser (circa 1970), while heavily contributing to improvements of the high-power diode laser (circa 1991)—today the world standard for a myriad of industrial applications. Nobel Prize winner.

GATES, BILL: Born 1955. American businessman, entrepreneur, and philanthropist: Co-founder of the Microsoft Corporation (1975), more popularly known as Microsoft. He is also credited with co-inventing the

Altair BASIC programming language (1975), the MS-DOS operating system (1981), the Microsoft Network (1995), and the Windows CE operating system (1996).
GATES, ELMER R. 1859-1923. American scientist, psychologist, chemist, and inventor: Best known for his invention of the foam fire extinguisher (1890), now used almost universally for Class A and Class B types of fires: fires caused by flammable solids (e.g., paper, textiles, wood) and liquids (e.g., gasoline, paint, diesel fuel).
GATLING, RICHARD J. 1818-1903. American gunsmith and inventor: Most famous for his invention of the first successful machine gun (1862), one known as the "Gatling gun" (a crank-operated, rapid-fire, multibarrel machine gun), he also invented a hemp-breaking machine (1850) and a steam plow (1857), as well as numerous other innovations.

Gatling Gun.

GAUSE, GEORGY F. 1910-1986. Russian scientist, biologist, ecologist, evolutionist, researcher, and experimentalist: Noted for his discovery of gramicidin S (1942)—that is, Soviet gramicidin—a powerful antibiotic that saved untold numbers of Russian soldiers after it was mass produced during World War II. He also proposed one of the fundamentals of modern ecology; namely, the competitive exclusion principle (1934), which stipulates that there can only be one result when two species competing for the same resources attempt to occupy the same ecological habitat or niche: one will survive while one will perish.
GAYLORD, NORMAN G. 1923-2007. American scientist, industrial chemist, researcher, educator, and inventor: Lead developer of siloxane-methacrylate, a rigid but gas-permeable material used in the production of contact lenses. By permitting oxygen to pass through the lens to the cornea, this innovation greatly accelerated contact lens technology.
GEIB, KARL-HERMANN. 1908-1949. German chemist: Invented the Girdler Sulfide Process (1943), also known as GS, an efficient and inexpensive method of producing heavy water (deuterium oxide), which is used in nuclear reactors to generate power, and also to produce radioactive elements such as plutonium.
GEIGER, HANS WILHELM. 1882-1945. German physicist, military officer, and inventor: Invented the Geiger Counter (1911), an instrument used to quickly detect and measure potential sources and levels of ionizing radiation. As radiation (in large doses) can be hazardous to human health, this device has saved countless lives, and is today used in a myriad of industries, from law enforcement, military, hospitals, mining, homeland security, and environmental research, to nuclear power plants, fire fighters, cancer patients, physics, emergency response,

and homeowners.[5]

GEIM, ANDRE. Born 1958. Russian physicist, researcher, and educator: Co-discovered graphene (2004), the thinnest material currently known to science. Measuring a mere one atom thick, this two-dimensional, pure carbon product is 200 times stronger than steel, which also makes it the strongest material in the world. Additionally, graphene is not only clear, light, strong, and pliable, it is also an excellent conductor of heat and electricity. This makes it suitable for a wide spectrum of uses in industries ranging from science, engineering, lighting, medicine, sports, apparel, and optics, to food, health, energy, nanotechnology, automotive, construction, and electronics. Nobel Prize winner.

GENKO, NESTOR K. 1839-1904. Russian scientist and forester: Created the first large-scale watershed protection forest belt system (1886-1903). Known as Genko's Forest Belt, this scientifically important forest belt acts as a windbreak that not only helps prevent soil erosion and improves microclimate, but also safeguards farms, crops, livestock, and other agricultural and ranching concerns. Located in Russia's Ulyanovsk Oblast, and now legally protected as a nature reserve, scientists and conservationists continue to study the effects of the forest belt on the local environment.

GEORGE II OF GREAT BRITAIN. 1683-1760. English and Irish king: Founded New York City's King's College (1754), which was subsequently renamed Columbia University (1896).

GERBER, CHRISTOPH. Born 1942. Swiss physicist, educator, and inventor: Co-invented the atomic force microscope (1986), or AFM, an instrument that employs a high-resolution imaging technique allowing it to scan almost any type of surface down to the atomic level. This microscope, which uses a technology known as scanning probe microscopy, visualizes the surface topography of materials at the nanoscale by which it can determine nearly any measurable force interaction. Due to its high resolution capabilities, versatility, and non-destructive imaging, the AFM has a broad degree of applications in such industries as molecular research, biotechnology, pharmacology, physical science, vision science, microelectronics, medicine, live cell imaging, mechanical property analysis, biology, and electrical property analysis, among many others.

GERKE, FRIEDRICH CLEMENS. 1801-1888. German engineer, telegraphist, writer, journalist, and musician: Improved and revised the original American Morse Code (1848), which is today known as the International Morse Code.

GERMER, EDMUND. 1901-1987. German physicist, light scientist, and inventor: Co-invented the fluorescent light (1926) and the high-pressure mercury-vapor lamp (1927). Germer is known as the "father of the fluorescent lamp."

GERSTENZANG, LEO. 1892-1961. Polish inventor: As individuals have been fitting cotton wadding to sticks (as cleaning tools) for generations, the inventor of the first Q-tip continues to be debated. It is my belief that the credit must go to Gerstenzang, however, for it was he who invented the first modern Q-tips (1923), originally called "Q-Tips Baby Gays"—the "Q" representing the word "quality." Baby Gays (i.e., "happy infants") was eventually dropped from the title, leaving the brand name known so well today. Gerstenzang was a representative of the American Jewish Joint Distribution Committee and the founder of the Leo Gerstenzang Infant Novelty Company (circa 1923).

GESTETNER, DAVID. 1854-1939. Hungarian manufacturer, industrialist, businessman, farmer, designer, and inventor: Invented the cyclostyle duplicating process, producing one of the first functional photocopiers: the Gestetner Automatic Cyclostyle Mimeograph (1879). He founded the Gestetner Cyclograph Company in London, England (1881), and is known as the "father of modern stencil duplicating."

GETTING, IVAN A. 1912-2003. American physicist, engineer, and inventor: Co-inventor and developer of the Global Positioning System (1973), better known as GPS. Owned by the U.S. government and operated by the United States Air Force, GPS is a space-based radio-navigation system that supplies precise data concerning one's location, speed, and time. GPS has been key in further developing technologies surrounding national security, the military, electronics, agriculture, emergency services, commercial products, meteorology, mapping, environmental protection, surveying, biology, package and food delivery, transportation, mining, seismology, sports, marine concerns, home appliances, public safety, railways, banking, space, recreation, travel, power grids, automobiles, construction, and aviation, to name but a few.

GIANNI, ALBERTO. 1891-1930. Italian salvage diver, soldier, and inventor: Invented the Torretta Butoscopica (circa late 1920s), or "exploration tower," an underwater observation chamber that was connected to the surface via cable, oxygen, and communications. Gianni's single-man dive chamber contributed to advancements in related underwater technologies.

GIBBON, JOHN HEYSHAM. 1903-1973. American physician, surgeon, educator, author, and inventor: Invented the heart-lung machine (1949), which allowed for the first successful open-heart bypass surgery. Gibbon's medical innovation has been responsible for saving countless lives.

GIEMSA, GUSTAV. 1867-1948. German chemist and bacteriologist: Invented the dye solution known as Giemsa Stain (1902), which, along with histology, is still used today in such disciplines as hematology, cytology, and bacteriology. Giemsa's histopathological staining method was originally intended to diagnosis malaria, but turned out to have other applications as well, such as the identification of such parasites as Plasmodium, Trypanosoma, and Chlamydia.

GIESL-GIESLINGEN, ADOLPH. 1903-1992. Austrian engineer, locomotive designer, educator, and inventor: Invented the Giesl Ejector (1951), which greatly improved locomotive engine performance. As a result, it was installed in railroad locomotives across Europe, aiding in the advancement of engine technologies.

GIESON, IRA VAN. 1866-1913. American scientist, psychiatrist, neurologist, histologist, hypnotist, neuropathologist, and educator: Invented Van Gieson's Stain (1889), a histological staining method for observing, studying, and analyzing connective tissue.

GIFFARD, HENRI. 1825-1882. French engineer and inventor: Invented both the water injector and a steam-powered, passenger-carrying blimp (1852)—a lighter-than-air, hydrogen-filled aircraft called the "Giffard Dirigible Airship." Giffard is known as the "Robert Fulton of air navigation," and is listed, by Victorian architect Gustave Eiffel, on the Eiffel Tower in Paris, France, as one of 72 important French scientists and engineers who lived and worked during the 18th and 19th Centuries.

Henri Giffard.

GILBERT, ALFRED CARLTON. 1884-1961. American athlete, magician, and toy inventor: Invented the still best-selling children's toy Erector Set (circa 1913), a miniature metal construction kit. He also helped popularize the toy model train set, American Flyer Trains (1946).

GILBERT, WILLIAM. 1544-1603. English physicist, physician, philosopher, educator, author, and inventor: Credited with inventing the first electroscope (circa 1600), or what he called a "Versorium." The device, which detected and measured electricity, has been replaced today by the more accurate instrument, the electrometer.

GILL, FOREST P. 1906-2005. American silkscreen printer: Credited with inventing the world's first bumper sticker (1934). Founder of Gill Studios, Inc. (1934).

GILLETTE, KING CAMP. 1855-1932. American inventor, businessman, and entrepreneur: Invented the first disposable (double-edge) safety razor (1904). It was not fully disposable, like many modern versions, but instead used a replaceable blade. His business enterprise, the Gillette Company (founded in 1901), manufactured millions of Gillette razors a day, after which the popular product was adopted by the U.S. military as standard issue for American soldiers during World War I.

GILLIAM, TERRY. Born 1940. American filmmaker, animator, comedian, and actor: Co-created and acted in the hit TV series and comedy troupe *Monty Python's Flying Circus* (1969).

GILLIGAN, VINCE. Born 1967. American producer, director, and screenwriter: Created the hit TV series *Breaking Bad* (2008).

GILMAN, DANIEL COIT. 1831-1908. American educator, philanthropist, and author: Founded Johns Hopkins University (1876), which was named after American businessman, benefactor, and philanthropist Johns Hopkins.
GLASER, DONALD A. 1926-2013. American scientist, physicist, neurobiologist, researcher, experimentalist, educator, and inventor: Invented the bubble chamber (1952), an instrument that detects and identifies charged particles (in collision) while measuring their interactions, movements, momentum, and decay. Still used in high-energy physics research for such work as the study of dark matter. Nobel Prize winner.
GLASS, JOSEPH. 1791-1867. English engineer, chimney sweep, construction builder, businessman, social activist, and inventor: Invented the first chimney-cleaning machine (1827), which included brushes and jointed (i.e., flexible) cane rods, surmounted by a large, stiff, circular brush made of whalebone.
GLOCK, GASTON. 1929-2023. Austrian businessman, entrepreneur, engineer, and inventor: Founder of the Glock firearms company (1963), whose weapons are popular with law enforcement agencies, the military, and consumers. Arguably its most popular model is the Glock 19 (1988).
GLUSHKO, VALENTIN P. 1908-1989. Ukrainian rocket scientist, engineer and designer: Co-developed (and launched) the first intercontinental ballistic missile (August 1957) as well as the first successful artificial satellite, Sputnik I (October 1957). Eventually attaining the position of director of the entire Soviet space program, he was also responsible for the invention of a myriad of advanced Soviet rocket engines, including the liquid-fueled RD-170 (1976), at that time the most powerful engine of its kind.
GOBYATO, LEONID N. 1875-1915. Russian combat engineer, military officer, and artilleryman: Invented the modern portable mortar (1915), a small, rudimentary, movable cannon that is muzzle-loaded with high-arching explosive shells.
GODDARD, ROBERT. 1882-1945. American scientist, physicist, aerospace engineer, astronomer, mathematician, educator, and inventor: Invented the liquid fuel rocket (1926), revolutionizing the field of space flight. Goddard is known as the "father of modern rocket propulsion."
GOEBEL, HEINRICH. 1818-1893. German repair mechanic, businessman, watchmaker, salesman, entrepreneur, and inventor: Credited by some (and highly disputed by many) with the invention of the incandescent light bulb (1854)—a device later referred to as the "Goebel Lamp"—nearly thirty years before Thomas Edison's creation of the incandescent light bulb (in 1880). Unfortunately for history, while Edison patented his invention, Goebel neglected to patent his, and, in the eyes of many—despite numerous claims, lawsuits, eyewitness testimonies, and court trials—the entire case remains unresolved. (Note: In the 1885-1892 case of Edison Electric Light Company vs. United

States Electric Lighting Company, Edison was re-verified as the original inventor of the incandescent light bulb, while the story of the Goebel lamp was relegated to that of nothing more than a Victorian "legend.")

GOLDBERG, EMANUEL. 1881-1970. Israeli scientist, physicist, and inventor: Invented the forerunner of the modern optical character recognition machine, or OCR device (1920s). His "Statistical Machine" determined patterns via a photoelectric cell.

GOLDEN, SAM. 1915-1997. American chemist, businessman, paintmaker, and inventor: Co-invented acrylic paint (circa 1946). Made of pigment suspended in acrylic polymer emulsion, it forever altered the art world by offering an inexpensive, quick-drying, water-resistant paint that could be used on almost any surface.

GOLDMARK, PETER CARL. 1906-1977. Hungarian physicist, engineer, researcher, and inventor: Invented the first commercially viable color television (1940), as well as the first long-playing (LP) phonograph record (1948), a vinyl disk that rotated at $33\ 1/3$ revolutions-per-minute (RPM)—replacing the earlier 78 RPM disk.

GOLDSTEIN, ISRAEL. 1896-1986. American rabbi and author: Co-founded Brandeis University (1948), named after Louis Dembitz Brandeis, American lawyer and associate justice of the Supreme Court.

GOLGI, CAMILLO. 1843-1926. Italian scientist, physician, biologist, histologist, pathologist, and educator: Invented Golgi's Method (1873). Also known as the "black reaction," the technique is used to stain brain tissue neurons with silver chromate, making the bodies of the cells appear black against the lighter background, disclosing the neurons along with their processes. Still in use today, the Golgi Method was an important step in accelerating our knowledge of the nervous system. Golgi also discovered the three forms of parasites and the three types of fever that are associated with malaria. Nobel Prize winner.

GOMORI, GYORGY. 1904-1957. Hungarian surgeon, histochemist, pathologist, and educator: Invented the Gomori Methenamine Silver Stain method (1946), or GMS, used in cytopathology to identify fungal organisms; he also developed the Gomori Trichrome Stain method (1950), used in myopathy to stain muscle tissue to evaluate mitochondrial disease.

GOMPERTZ, LEWIS. Circa 1783-1861. English animal rights activist, mechanical engineer, author, philosopher, and inventor: Co-founder of the Society for the Prevention of Cruelty to Animals, or SPCA (1824), now known as the Royal Society for the Prevention of Cruelty to Animals, or RSPCA.

GOODYEAR, CHARLES. 1800-1860. American engineer, chemist, businessman, author, and inventor: Invented the vulcanization process (heating rubber with sulfur) that opened up the

Charles Goodyear.

way for the commercialization of rubber (1839). In essence, vulcanization improves the properties of natural rubber, greatly increasing its elasticity, durability, and strength.

GORE, ROBERT W. 1937-2020. American scientist, engineer, businessman, entrepreneur, and inventor: Invented Gore-Tex (1969), a lightweight, windproof, breathable, waterproof fabric made from the expanded form of the polymer polytetrafluoroethylene (PTFE). Used primarily by outdoor and sports enthusiasts, his unique clothing represented a major development in the field of performance outerwear. Gore also invented Elixir guitar strings, a singular advancement in coated guitar-string technology.

GORYNIN, IGOR. 1926-2015. Russian scientist, metallurgist, educator, and inventor: Worked primarily with metal alloys, inventing numerous unique and advantageous materials, from high tensile strength aluminum alloys to weldable and corrosion resistant titanium alloys—products that have found applications in fields as diverse as bicycle technology and ship hull design.

GOSLING, JAMES. Born 1955. Canadian computer scientist: Invented the Java programming language (1994), a high performance, versatile, object-oriented, robust, network-centric, portable, fast, reusable, easy-to-use, secure, platform-independent, reliable, general-purpose code that can run on all platforms supported by Java. Still in wide use. Gosling is known as the "father of Java."

GOULD, RICHARD GORDON. 1920-2005. American scientist, physicist, educator, and inventor: The (disputed by some) inventor of the laser (1959). Also credited with coining the word "laser."

GOWER, RICHARD HALL. 1768-1833. English commander (Royal Navy), naval architect, businessman, educator, entrepreneur, merchant seaman, author, philosopher, and inventor: An ingenious ship-builder and designer, he invented the unsinkable Landguard Fort Lifeboat—among a number of other marine-related innovations—but is mainly remembered for his many improvements on the designs of ships' hulls, outfittings, sails, and rigging.

GRABOVSKY, BORIS. 1901-1966. Russian physicist, engineer television pioneer, and inventor: Invented the first fully electronic television (1928), which operated using horizontal and vertical scans. On July 26 of that year Grabovsky made history by being the first to transmit a moving image via wireless radio.

GRAHAM, GEORGE. 1674-1751. English mechanical engineer, scientific instrument maker, watchmaker, and inventor: Credited with numerous sweeping improvements in watch technology. He is best known, however, for inventing the orrery (1704), named after nobleman Charles Boyle, 4th Earl of Orrery (who commissioned the project). Also known as a planetarium, an orrery is a mechanical device representing our solar system and the planets' relative positions as they orbit the sun. The clockwork-operating apparatus is most often used as an educational model to demonstrate the movements of our planetary system.

GRAM, HANS CHRISTIAN. 1853-1938. Danish scientist, physician, pharmacologist, bacteriologist, microbiologist, author, lecturer, and educator: Invented the Gram Staining Method (1884), which helps identify and classify bacteria microscopically. Still in use today.

GRAMME, ZÉNOBE T. 1826-1901. Belgian electrical engineer, electrician, businessman, carpenter, and inventor: Invented the first direct-current (DC) dynamo (1871). Known as the Gramme Dynamo or the Gramme Machine, it transforms mechanical energy into electricity, and was the first generator of its kind to be used commercially—at the time mainly by industrial manufacturers and farmers. Gramme was also the inventor of the Gramme Ring and the Gramme Armature, as well as a host of other innovations.

GRÄTZEL, MICHAEL. Born 1944. German scientist, chemist, researcher, educator, author, lecturer, and inventor: Co-invented the dye-sensitized solar cell (1988), or DSSC. Known as the Grätzel Cell, the flexible, cost-efficient, biodegradable, easy to produce, extra thin-film cell turns sunlight into electricity using titanium dioxide ($TiO2$) particles coated with dye.

GREATHEAD, JAMES HENRY. 1844-1896. South African civil engineer, railway engineer, mechanical engineer, and inventor: Made improvements on the tunneling shield (circa 1869), a temporary protective cylindrical steel covering that allows a boring machine to excavate tunnels in soft ground, and more particularly, under rivers and other types of unstable watery soil. An early proponent of building an underwater tunnel from England to the European mainland (under the English Channel), he is best known for his work on England's underground and overhead railways. Greathead's design, called the "Greathead Shield," is essentially still in use today—though updated for use with modern equipment.

James H. Greathead.

GREBENNIKOV, VIKTOR. 1927-2001. Russian scientist, entomologist, author, paranormal investigator, and inventor: Famous for his work in the field of anti-gravity, and in particular what is called the "cavity structural effect." Considered "fringe science" by mainstream science, Grebennnikov asserted that biological structures (e.g., plant stalks, insect wings, flowers, etc.) can affect the human mind and body therapeutically—and even cause physical objects to both levitate and fly. Although potentially scientifically promising, his research has been both widely dismissed and suppressed.

GREENWOOD, CHESTER. 1858-1937. American businessman, entrepreneur, and inventor: Though he did not invent earmuffs, he improved on the design by patenting the idea for a strip of metal that passes over the top of the head and attaches to both ear coverings (1877). Known at the time as "ear protectors" or "ear mufflers," Greenwood's innovation is still in use today.

GREGORY, JAMES. 1638-1675. Scottish mathematician, astronomer, geometrist, optics engineer, mechanical engineer, and inventor: Invented the Gregorian telescope (1663), the first practical reflecting instrument of its kind. A student of trigonometry functions, he was also the first to describe the distinction between convergent and divergent infinite series, his most famous construction being what is called "Gregory's series for the arctangent function."

GRETSCH, FRIEDRICH W. 1856-1895. German industrialist, businessman, and entrepreneur: Founded the Gretsch Company (1883). A global seller of various musical instruments and accessories, it is best known today for its popular Gretsch drum line and guitar series. The company has been endorsed by numerous notable names, such as the Beatles, Eric Clapton, Phil Collins, Elvin Jones, Steve Ferrone, Hannah Ford-Welton, Michael Nesmith, Bono, Chet Atkins, Duane Norman, Bo Diddley, and Eddie Cochran.

GRIGGS, WILLIAM. 1832-1911. English illustrator, chromolithographer, and inventor: Invented the printing method known as photochromolithography (1868), a planographic technique for making spectacular color prints.

GROENING, MATT. Born 1954. American producer, cartoonist, writer, and animator: Created the hit TV series *The Simpsons* (1989).

GROVE, WILLIAM ROBERT. 1811-1896. Welsh scientist, physicist, chemist, naturalist, photographer, lecturer, judge, educator, and inventor: Invented the first fuel cell (1842), which used water from hydrogen and oxygen gas to create electricity. He is also responsible for inventing an improved wet-cell battery, a two-fluid electric cell called the "Grove Cell" or "Grove Battery" (1839), as well one of the first incandescent electric lights (1840). Lastly, among a number of other innovations and hypotheses, he predicted the theory of the conservation of energy (1842).

GUANELLA, GUSTAV. 1909-1982. Swiss engineer and inventor: Invented the Direct Sequence Spread Spectrum method (1942), or DSSS, a technique for transmitting clandestine messages. Guanella was the claimant of over 200 patents, and is also remembered for inventing the Guanella Balun (1944), a special transmission line transformer—balun being an acronym for "balanced line to unbalanced line."

GUERICKE, OTTO VON. 1602-1686. German scientist, physicist, engineer, natural philosopher, politician, mayor, and inventor: Invented the first air pump (1650), the first dasymeter (1650), and the first electric generator (1663), and was the first eyewitness of electroluminescence. (1672). He is credited by some with inventing the first manometer (1661)—an instrument used to measure pressure in a fluid.

GUILLAUME, MAXIME. Born 1888-circa mid 1900s. French engineer: Invented (patented) the world's first turbojet engine (circa 1921). Unfortunately, the technology required to build his motor did not exist at the time. Thus it was not until 16 years later (1937) that the first turbojet engine was constructed and bench-tested by Frank Whittle. The first actual flight of a turbojet engine took place two years after that (1939) under the guidance of Hans von Ohain.

GUILLOTIN, JOSEPH-IGNACE. 1738-1814. French physician, politician, and inventor: Credited with co-inventing the eponymous machine that bears his name: the Guillotine (1789). It would be more accurate, however, to say that Guillotin merely helped improve upon its design, as similar execution devices had been in use, in various forms, long before then. It is so ancient, in fact, that the original inventor of the Guillotine is not known. Though the Guillotine appears barbaric to modern enlightened societies, its invention at that time did have the benefit of ushering in a wave of new and more humanitarian forms of capital punishment.

GÜLDNER, CARL HUGO. 1866-1926. German mechanical engineer, author, and inventor: Invented the two-stroke diesel engine (1899). Co-founder of the companies Güldner and Lüdeke Maschinenfabrik Magdeburg (1897), and Güldner-Motoren (1904).

GUREVICH, MIKHAIL. 1893-1976. Russian engineer, aircraft designer, and inventor: In partnership with Artem Mikoyan, he co-invented the MiG-series Russian fighter aircraft beginning in 1939. Notable models included the MiG 1, the MiG 3, the MiG-15, the MiG-21, and the MiG-25. MiG is a Russian acronym for "Mikoyan and Gurevich."

GURNEY, GOLDSWORTHY. 1793-1875. Cornish mechanical scientist, surgeon, chemist, educator, lecturer, house-builder, farmer, pianist, piano-maker, and inventor: Invented an extra bright-burning oil lamp known as the Bude-light (1839), as well as the Gurney Stove (1856). A precursor to the modern home radiator, his "stove" is still used in various parts of Britain to this day. Additionally, he may have been the first to create an ammonia engine (circa 1820s). Gurney also experimented with and improved upon a great many other ideas, systems, and products, including the steam jet, steam engine, gas lighting, the blowpipe, limelight, airflow (ventilation), telegraphy, lighting lenses, electric conduction, lighthouses, and the much disputed blastpipe—the latter which stirred much controversy due to Gurney's claim as its originator.

GUSMÃO, BARTOLOMEU LOURENÇO DE. 1685-1724. Brazilian priest, mechanical engineer, aviation researcher, and inventor: Invented what many consider to be the world's first successful flying machine (circa 1709). His small yet complex lighter-than-air balloon, called the "Passarola Airship," was designed to resemble a bird, and was comprised of avian wings, a head, and tail, as well as magnet-filled metal globes, tubes, bellows, and a rudder. Gusmão made a successful test run before King João V of Portugal, who encouraged and supported the inventor. While the elaborate paper contraption ultimately proved impractical for human flight, his innovations aided the work of future flight scientists.

GUTENBERG, BENO. 1889-1960. German scientist, seismologist, educator, author, and inventor: Co-inventor of the Richter Magnitude Scale (1935), which records ground movement in order to analyze the size and intensity of earthquakes.

GUTENBERG, JOHANN. Circa 1398-1468. German craftsman, metalworker, goldsmith, blacksmith, and inventor: Invented the Gutenberg Printing Press (circa 1436), which used movable type, oil-based ink, and an even-pressure press. The first European invention of its kind, Gutenberg's printing press made, among many other things, books more readily and widely available, earning him the title of "father of the information age."

Johannes Gutenberg.

GUTHRIE, SAMUEL. 1782-1848. American scientist, industrial chemist, physician, army surgeon, explosives researcher, businessman, and inventor: Discovered chloroform (1831), which he called "chloric ether." (Though others independently demonstrated the fluid that same year, Guthrie is considered the original discoverer.) Known scientifically as trichloromethane, the clear non-flammable liquid is no longer used as an anaesthetic (due to its many health risks), yet it continues to have many applications in the modern era—from refrigeration and plastics, to cleaning and fumigation. Guthrie is also known for inventing a special firearms' percussion priming powder (1851), the "percussion pill," and a punch lock to light it (a forerunner of the percussion cap), superannuating the older flintlock style, muzzle-loading musket.

GUTIERREZ, JAY. 1962-2015. American medicine man, pilot, jewelry maker, businessman, and entrepreneur: Did not invent, but improved and modernized the scientific concept of natural radiation hormesis (the idea that small doses of radiation are therapeutic) by packaging radioactive stones and soil in useable and even wearable forms, such as necklaces, mudpacks, and water stones. Based on a science known to ancient peoples, according to numerous testimonials Gutierrez's hormesis stones have been used to successfully treat a wide variety of ailments, from tennis elbow to cancer. Founder of Night Hawk Minerals.[6]

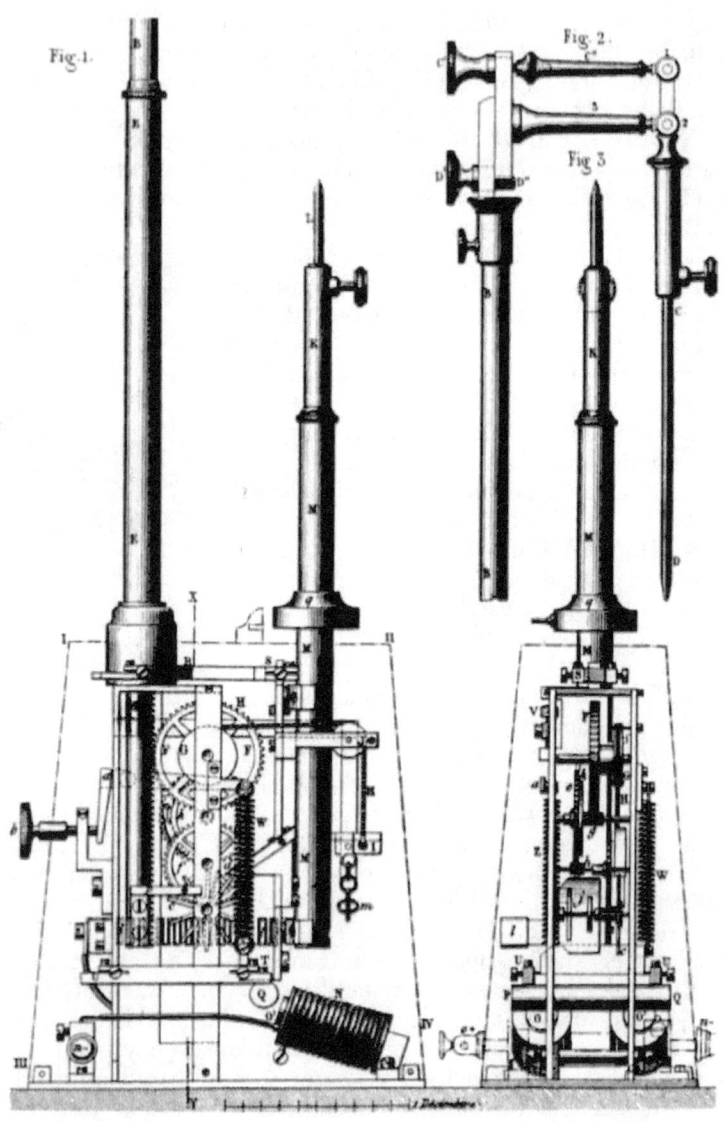

Patent drawing of Victor Serrin's arc light regulator, 1867.

HABER, FRITZ. 1868-1934. Prussian (now Polish) physical chemist, organic chemist, electrochemist, engineer, researcher, educator, author, and inventor: Invented the Haber-Bosch Process (1908), a high pressure, high temperature method of creating ammonia from hydrogen and nitrogen. The end product has numerous applications in fields as varied as agricultural fertilizers, textiles, explosives, refrigeration, plastics, and pharmaceuticals. Though ammonia, and its byproduct nitrogen fertilizer, have their issues (e.g., carbon dioxide emissions), because they now enable the farming industry to grow and harvest more food than ever before, many view the Haber-Bosch Process as one of the most important, though not always the most advantageous, developments of the modern era, even contributing to the current worldwide population explosion. Haber is also remembered for his pioneering work on nitrobenzene (1904), the hydrogen-oxygen fuel cell (1907), and the glass electrode (1909). Nobel Prize winner.

HADLEY, JOHN. 1682-1744. English physician, mathematician, astronomer, optics researcher, and inventor: Invented the forerunner of the sextant. Known as the octant (1730), it is a marine navigation instrument that can pinpoint a geographic position by measuring the altitude of the Sun (or a star) above the ocean's horizon. Hadley also made improvements on the reflecting telescope (1721), which greatly boosted its use and commerciality.

HAFFKINE, WALDEMAR. 1860-1930. Ukrainian zoologist, bacteriologist, and epidemiologist: Created the first vaccine for bubonic plague (1897) and the first effective low-side-effect vaccine for cholera (1892).

HAFNER, WILLIAM FREDERICK. 1870-1944. American toymaker and inventor: Invented a clockwork engine for toy cars (1901) and a clockwork train (1905), both which eventually led to the invention of American Flyer, a still popular collectible children's toy model train set.

HAGENS, GUNTHER VON. Born 1945. German-Polish scientist, physician, anatomist, educator, lecturer, businessman, and inventor: Invented a technique called plastination (1977), a method for preserving human tissue following death. Rather than preserving biological tissue in plastic (the traditional way), he conceived of the idea of injecting plastic into the (embalmed and dissected) tissue itself, which when hardened, enables one to view them in their original form. His educational though controversial public plastination exhibitions of the internal organs and muscles of human bodies has been viewed by tens of millions of people worldwide since they first opened in 1995. Founder of the Institute for Plastination, Heidelberg, Germany (1993).

HALL, CHARLES MARTIN. 1863-1914. American chemist, engineer, businessman, and inventor: Using electrolysis he developed an inexpensive technique, known as the Hall-Héroult Process, for producing aluminum (1886). This allowed for an explosion in commercial sector applications using aluminum, ranging from electronics, consumer goods, and medical equipment, to construction, transportation, and packaging. (The process derives its name from the fact that French inventor Paul-Louis-Toussaint Héroult conceived of the same method in the same year.)

HALL, FRANK H. 1841-1911. American inventor and superintendent of the Illinois School for the Blind: Invented the Hall Braille Writer (1892), the first Braille writing machine.

HALL, HOWARD TRACY. 1919-2008. American physical chemist, educator, researcher, and inventor: Invented the synthetic diamond (1954), revolutionizing not only the gemstone industry, but the electronics, medical, computer, mining, loudspeaker, water treatment, machining, and optics industries as well. Hall is known as the "father of the manmade diamond."

HALL, ROBERT N. 1919-2016. American physicist, engineer, and inventor: Invented the first semiconductor injection laser (1967), which gained popularity in science, medicine, optics, manufacturing, communications, and energy technologies due to its low cost, efficiency, and low-power needs. He also invented a unique diode rectifier, as well as a technique for the purification of germanium. Hall's innovations are still widely used today, in, for instance, fiber-optic communications, CD readers, TV remote controls, and laser printers, to name but a few common applications.

Andrew S. Hallidie.

HALL, SAMUEL. 1782-1863. English engineer and inventor: Best known for his invention of the surface condenser (1838). Used in the steam boilers of steam ship engines, it allowed for the recycling of fresh water, obsoleting the earlier method of using seawater—which tended to corrode metal engine parts. Though his condenser as a whole was a commercial failure, aspects of it are today still being used by the cooling industry. One of his earlier inventions, the textile lace gasser (1817), was quite successful, earning him a fortune. Tragically, he later squandered his wealth on unfruitful innovations, one of them being the surface condenser described above.

HALLIDIE, ANDREW SMITH. 1836-1900. English entrepreneur, construction engineer, bridge-builder, surveyor, wire rope maker, businessman, and inventor: Credited by some with inventing (or at least contributing to the invention of) the first cable car (circa 1873), a public transport vehicle that is pulled along a track by an underground cable system. One of the world's most notable examples of Hallidie's work are the celebrated cable cars of San Francisco, California, which he built expressly to safely traverse the city's famous hills.

HALSE, NICHOLAS. Died 1636. English governor and inventor: Invented the malt kiln (1635), which controlled germination by thoroughly drying out grain and hops prior to milling, as well as adding color, aroma, and flavor to beers and whiskeys.

HAMMING, RICHARD. 1915-1998. American computer scientist, mathematician, educator, and inventor: Invented Hamming Codes (1950), a system (originally intended for use with punch card readers) that finds and rectifies errors that occur during data transmission. Hamming Codes, in one version or another, are still used by the space industry to communicate with satellites and by the digital television industry to provide error-free TV signals, as well as by other technologies.

HAMMOND, JR., JOHN HAYS. 1888-1965. American engineer, businessman, radio wave pioneer, and inventor: Invented radio wave control (circa 1914), also known as remote-control or radio-control (RC). This world-altering innovation, along with his work on wavelength frequency modulation, auto-gyroscope stabilization, autopilot, unmanned aerial vehicles (UAVs), and unmanned combat aerial vehicles (UCAVs)—or what are now more popularly called drones—laid the groundwork for the development of all of today's radio-controlled technologies and devices, including wi-fi, Bluetooth, smartphones, smart TVs, countless household devices, fully automated aircraft, marine-craft, land vehicles, guided missile systems, robots, and model hobby toys (e.g., miniature RC cars, trucks, vans, buses, boats, airplanes, helicopters, jets, rockets, submarines, and trains). A holder of an estimated 400 to 800 patents, he was the founder of the Hammond Radio Research Laboratory, Gloucester, Massachusetts (circa 1910), and the architect of his now famous seaside estate, the medieval landmark known as Hammond Castle, Gloucester, Massachusetts (1926-1929).

Hammond is known as the "father of radio control."

HANSON, JAMES C. M. 1864-1943. Norwegian (later American) librarian: Co-invented the U.S. Library of Congress classification system (1899), or LC, which allows for the systematic organization of its current catalogue of some 40 million books. The LC is still used, not only by the Library of Congress, but also by numerous other libraries and academic institutions around the world.

HARA, MASAHIRO. Born 1957. Japanese engineer, author, and inventor: Co-invented the Quick Response Code (1994). Better known as a QR code, it is a type of gridded, square barcode that, when scanned, accesses whatever data or information its creator attaches to it. Examples: Websites, product information, contact information, payment processing, messaging, sales offers and discounts, etc.

HARGER, ROLLA NEIL. 1890-1983. American chemist, educator, and inventor: Invented the first usable in-the-field alcohol meter (1931), which he called the "Drunkometer." Today it is more commonly referred to by the brand name "Breathalyzer."

HARGREAVES, JAMES. 1721-1778. English hand-loom weaver, carpenter, and inventor: Invented the spinning jenny (circa 1764), a wool and cotton spinning frame with multiple spindles that was far more efficient than the single hand spinners used at the time. By both reducing costs and labor and by increasing productivity, the new spinning jenny technology greatly aided in forwarding the Industrial Revolution (1760-1840), helping move textile manufacturing from homes to factories.

HARINGTON, JOHN. 1561-1612. English author, poet, translator, courtier, and inventor: Invented the first modern flush toilet (circa 1596). The innovation, which incorporated a seat attached to a porcelain basin and an upper cistern filled with fresh water, flushed waste down a pipe into a cesspool below. A godson of Queen Elizabeth I, she was so impressed with Harington's creation that she had one installed in her palace at Richmond, Surrey, England. The flushable toilet was key to advancements in sanitation, disease control, urban development, and overall public health. For better or for worse, Harington himself will forever be memorialized in the nickname still used to this day for his invention: the "John."

HARRIS, WILLIAM SNOW. 1791-1867. English physician, electrician, researcher, author, and inventor: Though Benjamin Franklin invented the first lightning rod, it was Harris who invented and developed marine lightning rods (circa 1820)—specifically made for use on ships. Not only was his creation adopted by the British Royal Navy, it was also used successfully on the voyage of the HMS *Beagle*, saving both the ship, and its famous passenger Charles Darwin, from numerous dangerous lightning strikes. Harris' marine lightning conductor technology is still in use today and has unquestionably saved countless lives around the world over the past two centuries. For his efforts he has been nicknamed "Thunder and Lightning Harris."

HARRISON, JOHN. 1693-1776. English horologist, clockmaker, mechanic, carpenter, and inventor: Invented the first marine chronometer (1735), a precision clock used to accurately calculate longitude at sea—despite fluctuations in motion, temperature, air pressure, and humidity. Harrison, a self-taught clockmaker who fought for years to prove the validity and usefulness of his invention (the first variation which he called "H-1"), is today known for his contributions to understanding and resolving the many issues of reckoning marine longitude. Though largely replaced today by GPS technology, Harrison's hand-wound, non-electric marine chronometer continues to be considered a practical if not vital backup instrument by many boat owners, sailors, skippers, and maritime pilots—especially during power-outage emergencies, when it can be used for celestial navigation.

HARRISON, ROSS GRANVILLE. 1870-1959. American zoologist, anatomist, biologist, researcher, and educator: Developed the first workable animal tissue culture (1907) by producing frog nerve cells in a matrix of clotted lymph. His discovery of how to cultivate nerve cells helped advance organ transplant science while forming the basis for modern nerve physiology and neurology.

HARVARD, JOHN. 1607-1638. English Puritan, clergyman, and educator: A co-founder of Harvard College (1636), the first college established in the new American colonies. The school, now known as Harvard University, was named after him.

HASHIMOTO, KAZUO. Died 1995. Japanese telephonic engineer and inventor: Credited with over 1,000 patents, he is best known for his invention of the analog telephone answering machine (1958), the caller-ID system (1976), and the digital telephone answering machine (1983).

John Harvard.

All three of these particular devices forever altered both social communications and telephonic technology. Hashimoto is known as the "Thomas Edison of Japan."

HASSELBLAD, VICTOR. 1906-1978. Swedish engineer, photographer, businessman, ornithologist, author, and inventor: Invented the Hasselblad modular V-System (1940s), as well as the 1600F (1948)—a 6 x 6cm format focal-plane shutter SLR camera; the 500C camera (1957); the 500EL camera (1965); and the H6D camera series (2016). Hasselblad's innovations have been responsible for a great many advancements in camera technology, while his products are still viewed as the gold standard by many photographers. In 1969 his Hasselblad 500EL/70 camera rode aboard the Apollo 11 spacecraft, and was used by astronauts Neil Armstrong and Edwin "Buzz" Aldrin, Jr. to capture both the first photos of earth from the moon and the first photos of man on the moon.

HAWKING, STEPHEN. 1942-2018. English scientist, astrophysicist, cosmologist, theoretical physicist, mathematician, author, and educator: While not an inventor per se, Hawking made a number of significant scientific proposals and discoveries that have reshaped our worldview, among them: top-down cosmology, evidence of a multiuniverse, the gravitational singularity theorem, and perhaps above all, black hole mechanics, which includes his discovery that black holes emit radiation—a concept now known as "Hawking Radiation."

HAYDEN, FERDINAND VANDEVEER. 1829-1887. American geologist, physician (Union army), surveyor, educator, and explorer: Credited with inspiring the preservation of what would become Yellowstone National Park (1872)—America's first national park. This was largely due to his thorough surveying expedition into northwest Wyoming (1871), which brought back dramatic geological findings, photographs, and descriptions of the area. As one who I would consider the founder or "father of Yellowstone National Park," he has received numerous honors, including the naming of a massive sub-alpine plain in central Yellowstone after him: "Hayden Valley."[7]

HEILMEIER, GEORGE H. 1936-2014. American electrical engineer, computer scientist, DARPA director, and inventor: A pioneer in electro-optic effects, his research aided in the invention and development of liquid crystal displays (1964), or LCDs, the foundation of today's flat screen displays. Though LCDs are gradually being replaced by light emitting diode (LED) and organic light emitting diode (OLED) technology, due to their low cost LCDs are still used in some mobile phones, computers, and TVs.

HEIMLICH, HENRY J. 1920-2016. American surgeon: Credited with inventing the Heimlich Maneuver (1974), a simple and quick first-aid technique for dislodging choking blockages in the windpipe. The method, which works by wrapping one's arms around the victim from behind and applying pressure on his or her abdomen (just under the rib cage), is estimated to have saved over 100,000 lives.

HEINLEIN, ROBERT A. 1907-1988. American science fiction writer and engineer: Credited (though disputed by some) with conceiving the idea of the waterbed (1942), a plastic mattress filled with water—meant to be an alternative to traditional mattresses. A former naval officer, Heinlein is said to have thought of the concept during an illness in which he was bedridden.

HELL, JOZEF KAROL. 1713-1789. Slovakian miner, mining engineer, and inventor: Among his many mining-related inventions he is best remembered for his development of the water pillar (1749), an air-pumping machine that draws water out of mines. The basic technology of Hell's device is still in use today.

HELL, RUDOLF. 1901-2002. German engineer, author, and inventor: Invented the Hellschreiber Teleprinter System (1920s). Also known as the "Hell Printer," it is a dot-matrix impact teleprinter designed specifically for use with wired and wireless telecommunication. A prolific inventor, he also invented the Photoelectric Image Scanning Tube (1925), the Klischograph (1951), the first practical fax machine (1956), a color scanner (1963), and a computerized CRT type-setter (1965)—among other innovations. Hell is known as the "father of digital word processing."

HELMHOLTZ, HERMANN VON. 1821-1894. German polymath, physicist, physician, ophthalmologist, natural scientist, physiologist, mathematician, anatomist, educator, philosopher, psychologist, scholar, author, naturalist, researcher, and inventor: Most famously, he invented the ophthalmoscope (1851), a device used to medically examine the inner condition and workings of the eye. (Though slowly being replaced by newer technologies, the ophthalmoscope is still used.) Although he was involved in the development of dozens of other world-altering innovations, he is also chiefly remembered for his invention of the Helmholtz Pitch Notation System (1863) and the Helmholtz Resonator (1890), as well as many other accomplishments in the fields of physics, geometry, philosophy, optics, topology, electromagnetism, acoustics, thermodynamics, epistemology, mechanics, sensory fluid dynamics, perception, neurology, electrodynamics, and conservation of energy.

Hermann von Helmholtz.

HENDERSHOT, LESTER JENNINGS. 1898-1961. American inventor: Invented a fuel-less "miracle motor" known as the Hendershot Generator (1928); said to derive its power solely from "terrestrial magnetism"—that is, earth's magnetic fields. Hendershot demonstrated his invention to numerous individuals, including American aviator-hero Charles A. Lindbergh, as well as members of the U.S. Air Corps and the American press. Many were convinced of its authenticity, describing its performance as "impressive" and "uncanny." Others believed it was "bunk." Either way, Hendershot passed away before he could turn his limitless energy producing machine into a commercial reality. With little to no support from the scientific community, both he and his generator quietly disappeared from the pages of mainstream history, many assuming his was another case of invention suppression.

HENG, ZHANG. 78-139. Chinese polymath, scientist, mathematician, geographer, astronomer, astrologer, meteorologist, cartographer, engineer, geologist, machinist, scholar, author, poet, artist, and inventor: Invented the seismometer or seismograph (year 132), a copper instrument, which Heng called "Hou Feng Di Dong Yi," used to successfully measure earthquakes. He is also credited with inventing the first equatorial, hydraulic-powered armillary sphere, enabling the construction of accurate (for that time period) star maps.

HENRY, CHARLES H. 1937-2016. American physicist: Invented the quantum well laser (1972), a semiconductor laser that restricts charge carriers in a two-dimensional plane, producing high output powers and low threshold current. The efficiency and precision of the quantum well laser makes it more useful than other types of lasers in such fields as laser printing, telecommunications, medicine, and CD technology.

HENRY, JOSEPH. 1797-1878. American scientist, engineer, researcher, educator, and inventor: Invented the electromechanical relay (1835) and was the discoverer of self-inductance (1830s)—a varying current in a circuit that generates or induces an electromotive force in the same circuit. In honor of the latter breakthrough, the name "henry" was coined for the SI unit (*Système international d'unités*) of inductance. He also invented a precursor to the modern direct current (DC) motor, an early telegraph, and the first commercial electric product: an ore separator. The first secretary of the Smithsonian Institution (1846), Henry's many innovative contributions, including his work in ballistics, acoustics, aeronautics, and meteorology, have heavily influenced not only the field of electromagnetism, but science in general.

HERON OF ALEXANDRIA. Circa AD 10-70. Greek scientist, astronomer, engineer, geometer, mathematician, author, and inventor: Also known as Hero, he is credited by some with inventing the aeolipile. Also known as "Heron's Engine," it was the first steam engine.

HÉROULT, PAUL-LOUIS-TOUSSAINT. 1863-1914. French chemist and inventor: Invented the Héroult Furnace (1886), an electric arc furnace used to make steel, as well as the electrolytic method of producing aluminum (1886). Since the latter was conceived independently by American chemist Charles Martin Hall at the same time, it is known as the Hall-Héroult Process.

HERSCHEL, JOHN. 1792-1871. English polymath, astronomer, physicist, mathematician, optical scientist, chemist, photographer, botanist, author, and inventor: Discovered how hyposulfite of soda acts on insoluble silver salts (1819), invented a photographic process using sensitized paper (1839), took the first glass-plate photograph (1839), coined the word "photography" (1839), and invented the actinometer (1825)—an instrument that measures the intensity of incident radiation. Son of William Herschel below.

HERSCHEL, WILLIAM. 1738-1822. German astronomer, composer, and musician: Discovered the first planet using a telescope, Uranus (1781), along with its moons Titania and Oberon (1787); he also discovered Saturn's moons Enceladus and Mimas (1789). Lastly, he was the discoverer of infrared radiation (1800), used globally today in

William Herschel.

telecommunications, remote controls (electronics), medicine, astronomy, night vision equipment, meteorology, thermal imaging, heat and sun lamps, spas, research, computer ports, firefighting, light therapy, thermometers, fiber optic cables, health, and cooking, warming, drying, broiling, frying, grilling, and roasting foods. Father of John Herschel above.

HERTZ, HEINRICH R. 1857-1894. German physicist and educator: Discovered electromagnetic or radio waves (1887), today widely used in a variety of applications, from GPS, radar, television, satellites, and various types of communications, to microwave ovens, cell phones, astronomy, medicine, and food production. His experiments proved that heat and light are forms of electromagnetic radiations. The frequency unit "Hertz" (Hz), which measures cycles per second, was named after him. Must be considered the true "father of radar."

HERTZANO, EPHRAIM. 1912-1987. Romanian board game developer: Invented Rummikub (1940s), a worldwide popular board game that combines aspects of chess, dominoes, rummy, and mah-jongg.

HESSEL, LASSE L. 1940-2019. Danish physician, author, nutrition adviser, researcher, and inventor: Invented the female condom (mid 1980s), medical diagnostic and test kits, and diet and sex pills, as well as other types of developments—including innovations related to nature and the environment.

HEVESY, GEORGE DE. 1885-1966. Hungarian scientist, physical and organic chemist, educator, and author: Co-discoverer of the element hafnium (1923), introduced fluorescent X-rays, and co-invented the radioactive tracer (1913)—the latter which is used in such applications as imaging (i.e., nuclear medicine scans and positron emission tomography or PET). Nobel Prize winner.

HEWITT, PETER COOPER. 1861-1921. American electrical scientist, engineer, and inventor: Invented the mercury-vapor lamp (1901), a precursor to the fluorescent light. Mercury-vapor lights are used to this day to light roadways and large structures, from expressways and factories to gyms and sports arenas, providing added safety and visibility.

HEWLETT, WILLIAM R. 1913-2001. American businessman, entrepreneur, and engineer: Co-founded the Hewlett-Packard Company (1939), more popularly known as HP.

HICKMAN, RONALD PRICE. 1932-2011. South African automobile designer and inventor: Designed the classic cars known as the Lotus Elan (1962), the Lotus Europa (1966), and the Lotus Elan +2 (1967). He was also the inventor of the immensely popular portable workbench, the Black and Decker Workmate (1961).

HILL, ROWLAND. 1795-1879. English postal administrator, educator, and inventor: Invented the postage stamp (1837), greatly increasing postal efficiency and profitability.

HILLEMAN, MAURICE. 1919-2005. American scientist, microbiologist, vaccinologist, and immunologist: Invented some 40 anti-disease vaccines, including vaccines for rubella, streptococcus, pandemic flu, human cancer, chickenpox, pneumococcus, meningococcus, mumps, hepatitis A, hepatitis B, and measles, among many others. Known as the "father of modern vaccines," his work is estimated to save nearly 10 million lives every year.

HILLIS, DANNY. Born 1956. American computer scientist, entrepreneur, and inventor: Co-invented pinch-to-zoom technology (2005), a multitouch finger gesture that allows the user to zoom in and out of text, a video, a photo, etc., on the screen of an electronic device.

HINDLEY, HENRY. 1701-1771. English mechanic, clockmaker, businessman, entrepreneur, and inventor: Besides creating numerous scientific instruments, including precise watches and machine tools, he is best known for his turret (tower and church) clocks, the most famous timepiece which he built for York Minster (1750).

HINGS, DONALD. 1907-2004. English inventor: Credited by some with inventing the first walkie-talkie (1937), a portable radio signaling device—at the time known as the "packset."

HIPPARCHUS OF NICEA. Circa 190 BC-127 BC. Greek scientist, astronomer, and mathematician: Credited with inventing the first astrolabe (an early scientific device used to determine time and make astronomical measurements), and for founding the science of trigonometry, discovering the precession of the equinoxes, and for producing the first detailed star catalog in the West.

HISASHIGE, TANAKA. 1799-1881. Japanese businessman, entrepreneur, mechanical engineer, scholar, and inventor: Founded the Tanaka Engineering Works (1875), which later became the Toshiba Corporation (1975), more popularly known as Toshiba.

HOFF, MARCIAN E. Born 1937. American electrical engineer, computer scientist, and inventor: Co-inventor of the microprocessor (1971), an integrated circuit (IC) that revolutionized the electronics and communications industries. Today it is used in a myriad of applications, from computers, cell phones, and electronic games, to calculators, automobiles, and home appliances.

HOFFMANN, FELIX. 1868-1946. German chemist, pharmacist, and inventor: In an effort to produce new medicines he invented aspirin (1897) and independently developed heroin (1897). The former has eased the lives of millions while the latter has destroyed the lives of millions.

HOFMANN, ALBERT. 1906-2008. Swiss chemist and author: Invented the psychedelic drug LSD (1938). Known scientifically as lysergic acid diethylamide, it was safely and effectively utilized as a treatment by psychoanalysts for over a decade before it was made illegal due to public misuse.

HOLLERITH, HERMAN. 1860-1929. American scientist, mechanical engineer, statistician, educator, businessman, entrepreneur, and inventor: Invented a forerunner of the electronic computer: an automated tabulating machine (circa 1888) that used punch cards to process data. He was the founder of the Tabulating Machine Company (1896). This was eventually merged into the Computing-Tabulating-Recording Company (1911), which was later renamed the International Business Machines Corporation (1924), better known today as IBM.

Hollerith's Electric Tabulator.

HOLONYAK, JR., NICK. 1928-2022. American electrical engineer, educator, and inventor: Invented the first visible light emitting diode (1962), or LED, a semiconductor device that emits light when charged with an electric current. Up to 90 percent more energy-efficient than an incandescent light bulb, his revolutionary innovation finally made the long held dream of optical data communications a reality, and it is now used globally in countless everyday applications, from light bulbs, the Internet, smart phones, medical equipment, watches, street lights, task lighting, and vehicle headlights, to lasers, electronic displays (TVs, tablets, etc.), Christmas lights, robotics, calculators, and CD and DVD players. He was also the inventor of the silicon tunnel diode and the inventor of a component used in the light dimmer switch. Lastly, he was the co-inventor of both the quantum well laser diode (1977) and the transistor laser (2004). The many ways in which Holonyak's inventions have helped shape modern science are beyond counting.

HOLTER, NORMAN J. 1914-1983. American physicist, chemist, biophysicist, and inventor: Invented the Holter Monitor (1947), the first portable cardiac monitoring device. The non-invasive life-saving device can track heart rhythm abnormalities (arrhythmias), atrial fibrillation (AFib), and palpitations, as well as other heart irregularities, symptoms, and disorders. While newer heart monitoring technologies are beginning to appear, none have yet completely replaced the Holter Monitor due to its accuracy, effectiveness, and numerous capabilities.

HONDA, KOTARO. 1870-1954. Japanese scientist, physicist, metallurgist, engineer, educator, researcher, lecturer, and inventor: Invented KS magnetic steel (1917), a type of steel that, compared to tungsten steel, has three times the magnetic resistance. Due to this property it has a wide range of applications, from electric motors, fuel injectors, industrial pumps, electromagnetic actuators, transformers, valves, medical instruments, and industrial solenoids, to antilock braking systems, cargo ships, relays, electric generators, fuel pump laminations, injectors, and household appliances such as washing machines and refrigerators.

HONERKAMP, FRIEDRICH. Dates unknown. Nationality unknown. Credited with patenting the first toroidal fan (1930), a type of propeller with enclosed, curved, ring-shaped blades. The toroidal design improves fuel efficiency, reduces noise, reduces vibration, improves air and water flow, improves maneuverability, increases propeller rotation speed (which increases vehicle speed), cuts fuel consumption, increases water- and air-gripping capabilities, increases engine life, increases range, decreases emissions, increases load capacity, and reduces cutting damage (e.g., to underwater plant and animal life), all while increasing the strength and durability of the propeller. Toroidal props, which have revolutionized the science of fluid dynamics, have numerous modern applications, from helicopters, drones, and boats, to ships, wind turbines, submarines, and industrial and household fans.

HOOKE, ROBERT. 1635-1703. English polymath, experimental scientist, physicist, chemist, astronomer, geometrist, geologist, biologist, naturalist, paleontologist, surveyor, architect, artist, author, educator, philosopher, and inventor: Discovered the Law of Elasticity (1678), known as "Hooke's Law." A 17th-Century Renaissance Man and intellectual virtuoso, his wide-ranging interests and research took him into a diversity of fields involving watches, light, matter, fossils, gravity, earthquakes, combustion, pendulums, telescopes, comets, pre-Darwinian evolution, human memory, space science, cork, and gases—among others. He was the discoverer of both light diffraction and plant cells (he coined the word "cell"), and made improvements on such instruments as the hygrometer, barometer, and anemometer. Lastly, Hooke was the first to use a microscope to study fossils, and was the inventor of the iris diaphragm, the universal joint (the U-joint), the respirator, the compound microscope and illumination system, and the anchor escapement (used in watches and clocks).

HOPKINS, HAROLD H. 1918-1994. English optical physicist, mathematician, linguist, educator, lecturer, author, and inventor: His work on fiber optics and his invention of the rod lens system led to the realization of flexible endoscopes. Though his innovation did not replace rigid endoscopes (which are still used for some moderately invasive procedures), his long, thin, bendable endoscopes—with built-in rod lenses that magnify, illumine, and film the interior of the body—are today indispensable for highly invasive medical operations.

HOPKINS, JOHNS. 1795-1873. American businessman, entrepreneur, benefactor, and philanthropist: Johns Hopkins University and Johns Hopkins Hospital were named after him (1873).

HORNBY, FRANK. 1863-1936. English toy designer, toy inventor, politician, businessman, and entrepreneur: Invented Meccano (1901), a miniature toy construction kit comprised of a 600-piece set that he called "Mechanics Made Easy." The self-taught engineer was also the inventor of Hornby Model Railways (circa 1920)—toy clockwork train (and train accessories) construction kits, as well as the Dinky line (1931)—miniature zinc alloy toy cars, trucks, farm equipment, and railway station worker figures. Now considered collectible items by many, Hornby's creations were always extremely popular with boys, and still are to this day.

HORNER, WILLIAM GEORGE. 1786-1837. English mathematician, educator, and inventor: Invented the first modern zoetrope (1834), which he called the *Daedaleum*. This early cylindrical optical device possessed slits along its sides, while its inner wall was lined with sequential hand-drawn images of, for example, a person jumping rope. When a viewer peered through the slits and spun the cylinder, the illusion of motion was created, making the zoetrope, like the praxinoscope, thaumatrope, and stroboscope, a primitive forerunner of animation technology, and ultimately, the motion picture projector.

HOTZ, JIMMY. 1953-2023. American record producer, recording engineer, music producer, singer-songwriter, recording artist, computer programmer, computer technologist, music technologist, director, musical instrument designer, electronic music visionary, audiologist, videographer, musician, author, consultant, and inventor: Invented the popular computer software known as Hotz Box, the Hotz MIDI vest, the Hotz MIDI Translator, and the Atari Hotz Box. He also helped develop the award-winning software Zuma 3D Graphics.

HOUDAILLE, MAURICE. 1880-1953. French engineer and inventor: Credited with inventing the first hydraulic shock absorbers for motor vehicles (1906).

HOUDINI, HARRY. 1874-1926. Hungarian illusionist, magician, escape artist, stunt performer, film director, film producer, actor, aviator, and inventor: Among a number of inventions, arguably his most practical was an underwater dive suit that could be shed quickly in case of an emergency.

HOUDRY, EUGÈNE JULES. 1892-1962. French mechanical engineer, high performance fuel researcher, fuel oil engineer, businessman, soldier, and inventor: Credited with inventing the catalytic converter (1950s), a device that breaks down the toxic emissions produced by the internal combustion engine, thereby protecting both the environment and human health. Today it has largely been replaced by two-way and three-way catalytic converters—the latter which is able to render nearly 98 percent of toxic gases harmless.

Harry Houdini and his wife Beatrice.

HOUGHTON, JOEL. Born 1792. American inventor: Invented the first dishwasher (1850), a hand-cranked device that dampened then scrubbed dirty dishes. Though considered cumbersome and impractical by its users, Houghton's invention opened the door to innovations that eventually led to the modern electronic dishwasher.

HOUNSFIELD, GODFREY NEWBOLD. 1919-2004. English electrical engineer: Invented computer tomography (1972). CT, as it is more commonly known, is an imaging instrument that encircles a patient's body with a quickly rotating narrow beam of x-rays. The device's computer then processes these into cross-sectional images, aiding in medical diagnosis. CT technology has greatly advanced since its original development. Revolutionized medicine. Nobel Prize winner.

HOUSE, ROYAL EARL. 1814-1895. American electrical scientist and inventor: Invented the first printing telegraph (1844), an electrical instrument that could both print and transmit data (in the Roman alphabet), as well as acquire it via an electro-phonetic receiver. His invention allowed for great strides in the field of telegraphy.

HOUTEN, SR., CASPARUS VAN. 1770-1858. Dutch chocolate maker, chocolatier, businessman, entrepreneur, and inventor: Invented the hydraulic cocoa press (1828), which allowed for the creation of cocoa powder, cocoa butter, and chocolate milk. Casparus and his son, Dutch chemist Coenraad Johannes van Houten, later co-developed what is known as Dutch processing (1846), a technique for removing the natural bitterness found in cacao (by treating it with alkaline potassium carbonates). Van Houten Company chocolate products, originally created in Casparus' Amsterdam chocolate factory (founded in 1815), are still sold under the family name to this day.

HOWARD, HENRY TAYLOR. 1932-2002. American scientist, electrical engineer, radio engineer, businessman, entrepreneur, educator, and inventor: Credited with inventing satellite TV (1976), which is now gradually replacing cable TV technology.

HOWE, ELIAS. 1819-1867. American machinist, mechanical engineer, mechanic, and inventor: Invented the first practical sewing machine (1845). The mechanical device, which employed lockstitch technology, greatly exceeded the efficiency of hand-sewing, thereby revolutionizing the textile industry.

HUANG, JENSEN. Born 1963. Taiwanese electrical engineer, businessman, entrepreneur, and humanitarian: Co-founded the Nvidia Corporation (1993), an electronic graphics company, that among other products, is chiefly known for being a designer, producer, and seller of CPUs and graphics cards for applications ranging from computers, AI, and gaming, to robotics, video editing, and vehicles.

HUBBELL, HARVEY. 1857-1927. American industrialist, businessman, entrepreneur, and inventor: A specialist in inventing manufacturing equipment and electrical equipment, his numerous innovations include the thread rolling machine, automatic tapping machines, and stamping dies. He is best known, however, for inventing the pull chain light socket (1896) and the American electric wall plug and electric wall outlet or receptacle (1904), all three items used in millions of homes and buildings to this day.

HUBBLE, EDWIN P. 1889-1953. American space scientist, astronomer, and lecturer: His discoveries include the findings that the universe is expanding and that there are other galaxies besides the Milky Way. The Hubble Space Telescope was named after him.

HUGHES, DAVID EDWARD. 1829-1900. English physicist, electrician, educator, experimenter, and inventor: Invented the printing type telegraph (circa 1850) the carbon microphone (circa 1878), and a mobile receiver (circa 1879). He seems to have been the first to detect electromagnetic waves (1879), while metal detector technology later resulted from his work in the field of induction balance. Hughes coined the word "microphone."
HULL, CHUCK. Born 1939. American technology inventor, businessman, and entrepreneur: Credited with inventing the 3D printer (1983) as well as the stereolithographic file format STL (1987)—an acronym for Standard Triangle Language, or Standard Tessellation Language—that is typically used with 3D technology (design, printing, storage, etc.). He also co-founded 3D Systems (1986). Hull's innovations have impacted numerous fields, from education and electronics to aerospace and medicine. (Note: The stereolithography process was independently discovered around the same time by French inventors Olivier de Witte, Jean Claude André, and Alain Le Méhauté.)
HURTUBISE, TROY. 1963-2018. Canadian entrepreneur, conservationist, and inventor: Invented the Trojan Ballistics Suit of Armor (1996), a military suit that can protect against bullets; the Ursus suit (1996), a metal suit for protection against grizzly bear attacks; Angel Light, an instrument said to be able to see through objects; and Firepaste, a heat-proofing material.
HURWITZ, MITCHELL D. Born 1963. American producer, writer, and actor: Created the hit TV series *Arrested Development* (2003).
HUTCHISON, MILLER REESE. 1876-1944. American electrical engineer and inventor: Invented the first electric hearing aid (1898). Known as the Akouphone, it amplified weak audio signals using a carbon transmitter and an electric current. He was also the inventor of the klaxon horn (early 1900s)—both hand-powered and electric—an early turn-of-the-century horn used on bicycles and in cars, truck, trains, ships, and submarines.
HUYGENS, CHRISTIAAN. 1629-1695. Dutch scientist, physicist, mathematician, engineer, astronomer, optics scientist, clockmaker, author, artist, and inventor: Invented the pendulum clock (circa 1658) and the U-tube manometer (1661), and made improvements on the telescope through his novel lens grinding and polishing technique. Huygens, a co-founder of the French Academy of Sciences (1666), is also credited with describing the wave theory of light, and with discovering the true shape of the rings of Saturn. He is perhaps best remembered for the Huygens' Principle (1678), his explanation of how wave theory accounts for the laws of geometric optics: A plane light wave expands through space in every direction at the speed of light (depending on its velocity).
HYATT, JOHN WESLEY. 1837-1920. American chemist, industrialist, engineer, businessman, entrepreneur, and inventor:

John W. Hyatt.

Claimed, by some, to have formulated the process for manufacturing celluloid (1869)—though others generally attribute the discovery to Alexander Parkes, whose discovery occurred a few years earlier (1862). (Despite this, Hyatt, or possibly his brother Isaiah, is credited with coining the word "celluloid.") Hyatt was behind a number of other innovations as well, including the Hyatt Filter (1886), a water purification system (1880s), a sewing machine, a sugar cane mill, and a roller bearing.

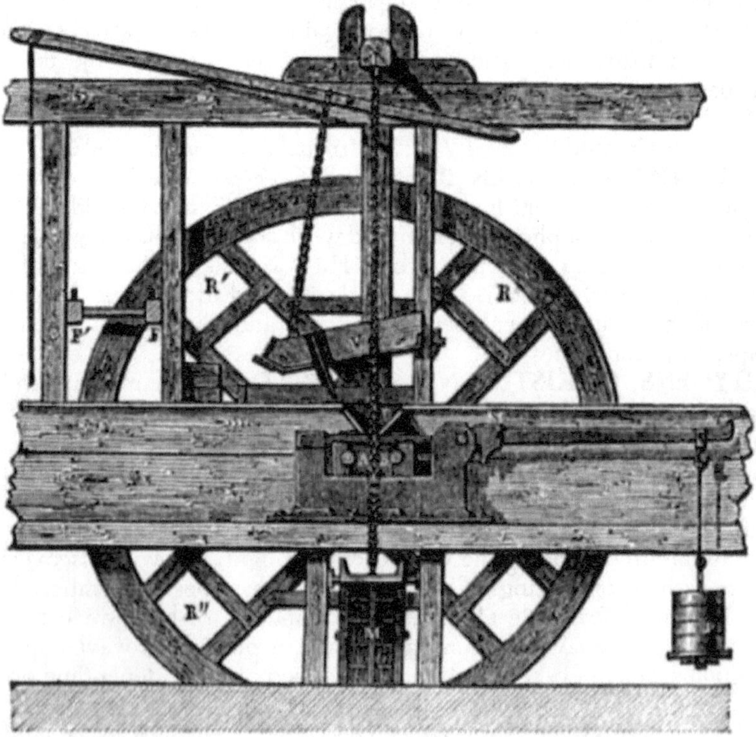

Patent drawing of an element of an ore preparing machine, late 1800s.

IBUKA, MASARU. 1908-1997. Japanese scientist, businessman, engineer, and entrepreneur: Co-founder of the Sony Group Corporation (1946), better known as Sony.

IDLE, ERIC. Born 1943. English actor, songwriter, musician, comedian, screenwriter, and playwright: Co-created and acted in the hit TV series and comedy troupe *Monty Python's Flying Circus* (1969).

ILIZAROV, GAVRIIL A. 1921-1992. Azerbaijani physician, orthopedic surgeon, and inventor: Invented the Ilizarov apparatus (1951), an external structural metal frame that fixates broken bones and fractures, allowing them to heal properly. This circular ringed fixator, which affixes to bone using adjustable threaded metal rods, transfixion wires, pins, and nuts, is also used to reshape and lengthen deformed limb bones. When the Ilizarov apparatus is used in a medical operation it is called the "Ilizarov Technique" or "Ilizarov Surgery," another one of his innovations. Though distraction osteogenesis was first described a century earlier by Bernhard von Langenbeck (1869), Ilizarov is credited with perfecting the bone lengthening medical procedure (1950s), which involves cutting a bone (osteotomy), then allowing new growth to develop in the opening.

ILON, BENGT ERLAND. 1923-2008. Swedish engineer: Invented the mecanum wheel (1972), an omnidirectional wheel that allows land-based vehicles to move in any direction. Also known as the Swedish Wheel, the Ilon Wheel, and the omniwheel, its rubberized, 45 degree-slanted rollers make it useful for machinery operating in small spaces, such as robots, wheelchairs, forklifts, and other types of transfer vehicles.

IMMINK, KORNELIS A. SCHOUHAMER. Born 1946. Dutch

scientist, digital pioneer, engineer, businessman, entrepreneur, author, educator, and inventor: Among his over 1,000 patents, he is credited with a long list of contributions related to the development of audio technology, video technology, and data recording technology. His work, in particular coding systems, continues to influence such applications as the compact disc, DVD, and Blu-ray.

IMURA, MAMORU. Born 1948. Japanese businessman, composer, and inventor: Invented the RFIQ (circa 2007). Also spelled RFIQin, it is a portable, waterless, automatic cooking device that uses induction heat, steam pressure, various sized pans, and electronic recipe cards to prepare a variety of foods.

INOUE, DAISUKE. Born 1940. Japanese businessman, musician, and inventor: Invented an improved version of the karaoke machine (1971), a device that allows amateur singers to sing along to taped music. The popular music system, which removes the original vocals, includes a microphone and speakers. The word karaoke derives from the Japanese words kara, "empty," and oke, "orchestra."

IRINYI, JÁNOS. 1817-1895. Hungarian chemist, engineer, author, and inventor: Invented the safety match (1836), a silent and non-explosive match that combines lead dioxide (rather than calcium chlorate) with phosphor in the match head. The ubiquitous safety match continues to be used for a wide variety of applications, from religious services, fireplaces, cooking, heating, and camping, to arts and crafts, emergency situations, lighting, fire prevention, and survival kits.

IWERKS, UB. 1901-1971. American cartoonist, animator, character designer, special visual effects innovator, film director, film producer, and inventor: Among many other endeavors, he refined (some say co-developed) the world beloved cartoon character Mickey Mouse (1928), improved the multiplane camera used for animation (circa 1937), tailored xerography to work with animation (circa early 1960s), and assisted in the development of numerous Disney theme park attractions. Iwerks is remembered for his many creative contributions to the Disney Company, to the film industry, and especially to the world of animation.

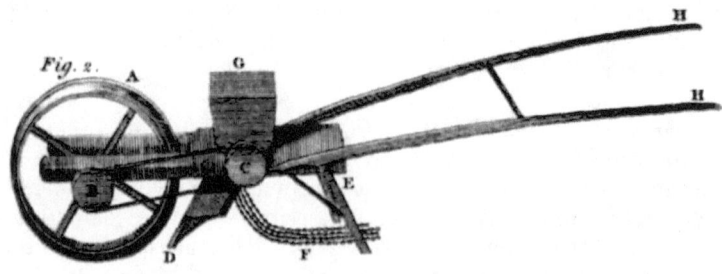

Patent drawing of Thomas Andrew Knight's Drill Machine, which furrowed and planted seeds simultaneously, 1804.

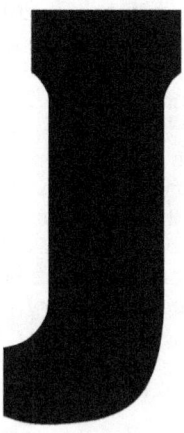

JACOBI, MORITZ HERMANN VON. 1801-1874. German physicist, engineer, architect, educator, and inventor: Credited with inventing the first electric motor (1834). He also invented electrotyping (1838), designed the first electric boat (1838), aided in the development of electric telegraph technology (1840s), and originated Jacobi's law (circa 1840), or maximum power theorem. He is also credited by some with inventing the first letter-typing telegraph machine (circa 1840s)—though this feat is usually attributed to Royal Earl House (1846).

JAENISCH, RUDOLF. Born 1942. Polish scientist, molecular biologist, geneticist, transgeneticist, epigeneticist, researcher, and educator: Created the first genetically modified (transgenic) mouse (1974). He was also an early researcher of stem-cells and master gene regulators, all of which has helped advance the study of various medical issues, such as neurological diseases, cancer, and epigenetic reprogramming. (Note: According to the science of transgenics, transgenesis involves artificially introducing one or more DNA sequences from a different plant or animal species. The result is a GMO, or genetically modified organism.)

JAMISON, ALCINOÜS BURTON. 1851-1938. American physician, proctologist, spiritualist, clairvoyant, educator, and author: A follower of the natural health concept of auto-intoxication (in which the body is said to create poisonous toxins—mainly in the bowel and digestive tract), he invented various

Alcinous B. Jamison.

internal cleansers by which his patients might avoid painful and dangerous surgeries. His, or similar, methods are still prescribed by naturopaths and are used by millions of sufferers around the world.

JANSKY, KARL G. 1905-1950. American physicist, astronomer, radio engineer, and inventor: Discovered radio waves (1932), or what he referred to as "star noise," and invented the Jansky antenna, a radio wave receiving beam antenna (1933)—the first (some say a precursor to the) radio telescope. Both helped launch the field of radio astronomy. In recognition of his discovery his surname was given to the unit of radio-wave emission strength: the "jansky." Jansky's work continues to have tremendous impact on everyday life in the 21st Century, influencing everything from GPS technology, medicine, energy, and satellite communications, to wi-fi, the oil industry, aerospace, and construction.

JANSSEN, ZACHARIAS. 1585-1632. Dutch optics scientist, glass maker, and inventor: Credited by some with inventing the first telescope (circa 1590), and by others with inventing the compound microscope (circa 1590).

JARVIK, ROBERT K. Born 1946. American biomedical scientist and engineer, medical pioneer, researcher, businessman, entrepreneur, and inventor: Credited with being the inventor of the world's first permanent artificial heart (1982). Known as the "Jarvik 7," it was first successfully implanted in a patient named Barney Clark (1982). Founder of Jarvik Research, Inc. (1987). Jarvik is known as the "father of the first successful artificial heart."

JATHO, KARL. 1873-1933. German aviator, airplane designer, builder, and pioneer, artist, and inventor: Credited by some, questioned by many, with performing the first manned, motorized, controlled airplane flight (November 1903) in a flying machine of his own design, soaring some 195 feet in distance and nearly 30 feet above the ground. This is said to have taken place one month before the famous flight of the Wright Brothers (December 1903). Jatho was the founder of the Hannoversche Flugzeugwerke Gesellschaft mit beschränkter Haftung (1913), roughly, the "Hannoversche Flugzeugwerke limited liability company." As the first known pilot to take off from a runway (in the beginning the Wright Brothers used a wooden railway track), I am bestowing on Jatho the title "father of the runway."

JAVAN, ALI. 1926-2016. Iranian physicist, educator, author, and inventor: Invented the gas helium-neon laser (1960), a type of laser that excites gas sealed in a tube, producing light. Additionally, he was the first to describe a technique for precisely measuring the speed of light. He also fostered the theory of the three-level maser, and inaugurated the field of high-resolution laser spectroscopy. Today Javan's gas lasers are used for countless applications, from Internet data transmission, holograms, scientific research, and telephony, to bar-code scanners, medicine, entertainment, and cutting and welding metal.

JEDLIK, ÁNYOS. 1800-1895. Hungarian polymath, physicist, mechanical engineer, electrical experimenter, author, educator, lecturer, clergyman, and inventor: Among many other innovations pertaining to electricity, electromagnetics, generators, and direct-current or DC motors (including commutators, rotators, and what he called a "tubular condenser"), he invented the Jedlik Dynamo (circa 1861), an early type of electrical generator that transforms mechanical energy into electricity. Some have called his dynamo the first electric motor. Nikola Tesla is said to have been influenced by him.

JEFFERSON, THOMAS. 1743-1826. American Founding Father, statesman, politician, neologist, author, and inventor: The third President of the United States of America, Jefferson is credited with numerous inventions—most created out of personal necessity. Among them: a lap desk, an improved iron plow, a revolving book stand, a rainwater catcher, a weight-powered clock, a folding footstool, a folding campstool, a folding ladder, a sundial, a swiveling chair, automatic doors, dumb waiters, a polygraph machine, a space-saving hanging bed (precursor to the Murphy Bed), a revolving clothes rack, a portable copying press, and a cipher wheel. The author of the world-altering Declaration of Independence, he was also a clever neologist, and was responsible for inventing over 100 new words, including the words: belittle, lengthily, monotonously, ottoman, Anglomania, electioneering, authentication, indecipherable, pedicure, and neologize. He led the team that codified the guidelines that inventors would use (and are still using) for patent applications. Founder of the University of Virginia (1825) and, more importantly, considered by many to be the primary founder of the U.S.A.

Thomas Jefferson.

JEFFREYS, ALEC JOHN. Born 1950. English geneticist, researcher, and lecturer: Invented the process known as DNA fingerprinting (1980s), which allows forensic authorities to profile and identify an individual by matching him or her with a genetic sample. Jeffrey's DNA typing technique, which can either convict or exonerate a suspect, has greatly impacted the modern world with a wide range of uses, from criminal investigations, medicine, anthropology, and biology, to paternity issues, population research, legal issues, and identification of missing persons.

JENKINS, CHARLES FRANCIS. 1867-1934. American electrical engineer, cinematography and TV scientist, businessman, stenographer, and inventor: Invented the Phantoscope (circa 1893), an early movie projector, and was one of the first to exhibit synchronized transmission of pictures and sound (1923)—though, at the time, in silhouette only. His company, Jenkins Television Corporation, launched America's first television broadcasting station (1925). The holder of over 400 patents,

Jenkins is one of the many early unsung pioneers of motion picture and television technologies.

JENNINGS, THOMAS L. 1791-1859. American clothing designer, tailor, businessman, and inventor: Invented a dry cleaning process known as dry scouring (1821), which laid the groundwork for modern dry cleaning.

JOBS, STEVE. 1955-2011. American entrepreneur, businessman, computer scientist, investor, financier, speaker, designer, film producer, and inventor: Co-founder (with Steve Wozniak and Ronald Wayne) of the technology company Apple Inc. (1976), as an inventor, co-inventor, and conceptualist, Jobs was directly or indirectly responsible for a host of innovative devices, software operating systems, and applications, including the following: the Apple I computer, the Apple II computer, the Macintosh computer, Pixar, NeXT, the Cube, the iPod, the MacBook, the iPhone, the iPad tablet, and the Bottom Line. His work in the field of personal computing not only revolutionized computer technology, it permanently changed the world.

JOEL, JR., AMOS EDWARD. 1918-2008. American electrical engineer, author, and inventor: Designed the first automatic telephone billing equipment (circa post 1945), directed the invention and development of the first electronic telephone switching systems (circa 1948), and greatly aided in the improvement of operator services (1960s). His many influential contributions to the fields of automatic message accounting (AMA) and telecommunications switching systems are still in use today.

JOHANSSON, CARL EDVARD. 1864-1943. Swedish mechanical engineer, businessman, entrepreneur, and inventor: Invented Jo-blocks (1896). Also known as "Johansson Gauges," these various-sized, hardened steel blocks are ground to the highest accuracy, allowing them to be used to produce precision dimensions. As such they are typically used by the military, automotive, aerospace, and manufacturing industries as instruments for measuring, calibrating, and setting angles with extreme accuracy. Jo-blocks greatly advanced the field of metrology.

JOHANSSON, JOHAN PETTER. 1853-1943. Swedish businessman, entrepreneur, and inventor: Invented the adjustable pipe or plumber wrench (1888) and made improvements on the adjustable spanner (circa 1891)—also known as a crescent wrench. Founder of the company Enköpings Mekaniska Verkstad (1887), which became the global company Bahco in 1916. Since that year Bahco has manufactured some 100 million spanners alone. Due to their amazing usefulness and practicality both his wrench and spanner are still produced and can be found in nearly every tool chest around the world.

JOHNSON, ERIC ARTHUR. Dates unknown. British engineer: Invented the computer finger-driven touchscreen interface (1965). This led to an explosion in touchscreen technology, which is now used in everything from cell phones and computer tablets to air traffic control computers and ATM machines. Inevitably it found its way into the hotel, medical, military, space, manufacturing, restaurant, retail, fast food, and theme park industries, as well as countless other modern commercial enterprises and applications.

JOHNSON, REYNOLD B. 1906-1998. American computer scientist and inventor: Led the team creation of the first commercial, moving head, hard disk drive (1956), or HDD. Known as the IBM Model 350, it weighed more than a ton, used fifty 24-inch diameter platters, could store only 5 MB of data, and had a search time—utilizing rotating magnetic disks—of around one second. Because internal hard drives did not exist at the time, all drives were inherently external, making him the inventor of the first external hard drive as well. Johnson is known as the "father of the hard disk drive."

JOLLY, PHILIPP VON. 1809-1884. German physicist, mathematician, mechanician, technologist, author, educator, and inventor: Invented the Jolly Balance (1864), a device that can fix the specific gravity—that is, weigh the relative density—of various objects, either in air and or when submerged underwater. He is also credited with inventing the Jolly Air Thermometer (1878), developing a unique eudiometer (1879), and improving the mercury air pump. Though the Jolly Balance itself has been obsolesed by more accurate technologies, Jolly's work in the fields of experimental physics, osmosis, and gravitation helped further new developments in metrology as well as other scientific fields.

JONES, FREDERICK MCKINLEY. 1893-1961. American engineer, mechanic, electrician, army officer, entrepreneur, and inventor: Invented (some say developed) the concept of portable refrigeration (1938). To begin with, his automatic mobile refrigeration units greatly aided U.S. military forces during World War II by transporting food, blood, and medicine. Perhaps more importantly, however, by making perishable foods accessible 365 days a year, his cooling system exponentially advanced refrigeration technology, changing the modern world as we know it. This self-taught innovator, who held some 60 patents and also made contributions to the film, theater, sound, recreational, and medical industries, co-founded the U.S. Thermo Control Company (1938), now known as Thermo King. Jones is known as the "father of refrigerated transportation."

JONES, ROBERT T. 1910-1999. American aerodynamicist, aeronautical engineer, flight engineer, and inventor: Among his many contributions to flight science he is chiefly known for his development of the delta wing concept for jet aircraft (1945), a triangular-shaped wing design that has largely replaced the straight wing design, and which is now standard on most modern high-speed and supersonic aircraft. Jones was also an ardent promoter of the oblique wing design.

JONES, SCOTT A. Born 1960. American entrepreneur, businessman, tech pioneer, investor, and inventor: Improved on voicemail technology (mid 1980s), and co-founded Boston Technology Corporation (2004), which specializes in a myriad of digital technology services. Jones is connected to other innovative companies, including Escient Technology, Precise Path Robotics, Gracenote, ChaCha, Art Technology Group, and the Eleven Fifty Academy.

JONES, TERRY. 1942-2020. English actor, director, comedian, historian, and writer: Co-created and acted in the hit TV series and comedy troupe *Monty Python's Flying Circus* (1969).

JONES, TOM PARRY. 1935-2013. Welsh chemist, businessman, entrepreneur, lecturer, aviator, and inventor: While he did not invent the alcohol meter, he did invent (or, according to some, co-invent with Bill Ducie) the first electronic alcohol meter (1967), as well as the first handheld electronic alcohol meter (1972)—devices more commonly known today by the brand name "Breathalyzer."

JORDANOFF, ASSEN. 1896-1967. Bulgarian engineer, aircraft pilot, aircraft designer, stunt pilot, businessman, entrepreneur, author, and inventor: Invented the world's first airbag (1957), a device responsible for saving the lives of millions of people since its inception some 70 years ago.

JOSEPHO, ANATOL. 1894-1980. Russian photographer and inventor: Invented the first coin-operated photo booth (1925). His small portable photo studio, which he called the "Photomaton," has found widespread popularity around the world, and has evolved over time to include a number of modern technologies, such as digital photography, video capabilities, photo editing, filters, virtual backgrounds, green screens, props, and social media-sharing functions.

JUDSON, WHITCOMB L. 1843-1909. American mechanical engineer, salesman, and inventor: Invented a forerunner to the modern zipper (1891), which he called the "Clasp Locker." He was also noted for his many contributions to railway technology, and was the founder of the Judson Pneumatic Street Railway (circa 1888). Whitcomb's metal zipper fastener made little commercial headway during his lifetime, and he did not live long enough to witness its incredible global success.

Electric train, circa 1898.

The device was later improved upon by subsequent innovators, most notably Swedish inventor Gideon Sundbäck—whose design is the one in use today. The B. F. Goodrich company coined the word "zipper" in 1923, 14 years after Whitcomb's passing. (Note: During the War Between the States Whitcomb was a Union soldier who fought at the Battles of Spring Hill and Franklin in the Fall of 1864, the latter where he was wounded.)[8]

JULIAN, PERCY LAVON. 1899-1975. American steroid chemist, researcher, educator, and entrepreneur: Invented a large scale synthesizing process for turning plants into compounds that could be used for medicinal applications (circa 1930s-1950s). His discovery of an industrial technique for mass producing drugs from flora, particularly steroids, lowered costs and made medications more widely available. Founder of Julian Laboratories (circa 1953), Julian Associates (1964), and the Julian Research Institute (1964).

JUN, MA. Lived circa 220-265. Chinese mechanical engineer and inventor: Invented the earliest known vehicle containing a differential gear (3rd Century), a mechanical device—later greatly improved upon by Onésiphore Pecqueur—that is still used in motorized vehicles and by various industries to this day.

JUNG, CARL. 1875-1961. Swiss scientist, psychoanalyst, psychiatrist, psychologist, psychotherapist, psychology pioneer, philosopher, illustrator, author, and essayist: Developed a number of important psychology-related ideas and concepts that are still in use today; among them: individuation, archetypes, the word association test, the collective unconscious, and introverted and extroverted personality types. His work inspired the invention of the Myers-Briggs Type Indicator (MBTI) as well as Alcoholics Anonymous.

JUREIT, JOHN CALVIN. 1918-2005. American engineer, cartographer, and inventor: Invented the gang nail plate (1955), a corrosion-resistant steel sheet with dozens of pointed nail-like prongs on one side that can be hammered into multiple wood surfaces simultaneously. The galvanized metal tie, which does away with the need for nails, glue, or screws, greatly strengthens the connection between such items as wooden joints, wooden beams, and wooden trusses. Jureit's invention revolutionized home construction by making houses stronger (e.g., in order to withstand strong winds) and by reducing building time as well as labor and production costs. The gang nail plate is said to have been a major factor in the explosive growth of affordable housing that began in the mid 20th Century.

Left: Samuel F. B. Morse's Key System, 1850s, a revolution in electric telegraphy.

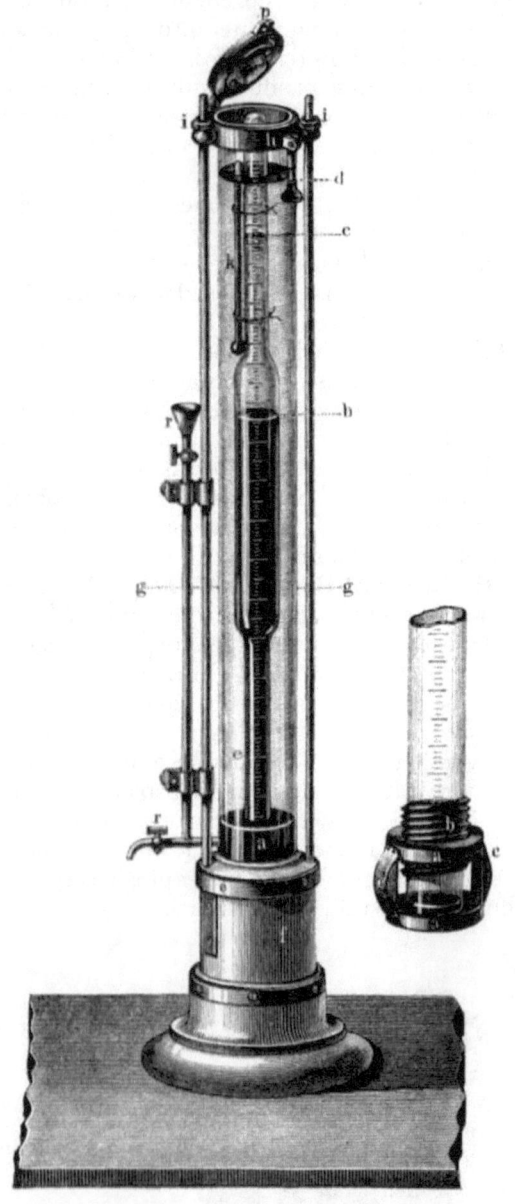

Patent drawing of an early glass measuring device, 1800s.

KAHN, BOB. Born 1938. American computer scientist and electrical engineer: Co-invented Internet protocol, or IP (1974), as well as Transmission Control Protocol, or TCP (1974)—also known as TCP/IP. Known as one of the "fathers of the Internet."

KAHNG, DAWON. 1931-1992. South Korean electrical engineer and inventor: Co-invented the metal oxide semiconductor field-effect transistor (1967), or MOSFET, a floating gate transistor (FGT) that is the foundation of non-volatile memory cells. Also known as a FGMOS, this vital electronic device is a key digital element used in a wide variety of applications, including flash memory, analog storage, neural networking, signal switching, integrated circuits, thin-film transistors, amplification, and digital potentiometers.

KALASHNIKOV, MIKHAIL T. 1919-2013. Russian military engineer, arms designer, politician, military officer, author, and inventor: Invented the AK-47, the "Kalashnikov" (1947), the most popular and widely used assault rifle in the world, with an estimated 120 million or more in circulation. According to some, his automatic rifle—officially named the "Avtomat Kalashnikova model 1947"—is perhaps not a purely original design, but more properly a practical blend of earlier weapon technologies. Its immense global popularity is due to its simplicity, accuracy, low cost, reliability,

Mikhail T. Kalashnikov holding his famous creation: an AK-47, circa 1990.

availability, adaptability, large caliber rounds, light weight, low maintenance, customization, durability, and ruggedness. For example, not only can the AK-47 withstand rust and extreme weather conditions, it can survive being frozen in ice, soaked in water, covered in snow, and buried in sand or mud. Used by the militaries of at least five dozen nations, it has been rightly called "the weapon of the century."

KAMEN, DEAN. Born 1951. American engineer, businessman, and inventor: Invented the Segway Human Transporter (circa 2001), or Segway HT—subsequently called the Segway Personal Transporter or Segway PT. The motorized, two-wheeled, self-balancing, electric vehicle transports orthostatic riders who steer using their body weight (that is, by leaning left or right). The Segway can attain a speed of 12-13 mph. Kamen is credited with a number of other innovations as well, from a portable infusion pump, an electric generator, and a robotic arm prosthesis, to a portable kidney dialysis machine, an insulin pump, and the iBOT (a powered mobility device or PMD).

KAMPRAD, INGVAR. 1926-2018. Swedish businessman, entrepreneur, author, and inventor: Founder of IKEA (1943), a global home furnishings retailer that specializes in ready-to-assemble furniture, as well as a wide variety of home decor and home goods.

KANNEL, THEOPHILUS VAN. 1841-1919. American engineer, businessman, entrepreneur, and inventor: Invented the revolving door (1888), which replaced conventional opening-and-closing doors while revolutionizing the architectural industry. To this day the revolving door continues to be an archetypal feature of modern city life, one particularly associated with skyscrapers.

KARAVODIN, V. V. Lived 19th Century-20th Century. Russian engineer, engine designer, and inventor: Credited with inventing the first valveless pulse engine (1906), a type of basic jet motor that lacks any moving parts—including mechanical valves. How it works: The repeated explosions inside the combustion chamber produce the thrust force, while oxygen is drawn back in as the exhaust gases are expelled. The incoming fresh air combines with the explosive fuel, which causes the cycle to repeat. While the valveless pulse engine offers higher speed potential, design simplicity, and greater fuel efficiency, it is gradually being replaced by turbojet technology.

KÁRMÁN, THEODORE VON. 1881-1963. Hungarian scientist, physicist, aerospace engineer and theoretician, mechanical engineer, astronautic engineer, author, and inventor: His many contributions to the flight industry include fluid flow theory, rocket motor design, aerodynamics, and his promotion of the swept-back wing design used in modern jets. He is best known, however, as the theoretical designer of supersonic flight, for which he has been named the "father of supersonic flight." Numerous scientific theories, concepts, and ideas have been named after him as well.

KAY, ALAN C. Born 1940. American computer scientist, computer programmer, educator, lecturer, composer, and inventor: A developer (or co-developer) of several computer programming languages, he invented the Dynabook (1968), a small children's laptop computer and the prototype for what would become the computer tablet. Founder of Viewpoints Research Institute (2001).

KAY, ANDREW. 1919-2014. American computer pioneer, businessman, entrepreneur, and inventor: Invented the first digital voltmeter (1954), or DVM, a device that measures the voltage in a circuit and displays the result numerically on a screen. Kay was also the creator of the famed Kaypro line of computers and was the founder of several companies, including Non-linear Systems (1952), the Kaypro Corporation (1984), and Kay Computers (circa 1995).

KEESHAN, BOB: 1927-2004. American producer, actor, and TV host: Created the hit TV series *Captain Kangaroo* (1955).

KÉGRESSE, ADOLPHE. 1879-1943. French mechanical engineer, military engineer and inventor: Invented the first half-track: The Kégresse Track (1911), an assembly consisting of flexible, tank-like, rubber or metal treads that could be fitted to a truck's rear axle, thereby increasing its traction, stability, and usability. Kégresse also invented the M-3 Half Track (1941), the dual-clutch transmission (1939), and the automobile sled (circa 1905), later known as the autochenile.

KEITH, CARL D. 1920-2008. American research chemist and inventor: Co-invented the three-way automotive catalytic converter (1973), a type of filter that reduces the pollution produced by internal combustion engines up to 98 percent. By cutting down on toxic emissions, it is estimated that the device has saved hundreds of thousands of lives, prevented millions of cases of throat and lung diseases, and improved the quality of life for billions of people around the world.

KELLOGG, JOHN HARVEY. 1852-1943. American physician, nutritionist, health food, therapy, and fitness innovator, researcher, businessman, lecturer, author, editor, and inventor: A vegetarian doctor, he is credited with being the first (some say among the first) to invent dry breakfast cereals, which launched the globally popular flaked cereal industry. Though the Aztecs (14th Century) and Incas (15th Century) were known to have created an early form of peanut butter, credit for the invention of the modern version must go to Kellogg (1895), whose now globally popular product was first marketed as an easily chewable "healthy protein" for people without teeth. A co-founder of the Battle Creek Sanitarium Health Food Company (1898), he was one of the first proponents of naturopathic medicine and instituted a wholistic health program for his patients called "Biologic Living." His medical

John H. Kellogg.

center, complete with a health spa and magnificent hotel, attracted visitors from around the world, including several individuals of note, such as Amelia Earhart, Thomas Edison, and Henry Ford.

KEMENY, JOHN G. 1926-1992. Hungarian mathematician, computer scientist, educator, author, and inventor: Co-developed the Beginner's All-purpose Symbolic Instruction Code (circa 1963), or BASIC, an elementary computer programming language that, due to its simplicity, versatility, and accessibility, helped open up computing to nontechnical students and to the general public.

KEMP, JOSEPH. 1841-1922. Canadian farmer, carpenter, and inventor: Though manure was already being used as a fertilizer at least 8,000 years ago, it was Kemp who invented the first known "practical" manure spreader (1874), a product that almost immediately revolutionized farming and food production. Founder of the J. S. Kemp Manufacturing Company (1900). Kemp is known as the "father of all manure spreaders."

KEMURDZHIAN, ALEXANDER. 1921-2003. Russian mechanical engineer: He and his space engineering team contributed to the invention of the first space exploration vehicle, the Soviet, robotic, battery-powered, lunar rover Lunokhod 1 (circa 1969). The first machine of its kind to land on a body outside earth's orbit, it reached the moon November 17, 1970, explored the surface for 11 lunar days (322 earth days), performed over two dozen soil analyses, and transmitted some 20,000 TV images before succumbing to low battery life on October 4, 1971. Originally configured to function for only 90 days, Kemurdzhian's successful invention paved the way for the rover technology that ensued over the following decades.

KENNARD, CHARLES W. 1857-1925. American businessman and inventor: In partnership with furniture maker E. C. Reiche, Kennard is credited with being the inventor of the popular Victorian paranormal parlor game which would later become known as the Ouija Board (1886). His company, the Kennard Novelty Company (founded in 1890), owned the patent for the Ouija Board (acquired in 1891) and would go on to be the "talking board" game's first manufacturer.

KEPLER, JOHANNES. 1571-1630. German scientist, mathematician, physicist, cosmologist, astronomer, astrologer, naturalist, theologian, musicologist, philosopher, educator, and inventor: Credited with numerous discoveries and developments in a myriad of disciplines, he is best known for his work in optics, which includes his invention of the Keplerian Telescope (1611) and research into the human eye. He is also celebrated for his still valid conception of the three laws of planetary motion (1609-

Johannes Kepler.

1619)—known as "Kepler's Laws." Kepler, whose work is still finding uses in science to this day, is known as the "father of modern optics."

KERIMOV, KERIM. 1917-2003. Azerbaijani space scientist, engineer, and rocket technologist: Leader of the Russian space program, behind which were a number of world "firsts." Among them, the first human to travel in space (Yuri Gagarin), the first woman to travel in space (Valentina Tereshkova), the first artificial satellite (Sputnik I), the first modular space station to orbit earth (Mir), the first spacewalk (1965), the first interplanetary probes (to Venus and Mars), the first robotic soft-landing on the moon (Soviet lunar rovers), and the first living being to be put into earth orbit (Laika the "space dog").

KERVOR, JACQUES EDWARD DE. 1928-2010. French industrial designer, toy designer, sculptor, artist, businessman, entrepreneur, and inventor: Invented or aided in the development of numerous innovations, from Delta faucets, televisions, and John Deere equipment, to the Ford Thunderbird, models for the Disney company, and diving apparatuses for Jacques Cousteau.

KETTERING, CHARLES F. 1876-1958. American engineer, technology innovator, businessman, and inventor: Invented the electric self-starter automobile ignition (circa 1912), the first electric cash register (circa 1904), the first commercially viable high-speed, two-cycle diesel engine (1930), a forerunner of the credit card (circa 1906), the first unmanned, propeller-driven guided, aerial missile (circa 1918), and a sophisticated high-compression automobile engine (1957). He also co-developed lead gasoline (early 1920s), Duco paint (1923), and the idea of air-cooled machinery and vehicle engines (early 1920s). Finally, his work on solar technology, electric lighting, and magnetics have helped improve the quality of human life in numerous industries, including automobiles, medicine, and pediatrics. Kettering was the co-founder of the Engineers Club of Dayton (1914), and founder of the Dayton Engineering Laboratories Company, or Delco (1909), the Dayton-Wright Airplane Company (1914), and the Kettering Foundation (1927). The city of Kettering, Ohio, is named after him.

KHAN, FAZLUR RAHMAN. 1929-1982. Bangladeshi civil engineer, structural engineer, and inventor: Invented the "bundled tube" structural system (1969) used in high-rise skyscrapers. This technology reduced costs, allowed for greater interior space, and significantly strengthened a building's infrastructure. Khan, whose creations include Chicago's Sears Tower (1973) and the John Hancock Center (1970), is known as the "father of tubular designs."

KHARITON, YULII BORISOVICH. 1904-1996. Russian scientist, nuclear physicist, explosives technologist, and munitions developer: Headed the Soviet atomic bomb program, and was thus a co-developer of the first Soviet atomic bomb (1949)—named the RDS-1. Khariton was the founder of the Chemical Physics' Laboratory of Explosives (1931), and is known as the "father of the Soviet atomic bomb." Like many Soviet inventors from this period, many of the details of his life were concealed for security reasons.

KHAZINI, ABD AL-RAHMAN AL-. Lived early 12th Century. Persian (of Greek origins) scientist, mathematician, doctor, astronomer, physicist, biologist, alchemist, philosopher, author, and inventor: Invented, developed, or discovered a host of scientific instruments and ideas. His work covered such areas as hydrostatic balance, astronomical tables, calendars, matter, scales, and gravitation. Al-Khazini, considered one of the greatest of medieval scholars and physicists, is responsible for such inventions as the dioptra and the triquetrum, as well as for improvements on the quadrant and astrolabe.

KHOJANDI, ABU-MAHMUD AL-. Circa 940-1000. Persian scientist, astronomer, mathematician, author, and inventor: Al-Khojandi developed one of the first (some say the first) accurate astronomical sextant (10th Century). His work in the fields of trigonometry, metrology, geometry, and instrument design helped forward science in a number of fields.

KHRENOV, KONSTANTIN K. 1894-1984. Russian electromechanical engineer, metallurgist, electrochemist, educator, lecturer, and inventor: Invented the process of underwater welding (1932). Also known as hyperbaric welding, this vital metallurgical binding technique is used in a variety of applications, from bridge maintenance and repair, dam maintenance and repair, underwater construction, underwater salvaging operations, pipeline maintenance and repair, to submarine maintenance and repair, oil rig maintenance and repair, environmental protection, dock maintenance and repair, ship maintenance and repair, and nuclear reactor maintenance and repair.

KIEFUSS, JOHANN. Circa 16th Century. German mechanical engineer and inventor: Credited by some with inventing the wheellock (circa 1517). Also known as a firelock, a wheellock is a steel friction wheel that rotates inside of a gun. This abrasion creates a spark that ignites gunpowder, which in turn fires a round of ammunition from the barrel. This, the first self-igniting firearm, was easy to use, dependable, and low-cost, revolutionizing gun technology and paving the way for the next major firearms development: the flintlock.

KIEPACH, MARCEL. 1894-1915. Croatian nobleman, economist, electrical engineer, and inventor: Invented a maritime compass (1910) that operated accurately despite outside magnetic or metallic influences. He is also credited with inventing an electric generator (1912), a type of dynamo that powered the lighting for various types of vehicles, as well as a gas pressure-powered electric switch that was used in the construction of X-ray machines. For his work and contributions in the fields of electrical science, acoustics, and magnetics, Kiepach has been nicknamed the "Tesla of Križevci"—the Croatian town where he was born.

KIETZ, ERHARD. 1909-1982. German scientist, physicist, author, musician, and inventor: Made numerous contributions in the fields of TV technology and sound recording technology.

KILBY, JACK. 1923-2005. American electrical engineer and inventor: Co-invented the first microchip or integrated circuit (circa 1959), opening the door for the development of the microprocessor and, in turn, modern day high-speed computing and telecommunications. To truly appreciate this momentous achievement, one must imagine the non-digital world prior to 1959, before computers, cell phones, the Internet, and social media. Kilby is also credited with inventing (some say co-inventing) the pocket calculator (1967). Nobel Prize winner.

KINDĪ, ABU YUSUF AL-. 801-873. Iraqi polymath, scientist, physician, chemist, ophthalmologist, mathematician, astronomer, cosmologist, pharmacist, meteorologist, geographer, philosopher, logician, scholar, author, musician, and inventor: Among countless other items, he invented several cipher-breaking methods, a medication scale for doctors, and numerous recipes for aromatic scents. Besides the professions listed above, he also made contributions in the fields of theology, chemistry, music therapy, astrology, psychology, metaphysics, eschatology, pharmacology, tidal science, thanatology, optics, cryptanalysis, and epistemology. Al-Kindi, whose name is sometimes Latinized as Alkindus in the West, is said to have authored some 270 books. To this day he is variously known as the "philosopher of the Arabs," the "father of cryptography," and the "father of perfumery."

KIPP, PETRUS JACOBUS. 1808-1864. Dutch scientist, pharmacist, chemist, apothecary, and scientific instrument designer and maker: Invented "Kipp's Gas Generator" (circa 1844), a glass device used to create and mix small quantities of gases—such as hydrogen sulphide and carbon dioxide. The glassware apparatus was considered standard equipment in laboratories around the world for over 100 years before it was replaced by more convenient pre-prepared gases.

KIRLIAN, SEMYON D. 1898-1978. Russian scientist, electrophotographist, and inventor: Invented electrophotography (1939), a form of high voltage photography that uses film sheets, metal plates, and an electric charge to record an object's corona discharge on film. His work laid the foundation for what is called "Kirlian photography." Note: Paranormal researchers and spiritualists assert that Kirlian photos show evidence of the aura, a type of invisible plasma or the "spiritual emanation," of a living being or even inanimate objects (in people or animals the color of the discharge—be it green, red,

Kirlian photo of a leaf.

blue, yellow, white, purple, etc.,—is thought to be an indication of one's present health condition, mood, or level of spiritual evolution). Mainstream scientists, however, claim that the corona discharge seen in Kirlian photographs is nothing more than the result of normal physical processes and does not violate known scientific laws.

KIRSCH, STEVE T. Born 1956. American electrical engineer, computer scientist, entrepreneur, and inventor: One of two independent inventors of the optical computer mouse (1980), a manual input device that permits one to operate and manipulate the cursor on a computer screen. Kirsch is the founder of numerous companies, including Infoseek, Abaca, OneID, and Frame Technology Group.

KLATTE, FRIEDRICH "FRITZ" H. A. 1880-1934. German chemist, pharmacist, and inventor: Discovered polyvinyl acetate (1912), or PVA, a vinyl polymer, or adhesive substance, that is today used in a myriad of applications, from 3D printing, book binding, construction, and lickable envelopes, to chewing gum, cigarette paper, food packaging, and wood glue. Klatte's pioneering work in the field of plastics chemistry led to the development of the first thermoplastic polymers, and in turn to the discovery of polyvinyl chloride (PVC). Though not the inventor of PVC, he was a co-inventor of the polymerization process (involving sunlight) that eventually led to PVC's commercial production, revolutionizing modern society in thousands of different ways.

KLEIN, YVES. 1928-1962. French conceptual artist, painter, sculptor, composer, actor, and photographer: A member of the movement known as Parisian Nouveau Réalisme, he invented an acclaimed variation of the color ultramarine blue that he called "International Klein Blue" (1960).

KNIGHT, THOMAS. Dates unknown. American computer scientist, biologist, educator, and inventor: Co-developed ARPANET (1969), the forerunner of the Internet. After Andrew D. Endy and Adam P. Arkin proposed standardizing biological parts (1999), Knight developed the idea of the BioBrick (2003), a type of manmade DNA sequence that aids researchers in the study of biology as well as the construction of bioparts—an innovation that led to the creation of the MIT Registry of Standard Biological Parts that same year. Knight, a co-founder of both Ginkgo Bioworks and iGEM (an international science competition), is known as the "godfather of synthetic biology."

KNUNYANTS, IVAN L. 1906-1990. Armenian scientist, chemist, engineer, and inventor: Invented polycaprolactam (early 1930s)—also known as capron, polyamide-6, or nylon-6—a manmade polymer used for a variety of industrial purposes. The founder of the Soviet fluorocarbon chemistry school and a pioneer in elemento-organic chemistry, he was both a major contributor to the Soviet chemical weapons program and the inventor of several chemotherapeutic drugs.

KOCH, ROBERT. 1843-1910. German scientist, physician, chemist, biologist, microbiologist, bacteriologist, educator, author, and inventor: The discoverer of tubercle bacillus (1882), a bacterium that causes tuberculosis in humans, he invented new techniques for growing pure bacteria cultures on solid media, as well as new bacteria staining methods. His serological and bacteriological work with pathogenic bacteria-borne ailments, like anthrax, tuberculosis, malaria, typhus, and cholera, allowed scientists to procure clean, untainted bacteria cultures for study, identification, and cure development. His laboratory standards, called "Koch's Postulates" (1884), were quite stringent and helped both prevent erroneous test results and advance scientific research into infectious diseases caused by bacteria, thus saving countless lives. Koch is known as the "father of microbiology." Nobel Prize winner.

Robert Koch.

KOLFF, WILLEM JOHAN. 1911-2009. Dutch physician, therapist, educator, and inventor: Invented the first functional artificial kidney or hemodialysis machine (1943), the first artificial heart (1957), and the first artificial wearable kidney (1975), and co-invented the intra-aortic balloon pump (1961). He was the first president of the American Society for Artificial Internal Organs (1955) and was the founder of Vital Assist (1970). His pioneering work with manmade organs and limbs revolutionized medical science and healthcare. Kolff is known as the "father of artificial organs."

KOMPFNER, RUDOLF. 1909-1977. Austrian scientist, physicist, electrical engineer, architect, educator, author, and inventor: Invented the traveling-wave tube (circa 1942), or TWT, a device that boosts radio frequency (RF) signals. This reliable form of microwave signal amplification has led to great advancements in the arenas of military technology, radar technology, and satellite communications.

KONSTANTINOV, KONSTANTIN. 1818-1871. Russian scientist, artillery officer, rocket and missile designer, author, and inventor: Invented a flight speed measuring instrument (1844), along with a myriad of sophisticated rocket and missile designs. His work in aeronautics, rocketry, and weapons systems opened up new vistas in the fields of both military science and space science.

KOROLEV, SERGEI P. 1907-1966. Russian rocket scientist and designer: Co-developed (and launched) the first intercontinental ballistic missile (August 1957), as well as the first successful artificial satellite, Sputnik I (October 1957). He also managed the design, build, and launch of the Soyuz, Voskhod, and Vostok crewed spacecraft, and the unmanned spacecraft series known as Kosmos, Molniya, and Zond. The lead intellectual power behind the Soviet spaceflight program, he well earned the Soviet nickname, "Chief Designer."

KOROTKOV, NIKOLAI. 1874-1920. Russian medical scientist, physician, surgeon, and inventor: Invented a technique for measuring arterial blood pressure called the auscultatory method (1905). Simply put, it employs a stethoscope and an inflating and deflating blood pressure cuff to detect the change in audible systolic and diastolic sounds as blood is squeezed and released through the compressed arteries.

KORSAKOV, SEMYON N. 1787-1853. Russian physician, homeopathic doctor, mathematician, photographer, and inventor: Proposed, designed, and created the first machines capable of computer-like data analysis (circa 1832). His devices, some of which bore the names "linear homeoscope," "flat homeoscope," and "ideoscope," used an assortment of data tables, wooden frames, pins, wooden lathes, and punch cards to store and retrieve information. His work and research laid the foundation for modern computer science.

KOSHKIN, MIKHAIL I. 1898-1940. Russia mechanical engineer, tank designer, confectioner, and inventor: Among his many tank designs, Koshkin invented what is considered by many to be the most perfectly conceived, armed, self-propelled, military fighting machine ever made: the T-34 medium tank (1930s), which played a major role in World War II (1939-1945).

KOSTOVIĆ, OGNJESLAV S. 1851-1916. Serbian scientist, mechanical and chemical engineer, military officer, and inventor: An inventor of a myriad of innovations related to ships, balloons, submarines, dirigibles, sea planes, and other aircraft, he is today best known for his invention of arborite (1906), a sturdy, manmade, plywood-like material that some consider the first form of plastic.

KOTELNIKOV, GLEB. 1872-1944. Russian mechanical engineer, military officer, composer, musician, actor, author, and inventor: Invented the RK-1 (1911), known more commonly as the knapsack parachute (a hard case and later a soft pack design), as well as the drogue parachute (1912)—a drag chute that helps slow down a falling or fast-moving aircraft, land vehicle, or skydiver. Kotelnikov's work in the field of parachute science not only saved lives, but also paved the way for the further development and refinement of both parachute design and the sport of parachuting.

KOUWENHOVEN, WILLIAM B. 1886-1975. American electrical engineer, medical science researcher, and inventor: Credited with inventing the world's first external cardiac defibrillator (1930). Kouwenhoven is known as the "father of cardiopulmonary resuscitation."

KRESS, WILHELM. 1836-1913. Germanic aviation scientist, author, and inventor: Invented a variety of rubber band-powered model aircraft (1877), the first aircraft control stick (1900), and what appears to be the first modern delta-flying hang glider (1877). Above all, Kress is best known for inventing and flying the world's first sea plane (circa 1901). The *Drachenflieger* ("Dragon-flier"), as he accurately called it, was a gas-powered amphibious aircraft with a 60 foot wingspan and numerous membranous airfoils, giving it a dragonfly-like appearance. Though difficult to

Kress' "Dragon-Flier."

handle, the unwieldy but beautiful aircraft laid the foundation for the future of sea plane technology.

KROEMER, HERBET. 1928-2024. German scientist, physicist, electrical engineer, educator, and inventor: He was the first person to describe the idea of a quantum well (1963), a microscopically thin layer (typically about 40 atoms thick) of a semiconductor material that restricts particles to one dimension, but which allows them to move freely in the other two dimensions; used mainly in electronics for such devices as lasers, microwave receivers, computer chips, CD players, and LEDs. Nobel Prize winner.

KROLL, WILLIAM JUSTIN. 1889-1973. Luxembourgian scientist, metallurgist, chemist, and inventor: Invented the Kroll process (1940), a method for extracting titanium from titanium tetrachloride—a chemical material made from titanium ores. In 2001 the Kroll process replaced most other extraction techniques, helping to further refine and develop the ever-expanding field of metallurgy. Titanium itself, a lightweight, non-corrosive, malleable, nearly indestructible metal, is used around the globe for everything from glass-making, the medical field, paints, the automotive industry, paper, bicycles, sunscreen, and aerospace science, to cookware, golf clubs, the military, jewelry-making, computers, dental implants, energy technologies, marine science, power plants, and plastics.

KRUPA, ALFRED JOSEPH. 1915-1989. Polish artist, painter, photographer, sportsman, educator, and inventor: Invented the first suitcase with wheels (circa early 1950s), as well as a lightweight folding canvas catamaran (circa early 1950s), a glass-bottom boat (1957), compressed air-filled boxing gloves, and an improved variant of da Vinci's invention, water-walking skis (1951), among a number of other innovations. Krupa is also said to have been the first person to paint a picture underwater using oil paints (1950).

KRYLOV, ALEKSEI N. 1863-1945. Russian mechanical engineer, shipbuilder, mathematician, naval scientist, military official, educator, author, essayist, and inventor: Invented the gyroscope stabilizer (1909) for use on ships. Krylov is known as the "father of Russian shipbuilding."

KUCHER, ANDREW A. Dates unknown. Nationality unknown. At the time Ford Motor Company vice president for engineering and research: Mastermind behind the Levacar Mach 1 (1950s), a single-occupant, wheelless hovercraft vehicle that rode on a thin cushion of air (generated by an onboard air compressor), with its forward motion powered by a heater fan. First publicly displayed in 1958 at the Ford Rotunda in Dearborn, Michigan, the small, sleek, finned, space-age looking concept car promised to revolutionize both the car industry and transportation. Despite worldwide enthusiasm over the Levacar, the entire project was scrapped just four years later (early 1962), allegedly a victim of design problems (for one thing, the air pressure provided could not lift the vehicle high enough off the road's surface to avoid bumps, potholes, and stones). Kucher's "dream machine" was relegated to a back storeroom, where it was later consumed by fire (late 1962), utterly evaporating from the pages of automotive history. Was the disappearance of the Levacar program as straightforward as it appears? Among the many conspiracy theories that abound, the primary assertion is that it may have been seen as a threat by conventional car manufacturers—including by Ford itself. (Note: The project was cancelled by then company CEO Henry Ford II.) While today most accept the mainstream explanation, to many others the truth remains open.

KULIBIN, IVAN. 1735-1818. Russian mechanical engineer and inventor: Founded the science of bridge construction; proposed various manmade mechanical limbs or prosthetics (circa 1790); invented a spotlight lamp with reflecting mirrors (1779), the first egg-shaped clock (1764), the screw elevator (circa 1793), a water current-powered barge (circa 1782), the optical color light telegraph (1794), and a self-powered carriage (1791), a forerunner of the bicycle, or even the automobile. Kulibin, a self-taught inventor whose work and innovations led to advances in many scientific fields, is known as the "Russian Archimedes."

Kulibin bridge.

KUO, SHEN. 1031-1095. Chinese polymath, scientist, physicist, engineer, mathematician, geologist, optics scientist, economist, pharmacist, cartographer, anthropologist, orographer, lawyer, astronomer, geographer, horologist, botanist, philosopher, government official, author, poet, and inventor: Invented an armillary sphere, a water clock or clepsydra, and a sundial pin or gnomon. Known for his copious, "ahead-of-his-time" writings on fossils, music, atmospheric science, art, hydraulics, medicine, anatomy, calendars, waterways, metaphysics, astronomy, natural history, metallurgy, and politics, among a myriad of other diverse subjects.

KURCHATOV, IGOR V. 1903-1960. Russian nuclear physicist and inventor: As director of the Soviet nuclear program (1943) he was responsible for inventing the first Soviet atomic bomb (1948), the first nuclear power plant (circa 1954), the first Soviet thermonuclear bomb (circa 1953), and Europe's first nuclear reactor (1946). Kurchatov, who also oversaw the development of electric power stations (1954) and the Lenin (1957)—a nuclear-powered icebreaking ship, is known as the "father of the Soviet atomic bomb."

KURTZ, THOMAS E. 1928-2024. American mathematician, computer scientist, educator, author, and inventor: Co-developed the Beginner's All-purpose Symbolic Instruction Code (circa 1963), or BASIC, an elementary computer programming language that, due to its simplicity, versatility, and accessibility, helped open up computing to nontechnical students and to the general public.

KURZWEIL, RAYMOND. Born 1948. American computer scientist, futurist, businessman, entrepreneur, author, and inventor: The founder of Kurzweil Computer Products, Inc. (1974), he launched a number of revolutionary products, including musical keyboard synthesizers, a text-to-speech synthesizer, a flatbed scanner, a reading machine for the vison-impaired, and speech recognition computer software. He is also credited with inventing the first omni-font OCR (optical character recognition) software (1974).[9]

KUTARAGI, KEN. Born 1950. Japanese computer scientist, electrical technologist and engineer, businessman, educator, and inventor: Led the development and invention of the popular video game console PlayStation (1994).

KYAN, JOHN HOWARD. 1774-1850. Irish inventor: Invented a technique (using bichloride of mercury) for preserving wood known as Kyanization (circa 1820s). Though at first quite successful in preserving such items as wooden dock pilings, wooden mine support beams, and wooden railroad ties (and even non-wood products such as paper, cloth, and canvas), Kyanization was eventually replaced with other less corrosive preserving processes, such as those using creosote.

Double-acting, single cylinder, 2 cycle diesel engine, 2000 horsepower, 1913.

Drawing of one of James Watt's early steam engines, 1767.

LACHINOV, DMITRY. 1842-1902. Russian scientist, physicist, electrical engineer, meteorologist, author, and inventor: Proposed, improved, or created a number of innovations, including such products as electrolyzers, photometers, and dynamometers, among others. His work aided in the growth and development of the field of electronics.

LAENNEC, RENÉ. 1781-1826. French physician, anatomist, academician, educator, lecturer, author, and inventor: Invented the stethoscope (1816), an auscultatory device that is used to listen to sounds produced inside a patient's body in order to determine his or her health; or more often, the health of the heart and lungs specifically. His original stethoscope (a plain hollowed out wooden tube) and the introduction of what he called "mediate auscultation," was revolutionary: Before its invention physicians simply laid their ear on a sufferer's bare chest (known as "immediate auscultation")—an often uncomfortable, inconvenient, and embarrassing form of diagnosis for both patient and doctor. Many of the results of Laennec's work and clinical innovations are still used in the health field, such as certain medical terminologies and the categorizations and summarizations of various diseases. In his honor a number of ailments and conditions have been named after him. Laennec is known as the "father of medical auscultation."

René Laennec.

LAKHOVSKY, GEORGES. 1869-1942. Russian scientist, engineer, philosopher, author, and inventor: Invented the Multi-wave Oscillator (early 1920s), also known as the "Lakhovsky Coil," a complex electrical device (employing a Tesla Coil) that he used to treat various diseases and health issues—primarily cancer. While mainstream medicine generally disputes Lakhovsky's ideas, viewing his invention as of "dubious value," many others in the health field have embraced his "secret of life"; namely that the cells of all living beings are created and maintained by natural radiation, and are thus miniature electromagnetic transmitter-receivers whose health depends on a precise balance of natural energies that derive from sunlight, outer space, terrestrial radioactive rocks and soil, natural foods, pure water, stress management, etc. His Multi-wave Oscillator, a type of cellular regenerator, was intended to help rebalance the bodily disequilibrium (and subsequent diseases) caused by our modern lifestyle. According to some early sources, Lakhovsky's machine (which was alleged to have had a very high rate of success curing cancer) was eventually banned and his ideas were suppressed by the medical establishment.

LAM, SIMON S. Born 1947. Chinese computer scientist, engineer, educator, and inventor: Invented both Secure Sockets (1991), the idea of a private protected online channel for communication between two devices, and Secure Socket Layers or SSL, the encryption code that allows for this type of secure communication. He also invented Secure Network Programming (1993), or SNP, an application programming interface (API). Lam's innovations, which revolutionized e-commerce technology, are used primarily to protect credit card, banking, and login information while an individual is shopping online or browsing the Internet. Every Website address (URL) that includes an "s" (e.g., https://) has been secured using Lam's SSL technology.

LAMBOT, JOSEPH-LOUIS. 1814-1887. French scientist, agriculturist, and inventor: Invented ferrocement (1848), a nonflammable, rust-resistant, wind-resistant, and earthquake-resistant building material produced by combining cement, wire, water, and sand, making it ideal for skyscraper, house, bridge, road, seawall, and dam construction (among many other types of structures). This durable composite paved the way for the development of rebar ("reinforcing bar"), also known as reinforced concrete.

LAND, EDWIN HERBERT. 1909-1991. American scientist, physicist, photographer, businessman, entrepreneur, and inventor: Invented polarizing filters as well as the Polaroid Land Camera (1948)—which could develop and print out a finished color photograph in one minute or so. He co-founded Land-Wheelwright Laboratories (1932), and founded both the Polaroid Corporation (1937) and the Rowland Institute of Science (1960). Land's world-altering polarizing technology continues to be widely used in, for example, sunglasses, photography filters, microscopes, 3D movie glasses, and liquid crystal displays.

LANGLEY, SAMUEL P. 1834-1906. American astrophysicist, engineer, astronomer, solar researcher, pilot, aircraft designer, draftsman, educator, author, and inventor: Invented the bolometer (circa 1878-1880), a device that measures radiation, allowing for the determination of minute temperature changes. He also designed and launched the world's first successful flight of an unmanned heavier-than-air airplane (1896), and aided in the refinement of American time scheduling by improving the system of Standard Time Zones. Secretary of the Smithsonian Institution.

Samuel P. Langley.

LANGMUIR, IRVING. 1881-1957. American scientist, metallurgical engineer, physicist, chemist, educator, and inventor: Invented the gas-filled incandescent light bulb (1913), as well as atomic hydrogen welding (1920s). AHW, as it is also known, has numerous advantages over traditional gas welding—such as a more durable weld, a wider variety of applications, and better protection from atmospheric contamination. Recently, however, AHW is slowly being replaced by newer welding technologies, such as shielded metal arc welding. Nobel Prize winner.

LARSEN, NORMAN BERNARD. 1923-1970. American industrial chemist, businessman, entrepreneur, and inventor: Invented the multipurpose product WD-40 (1953), an acronym for "Water Displacement 40th Formula." The brand name is derived from the fact that Larsen created and worked with 39 unsatisfactory formulas before attaining success on his 40th attempt. The worldwide enthusiasm for this popular penetrating oil is due to its multiplicity of uses, including lubrication, rust protection, degreasing, and moisture displacement. Larsen, said to have been self-taught, also delved into the areas of weaponry, wood preservation, and therapeutic creams. Note: The WD-40 attribution is disputed by some due to confusion between the man highlighted here, Norman Bernard Larsen (also known as Norm Larsen), and a man named Iver Norman Larsen, who was born around the same time period.

LATIMER, LEWIS H. 1848-1928. American engineer, draftsman, patent authority, poet, musician, playwright, author, and inventor: Helped Alexander Graham Bell develop the telephone (1876), worked with Thomas Edison on making improvements to the incandescent light bulb (1884), and assisted Hiram S. Maxim in upgrading the carbon filament. Among his own inventions was an enhanced railroad car toilet (1874), a technique for mass-producing carbon filaments (1882), and a precursor to the air conditioner.

LAURER, GEORGE J. 1925-2019. American engineer and inventor: Invented the Universal Product Code (1973). More commonly known as a UPC code, this 12-digit numerical barcode allows supply chains to identify and inventory products, an idea that revolutionized the manufacturing industry.

LAVAL, CARL GUSTAV PATRIK DE. 1845-1913. Swedish scientist, electrical engineer, politician, businessman, and inventor: Invented the centrifugal cream separator (1878), the steam turbine (1882), the reversible steam turbine (1893), and the milking machine (1894). Power plants were eventually able to operate using de Laval's turbines (1896), developments that would lay the foundation for today's high-speed turbine industry. A variation of his speed-increasing steam jet nozzle, the De Laval Nozzle, is used today by the aerospace industry. Among the 37 companies de Laval founded (and/or co-founded) was the famous AB Separator (1883), later renamed Alfa Laval.

LAWES, JOHN BENNET. 1814-1900. English agronomist, farmer, chemist, fertilizer pioneer, businessman, entrepreneur, and inventor: Invented superphosphate (1842), the world's first chemical fertilizer, which he formed by treating phosphate ore with sulfuric acid. Lawes founded (some say co-founded) the Rothamsted Experimental Station (1843), an agricultural science center that utilizes laboratory research and experimental cropland to study and improve food production in combination with environmental conservation.

LAWRENCE, ERNEST ORLANDO. 1901-1958. American scientist, physicist, educator, author, and inventor: Invented the Cyclotron (1929), a high-velocity nuclear particle accelerator that can operate at low voltages. Important analysis resulted from his invention, including advancements in the research and development of ontology, mesons, radioactive isotopes, time measurement, the atomic bomb (the Manhattan Project), and antiparticles. Nobel Prize winner.

LEAR, NORMAN. 1922-2023. American producer and screenwriter: Created the hit TV series *All in the Family* (1971), as well as *Maude* (1972), *Sanford and Son* (1972), *One Day at a Time* (1975), and *The Jeffersons* (1975).

LEBEDENKO, NIKOLAI. 1879-1948. Russian mechanical engineer, military tank designer, and inventor: Lead designer (along with his team) of the "Tsar Tank" (circa 1916). Also known as the Lebedenko Tank, this imposing but largely impractical combat machine was the largest armored gun transport ever created up to that time. Despite the fact that design issues prevented the unwieldy vehicle from being used in combat (eventually the project was discontinued and the vehicles themselves were sold for scrap metal), the Tsar Tank's many innovations helped pave the way for the evolution of military tank technology.

LEBEDEV, SERGEI V. 1874-1934. Russian chemist, engineer, educator, and inventor: Invented polybutadiene (1910), a durable type of synthetic rubber that proved to be easily mass-producible, and in turn, commercially successful on a grand scale. Indeed, Lebedev's innovation is today only gaining in usefulness and popularity, and is now used in an astonishingly wide array of products, from construction, gaskets, vehicle tires, adhesives, plastics, and sealants, to hoses, wire insulation, asphalt, footwear, conveyor belts, and railway padding.

LEE, ROBERT EDWARD. 1807-1870. American military engineer, U.S. army colonel, C.S. commander of the Army of Northern Virginia, Confederate States army general-in-chief, and president of Washington College—which was renamed Washington and Lee University in his honor (1871).[10]

Robert E. Lee.

LEE, WILLIAM. 1563-1614. English cleric, businessman, entrepreneur, and inventor: Invented the world's first stocking frame (1589), a framework knitting machine that could semi-automatically produce fabrics, and in particular, cotton stockings. Lee's invention spurred the growth of both textile production technology and, in turn, the Industrial Revolution (1760-1840).

LEEUWENHOEK, ANTONIE VAN. 1632-1723. Dutch scientist, physicist, microbiologist, microscopist, businessman, and inventor: Advanced microscope technology by grinding high quality (usually very small) glass lenses and constructing his own microscopes around them. His lens developments allowed him to become the first to observe microscopic organisms, such as bacteria, protozoa, and spermatozoa. Using his microscopes he was also the first to view and accurately describe not only red blood cells, but the life cycles of many types of insects, and larger animals as well, such as eels and freshwater mussels. Leeuwenhoek—whose work both overturned the then popular theory of spontaneous generation while greatly progressing the field of microscopy—is known as the "father of microbiology."

LEMELSON, JEROME H. 1923-1997. American aeronautical engineer, industrial engineer, and inventor: As an independent inventor he was involved in the research and development of numerous innovations in a diversity of fields, including integrated circuitry, word processing, lasers, medicine, toys, fax machines, facial recognition, automated industrial robotics, telephony, games, data recording, video recording, audio recording, semiconductors, computers, and biology. Awarded some 600 patents, he was a co-founder of the Lemuelson Foundation (1992).

LENOIR, JEAN-JOSEPH ETIENNE. 1822-1900. Belgian automotive engineer, naval architect, businessman, and inventor: Though he did not invent the first internal combustion engine, he did invent the first practicable and successful one (1860). After converting it from steam-power to liquid fuel-power, he installed the one-cylinder engine in his "Hippomobile," making him the inventor of the first automobile to use an internal combustion motor (1862). Lenoir is also credited with other important innovations, such as improved train braking systems

(1855), an automated telegraph (1865), and the world's first motorboat (1886). Based on his inventions subsequent inventors made numerous advancements in the fields of engine technology, as well as applying his principles to water pumps, power plants, and industrial machine tools.

LENTINI, GIACOMO DA. Lived 13th Century. Italian poet: Credited with having invented the sonnet (circa 1200s). The most familiar early form of poetry, a sonnet typically contains 14 lines with ten syllables per line. Written in iambic pentameter and employing various rhyming patterns, arguably the most famous example is William Shakespeare's poem Sonnet 18.

LERDORF, RASMUS. Born 1968. Greenlandic computer scientist and computer programmer: Invented PHP (1994), an acronym for "Personal Home Page," a simple, versatile, widely used hypertext preprocessing, open-source scripting language created for dynamic Website development. Lerdorf is known as the "father of PHP."

LETOURNEAU, ROBERT G. 1888-1969. American businessman, engineer, and inventor: Though he was primarily known for inventing massive, revolutionary, earthmoving and earth-digging equipment and heavy construction equipment, he was actually the holder of some 300 patents that covered a variety of innovations. For example, not only was he the inventor of the first portable crane, an all-terrain land train, the precursor to the modern bulldozer (1953), and a series a powerful earthmovers known collectively as "Carryall Scrapers" (1931), he also invented the "Scorpion" (1956)—the first portable offshore drilling platform, electric wheels, and the first log-stacking machine (1955). Among numerous other inventions and improvements related to the earthmoving and construction industries, he also headed the research and development of prefabricated concrete homes using a machine that could pour and mold an entire house in one step. Founded the R. G. LeTourneau Company (1929) and co-founded LeTourneau University (1946). Due to his strong Christian convictions and his charitable nature Letourneau was known as "God's businessman." The technologies he introduced are still in wide use today, with effects that reach around the globe.

LIBBY, WILLARD FRANK. 1908-1980. American physical chemist, radio chemist, author, and educator: Invented radiocarbon dating (circa 1946), a technique used to determine the approximate age of organic objects (i.e., bone, wood, and plant fossils) by measuring their content of carbon-14 atoms. Ever since its development radiocarbon dating has been widely used in such fields as archaeology, biology, anthropology, paleontology, climatology, forensic science, oceanography, and environmental science, to estimate the age of carbon-based artifacts—one of the more famous being the Dead Sea Scrolls. Libby was a member of the U.S. Atomic Energy Commission (1954). Nobel Prize winner.

LICKLIDER, JOSEPH CARL ROBNETT. 1915-1990. American

computer scientist and psychologist: Credited by some with inventing, or at least conceiving, cloud computer technology (1962)—or what he termed an "Intergalactic Computer Network."

LIE, HÅKON WIUM. Born 1965. Norwegian technologist and computer scientist: Invented CSS (1994), an acronym for "cascading style sheets," a computer markup language, or World Wide Web coding system, that is composed of a series of rules, allowing online content, such as color, fonts, and spacing, to be typographically styled.

LIEBIG, JUSTUS VON. 1803-1873. German scientist, chemist, biochemist, agronomist, educator, lecturer, author, editor, and inventor: Invented "chemical manure" (1845), a type of nitrogen-enriched fertilizer (1845). His research, work, and theories, much which is still used and studied to this day, opened up avenues for new developments in numerous fields, such as agriculture, biology, and in particular, organic chemistry.

LIGHT, EDWARD. 1747-1832. English composer, musician, organist, guitarist, educator, author, and inventor: Invented the harp lute or dital harp (1795), the harp guitar (1798), and the harp lyre (1816). His innovations combined various instruments with the world's oldest known stringed instrument, the harp, which introduced new sounds and new playing techniques to composers and performers.

LIK, HON. Born 1951. Chinese pharmacist and inventor: Invented the electronic cigarette (2001), also known as a "vape," which simulates smoking a tobacco-based cigarette. Lik's e-cigarette thus introduced the practice of vaping (using a battery-powered device containing an atomizer), wherein one inhales vapors instead of smoke. Though originally intended to provide a healthier alternative to ordinary cigarettes, the medical establishment asserts that vaping poses numerous health risks that are nearly as serious as those that derive from regular smoking.

LILIENTHAL, OTTO. 1848-1896. German mechanical engineer, aviator, businessman, entrepreneur, lecturer, author, and inventor: Invented the first successful hang glider (1891), proving the theory of unpowered heavier-than-air flight. He patented ideas in other fields as well, including those involving children's toys, steam engines, and mining. Lilienthal's work and research in aviation science inspired the Wright brothers and led to the development of the first manned aircraft.

LIN, YUTANG. 1895-1976. Chinese author, philosopher, poet, historian, linguist, translator, scholar, educator, and inventor: Invented the MingKwai typewriter (mid 1940s). The first electromechanical Chinese language typewriter, it opened the technological door to Chinese typing while anticipating various elements of the modern computer.

LINDBERGH, CHARLES A. 1902-1974. American hero, aviator, explorer, engineer, military pilot, author, and inventor: Co-invented the organ perfusion pump (1935). Technically known as the

"Carrel-Lindbergh Perfusion Pump," the hand-blown glass apparatus pumped oxygenated blood (natural or synthetic) into organs that had been removed from the body, preserving them for weeks at a time. Lindbergh's device was a forerunner of the heart-lung machine, which allows the heart to be stopped during serious medical operations.

Charles A. Lindbergh.

LINDQVIST, FRANS WILHELM. 1862-1931. Swedish businessman, entrepreneur, and inventor: Invented the Primus Stove (1892), the first compressed air, soot-free kerosene stove. Founded the company Primus (1892).

LINNÉ, CARL VON. 1707-1778. Swedish scientist, physician, naturalist, zoologist, ornithologist, botanist, mycologist, explorer, educator, lecturer, author, and inventor: Often known today by his Latinized name, Carolus Linnaeus, he invented the field of modern taxonomy (circa 1735), a binomial nomenclature system that enables the classification of all living organisms into logical biological hierarchical groupings (e.g., phylum, class, order, family, genus, species, etc.). While Linne's complex Latin (and Greek) naming system remains essentially intact, due to the discovery and identification of thousands of new plants and animals since his time, modern biologists have added numerous categories to his original phylogenetic trees. Linne, a co-founder of the Royal Swedish Academy of Sciences (1732), and a man with many interests besides taxonomy—among them Swedish patriotism, physiology, geology, pathology, the social sciences, entomology, theology, pharmacology, and embryology—is known as the "father of taxonomy."

LIPPERSHEY, HANS. 1570-1619. German optics scientist, eyeglass maker, businessman, entrepreneur, and inventor: Described (and possibly invented) the first telescope (1608), as well as the first binoculars (1608). He is also credited by some with inventing the first compound microscope. Due to a lack of concrete historical evidence, however—not to mention the vigorous claims of other inventors who lived at the same time, these assertions are highly disputed.

LIPPMANN, JONAS FERDINAND GABRIEL. 1845-1921. French scientist, physicist, scholar, horologist, educator, photographer, and inventor: Among numerous other innovations, he invented the Lippmann Capillary Electrometer (1873)—an electrocardiographical instrument used in medicine; the Lippmann Plate (1891)—a unique color image development process used in photography; and the Coelostat (1895)—a telescopic device used in astronomy. Nobel Prize winner.

LIVENS, WILLIAM HOWARD. 1889-1964. English military officer, weapons engineer, photographer, spiritualist, and inventor: Among other innovations he is best known for launching the science of chemical warfare (1915) through the invention of numerous deadly military arms—most notably the Livens Projector (circa 1916), which could propel large toxic gas-filled canisters onto the enemy, even at great distances. His gas, oil, and flame-based weapons were used against the Germans in World War I with devastating results, and have formed the basis for new developments in modern chemical warfare.

LJUNGSTRÖM, BIRGER. 1872-1948. Swedish mechanical engineer, designer, bicycle maker, businessman, entrepreneur, and inventor: Co-invented (with his brother, below) the Ljungström Turbine (1908), a double rotation steam turbine, as well as the Ljungström Regenerative Air Preheater (circa 1920), both which were successfully used in power stations. His innovations greatly impacted numerous disciplines, including mining, the dairy industry, electricity production, and ship and locomotive engine design, and are used globally to this day.

LJUNGSTRÖM, FREDRIK. 1875-1964. Swedish mechanical engineer, designer, bicycle maker, businessman, entrepreneur, and inventor: Co-invented (with his brother, above) the Ljungström Turbine (1908), a double rotation steam turbine, as well as the Ljungström Regenerative Air Preheater (circa 1920), both which were successfully used in power stations. His innovations greatly impacted numerous disciplines, including mining, the dairy industry, electricity production, and ship and locomotive engine design, and are used globally to this day.

LOCHER, EDUARD. 1840-1910. Swiss mechanical engineer, businessman, and inventor: Invented the Locher Rack Railway System (circa 1880s), a locking gear that securely fastens railroad car wheels to the rail track. His innovation made rail travel on steep and dangerous inclines possible. Locher used his apparatus on the world's steepest cogway railroad, Switzerland's Pilatus Railway Line (opened 1889), which he proposed, designed, and constructed. The railway is in operation to this day.

LODYGIN, ALEXANDER NIKOLAYEVICH. 1847-1923. Russian scientist, physicist, engineer, and inventor: Credited by some with being the inventor of (or more likely, one of the developers of) the incandescent lamp (1872). His innovation, which used a vacuum tube and carbon wire, seems to have been a commercialized improvement on earlier designs. As of yet there is no uncontestable historical consensus.

LOMBARD-GÉRIN, LOUIS. 1848-1918. French scientist, engineer, carpenter, and inventor: Credited with developing and improving the trolleybus (circa 1900), a trackless, public transport vehicle that draws power from four trolley wheels mounted on the end of trolley poles, that are, in turn, attached to two electrified overhead cables.

LOMONOSOV, MIKHAIL. 1711-1765. Russian polymath, scientist, chemist, physicist, mechanical engineer, historian, astronomer, metallurgist, geologist, glass-maker, geographer, educator, grammarian, linguist, author, philosopher, poet, and inventor: His work in the area of atmospheric electricity aided in the development and invention of the

Mikhail Lomonosov.

lighting rod (circa early 1750s), quite independently of Benjamin Franklin. Lomonosov was also responsible for developing night vision, as well as an underwater viewing device. Among dozens of other contributions he made across a wide spectrum of disciplines, he improved the telescope, discovered an atmosphere on Venus, and developed a helicopter-like device (circa 1755) that was powered by the world's first known coaxial rotor (still used in some aircraft today). He was the first to propose the idea of establishing a school in Moscow (circa 1750s)—which later became known as Moscow State University. His great name, stature, and reputation inspired a host of posthumous eponymous tributes, such as the city of Lomonosov, Russia, which was named after him in 1948, and a moon crater that was named after him in 1961. A member of the Royal Swedish Academy of Sciences who was recognized for his unique glass mosaics, his innovations and research paved the way for numerous advances in countless scientific fields.

LOMONOSOV, YURY V. 1875-1952. Russian railroad engineer, educator, author, and inventor: Invented the first successful diesel internal combustion locomotive engine (circa 1923). It replaced steam engine-powered locomotives, thereby revolutionizing the railroad industry.

LORAN, ALEKSANDR G. 1849-1911. Russian chemist, businessman, entrepreneur, educator, and inventor: Invented firefighting foam (1902), as well as the first foam fire extinguisher (1904). The latter device spreads a blanket of thick, water- and air-saturated, bubble-packed lather over a fire, smothering it more quickly and efficiently, and with less damage to property, than water.

LOSEV, OLEG. 1903-1942. Russian scientist, physicist, engineer, and inventor: Largely self-taught, he seems to have been the first to scientifically describe the light-emitting diode (1924), better known as the LED. Because of this, some credit him with actually inventing the LED. Beyond dispute is the fact that Losev's research aided in the process of the commercialization of electroluminescence technology.

LOUD, JOHN J. 1844-1916. American lawyer, leather worker, and inventor: Invented the world's first ballpoint pen (1888), a writing implement that allowed ink to flow smoothly and consistently from the reservoir via a tiny rotating ball bearing located in a socket at the tip. While Loud's ball-and-socket device was ingenious for the time (replacing both the pencil and the fountain pen for many applications), due to several issues it was not yet commercially viable. However, it did lay the groundwork for later developments and improvements, which eventually led to the modern ballpoint pen—today the most popular and commonly used writing tool in the world.

LOUIS, ANTOINE. 1723-1792. French surgeon, physiologist, historian, educator, author, and inventor: Invented and/or improved numerous medical devices. However, he is best known for being the co-inventor of what might best be called the modern Guillotine (1789), as similar execution devices had been in use, in various forms, long before then.

LOW, ARCHIBALD M. 1882-1956. English scientist, physicist, mechanical engineer, military official, futurist, author, and inventor: Considered one of the world's most inventive but largely unheralded inventors, among many other innovations he developed the "TeleVista" (1914), a precursor to the modern television. Additionally, while he co-invented the first electric-guided rocket (1917), he was the inventor of the world's first drone (1916), the first electric-powered gyroscope (circa 1917), and the first rocket-driven motorcycle (circa 1940s). Founder of the Low Engineering Company (circa 1919) and co-founder of the British Interplanetary Society (1933), Low is known as the "father of radio guidance systems."

LOWE, EDWARD. 1920-1995. American businessman, entrepreneur, salesman, and inventor: Invented cat litter (1947). Designed using a type of water-absorbing clay (rather than water-repelling sand, the commonly used material at the time), his product, "Kitty Litter," launched an entirely new field and revolutionized the pet industry.

LOZINO-LOZINSKIY, GLEB. 1909-2001. Ukrainian scientist, aerospace engineer, and inventor: Worked on Russian's MIG fighter jet project, and led the Spiral project (circa 1965), a type of mini space shuttle. While he was head of NPO Molniya (founded 1976), Moscow, Russia, he began work on the Buran (circa late 1970s)—a Soviet-style, reusable orbiting spacecraft, which succeeded in making the world's first unmanned automated earth landing (1988). Lozino-Lozinskiy is also credited with heading the team that invented the MAKS project (1990s), a Russian acronym for "Multi-purpose Reusable Aerospace System."

LUDWIG, THEOBALD. 1889-1918. American businessman, entrepreneur, performing musician, percussionist, drum manufacturer, and inventor: Co-founded (with his brother William, below) the Ludwig Drum Company (1909), world famous maker of drums, percussion instruments, drum hardware, and various percussion accessories. Notable artists and bands have long used Ludwig drums, helping catapult the company to global success. Such individuals and groups include: Fleetwood Mac, the Beatles, the Doors, Black Sabbath, Buddy Rich, the Carpenters, the Foo Fighters, the Yardbirds, Jimi Hendrix, the Tonight Show Band, Aerosmith, Cream, Emerson, Lake and Palmer, Todd Rundgren, Grand Funk Railroad, REO Speedwagon, Max Roach, Led Zepplin, and Pink Floyd.

LUDWIG, WILLIAM F. 1879-1973. German businessman, entrepreneur, performing musician, percussionist, drum manufacturer, and inventor: Co-founded (with his brother Theobold, above) the Ludwig Drum Company (1909), world famous maker of drums, percussion instruments, drum hardware, and various percussion accessories. Notable artists and bands have long used Ludwig drums, helping catapult the company to global success. Such individuals and groups include: Fleetwood Mac, the Beatles, the Doors, Black Sabbath, Buddy Rich, the Carpenters, the Foo Fighters, the Yardbirds, Jimi Hendrix, the Tonight Show Band, Aerosmith, Cream, Emerson, Lake and Palmer, Todd Rundgren, Grand Funk Railroad, REO Speedwagon, Max Roach, Led Zepplin, and Pink Floyd.

ŁUKASIEWICZ, IGNACY J. 1822-1882. Polish pharmacist, humanitarian, politician, political activist, and inventor: Invented (some say co-invented) the first known kerosene lamp (1853), which, along with the purchase of rich oil fields and the opening of several oil refineries, marked the advent of the petroleum industry in Poland. The onetime head of Poland's National Petroleum Society (1880), as well as a generous humanitarian and benefactor, Łukasiewicz is known as the "father of Poland's oil industry."

LUMIÈRE, AUGUSTE M. 1862-1954. French engineer, businessman, filmmaker, film producer, director, actor, and inventor: Based on an earlier design by French inventor Léon Guillaume Bouly (1892), he (along with his brother below) is credited with co-developing and co-inventing the Cinématographe (1895)—from which derives the word "cinema." With this device, the world's first movie film projector, they produced the first known motion picture, *La Sortie des Ouvriers de L'usine Lumière* (1895).

LUMIÈRE, LOUIS. 1864-1948. French engineer, businessman, entrepreneur, filmmaker, film producer, director, photographer, and inventor: Based on an earlier design by French inventor Léon Guillaume Bouly (1892), he (along with his brother above) is credited with co-developing and co-inventing the Cinématographe (1895)—from which derives the word "cinema." With this device, the world's first movie film projector, they produced the first known motion picture, *La Sortie des Ouvriers de L'usine Lumière* (1895).

Louis Lumière.

LUN, CAI. Circa 55-121. Chinese government official: Said to have been the inventor of the first known form of paper (105), which he produced by combining tree bark (cellulose), vegetable waste, and cloth. Due to its low cost, durability, and superior writing surface, it quickly became the standard recording material, not only across China, but eventually across the rest of the world. Today there are thousands of different types of paper available, all which are divided into seven basic categories: printing paper, writing paper, drawing paper, wrapping paper, handmade paper, blotting paper, and speciality paper.

LUPPIS, GIOVANNI. 1813-1875. Croatian naval official and inventor: Proposed and experimented with the first self-propelled torpedo (circa 1850s). His designs, though innovative, were impractical and thus essentially unsuccessful. However, they did pave the way for developments that led to modern self-propelled torpedoes.

LYON, GEORGE ALBERT. 1882-1961. American industrialist, businessman, entrepreneur, manufacturer, and inventor: Credited with inventing the modern automobile bumper (1923). Lyon's innovation was a great advancement over earlier bumper designs (that date as far back as 1897), which were essentially decorative rather than practical. Founder of Lyon, Inc. (1930), known as "the world's largest producer of wheel accessories"—which included the popular tire product, snap-on metal "Lyon Whitewalls."

LYON, GUSTAVE. 1857-1936. French civil engineer, instrument maker, acoustician, military official, and inventor: Invented a number of musical instruments, including new types of pianos, orchestral percussion instruments, and harps. He is best known, however, for his work in the field of acoustics, and was responsible for designing and/or improving the sound dynamics of a number of notable concert halls in France.

LYON, RICHARD F. Born 1952. American electrical engineer, scientist, educator, author, and inventor: One of two independent inventors of the optical computer mouse (1980), a manual input device that permits one to operate and manipulate the cursor on a computer screen. Lyon is the co-inventor of numerous other innovations in such fields as auditory science, GPS, Ethernet, digital memory, and photography.

LYULKA, ARKHIP. 1908-1984. Russian scientist, mechanical engineer, and inventor: Invented the first turbofan engine (circa 1940), an engine technology that is used in nearly all modern jet aircraft to this day.

Two reversible, 6 cylinder, 4 cycle, 120 horsepower engines, designed for paddlewheel water vessels, 1913.

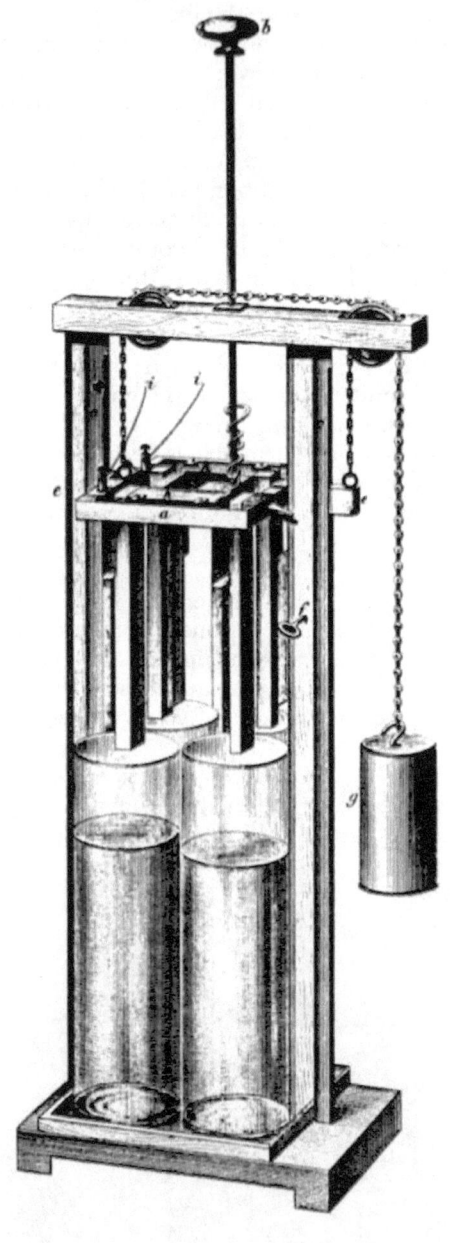

Patent drawing of Robert Bunsen's chromic acid battery cells, circa 1875.

MACH, ERNST. 1838-1916. Austrian scientist, physicist, mathematician, philosopher, lecturer, writer, and educator: Discovered the phenomenon of the "Mach Wave," which describes the conical supersonic shock wave that surrounds a high-speed projectile from the tip down (the "Mach Cone"). When an object, like an aircraft, travels faster than the speed of sound in air (767 mph—or about 1,100 feet per second, or about one mile in just under 5 seconds), it produces a shock wave, resulting in a sonic boom. The velocity of an object in relation to the speed of sound is known as the "Mach Number"; thus an object surpassing the speed of sound is said to be traveling at Mach 1, while an object traveling over Mach 5 is said to be traveling at hypersonic speed. His "Mach's Principle"—a cosmological hypothesis stating that the mass of an object is the result of universal gravitational pull rather than a property of the object itself—influenced the development of Albert Einstein's theory of relativity.

Ernst Mach.

MACINTOSH, CHARLES. 1766-1843. Scottish scientist, chemist, entrepreneur, businessman, and inventor: Using water-resistant, rubberized, flexible materials that he developed via his knowledge of chemistry, he invented the first modern-era waterproof raincoat (1824), today's version which is known as the "Macintosh," the "Mackintosh," or simply a "Mack."

MADISON, JAMES. 1751-1836. American Founding Father, fifth U.S. secretary of state (1801), fourth president of the United States (1808), Virginia House representative, politician, military officer, coauthor of the U.S. Constitution, the Bill of Rights, and the Federalist Papers: Proposed the concept of a congressional book depository (1783), which later became the Library of Congress (1800). Madison is known as the "father of the Constitution."

James Madison.

MAIMAN, THEODORE H. 1927-2007. American scientist, physicist, and electrical engineer: Invented the world's first laser (1960), revolutionizing every modern technology, from medicine, communications, scientific research, and manufacturing, to construction, electronics, the military, and all industrial fields.

MAJAN, AHMED. Born 1963. Emerati engineer, military officer, and inventor: Invented an "intelligent" saddle—a racehorse saddle with built-in electronics that track, weigh, film, and measure various metrics of horse and rider. Majan's many other innovative contributions, as just a few examples, involve such fields as GPS, remote control, and bicycles.

MAKAROV, ALEKSANDR. Born 1966. Russian physicist and inventor: Co-invented the Orbitrap (1999), a type of trap mass analyzer that is used in mass spectrometry, with applications in such fields as biology, scientific research, health and medicine, food, environmental research, forensics, beverages, and pharmaceuticals.

MAKAROV, STEPAN O. 1849-1904. Russian oceanographer, naval architect, naval engineer, naval officer, author, explorer, and inventor: Designed and invented the Ermak (1897), the world's first authentic icebreaker. With her thick, super-strengthened hull, which allowed her to easily break up and fragmentize pack ice, the Ermak served mainly in military capacities throughout World War I and World War II, not being decommissioned until 1963. Her design opened the door to numerous advancements in icebreaker ship technology. Makarov was also responsible for innovations involving armor-piercing shells, torpedo boats, and mine-laying ships.

MAKEEV, VICTOR. 1924-1985. Russian scientist, aerospace engineer, and inventor: Invented the submarine-launched ballistic missile (1955), or SLBM. The first Soviet missile deployments of this type were made from diesel-powered subs, but were transferred to nuclear-powered subs soon thereafter, escalating the global arms race and energizing America's Polaris missile program (1960).

MAKSUTOV, DMITRI DMITRIEVICH. 1896-1964. Russian physicist, military engineer, astronomer, optics scientist, author, and inventor: Invented a variety of scientific instruments, most notably the the Maksutov telescope (1941), which relied on his meniscus corrector lens system (1941). Maksutov also invented the aplanatic optical system (1923) as well as the photogastrograph (1928), and made countless improvements to optical lenses, optical mirrors, and optical surfaces.

MALACHOWSKY, CHRIS. Born 1959. American electrical engineer and businessman: Co-founded the Nvidia Corporation (1993), an electronic graphics company, that among other products, is chiefly known for being a designer, producer, and seller of CPUs and graphics cards for applications ranging from computers, AI, and gaming, to robotics, video editing, and vehicles.

MALYUTIN, SERGEI V. 1859-1937. Russian modernist artist, painter, architect, scenographer, and illustrator: Credited by many with inventing and painting the first Matryoshka Doll (1890)—the first actual carving of one which was later created by Russian wood carver Vasilii Zvyozdochkin—based on Malyutin's artwork. More popularly known as Babushka ("Grandmother") Dolls or Russian nesting dolls, this set of hollowed out, painted wooden figures, which gradually decrease in size, "nest" inside one another, with the last or innermost doll (usually depicted as a baby), being the smallest. The word *matryoshka* (from the Latin word *mater*, "mother") translates as "little matron" in Russian. Etymology betrays the toy's primal foundations: The Matryoshka Doll, a universally recognized symbol of traditional Russian culture, is a fertility symbol whose archetypal origins date back to prehistoric times, when, as overtly chronicled in the archaeological record, reverence for the Great Mother and the religious practice of goddess-worship (thealogy) were at their zenith. Malyutin is known as the "father of modern Russian art."[11]

MAMYRIN, BORIS A. 1919-2007. Russian scientist, physicist, and inventor: Invented an ion mirror known as the reflectron (1974), an important part of time-of-flight analyzers that are used in the field of mass spectrometry.

MANBY, GEORGE WILLIAM. 1765-1854. English military officer, author, illustrator, lecturer, and inventor: Invented the first hand-held fire extinguisher (1813), the first hand-held harpoon gun (circa 1820), and a nonsubmersible lifeboat. His Manby Mortar (circa 1807), an artillery weapon retrofitted to shoot a rope line from shore out to a sinking ship, has saved countless lives since its invention. Manby also founded what would later become England's Royal National Lifeboat Institution (1824).

MANTOUX, CHARLES. 1877-1947. French physician and medical innovator: Invented the Mantoux Test (1907), or tuberculin skin test, which screens for tuberculosis.

MARCONI, GUGLIELMO. 1874-1937. Italian scientist, physicist, electrical engineer, businessman, entrepreneur, politician, military officer, and inventor: Received the first patent for a wireless radio telegraphy system (1896). Among countless other innovations, he was also the inventor of syntonic telegraphy (1900), a magnetic wireless

Guglielmo Marconi.

receiver (1902), and a horizontal directional aerial (1905). Founded the Wireless Telegraph and Signal Company Limited (1907); his wireless devices were instrumental in saving lives during the Titanic disaster (1912), adding to his fame. Marconi's work with wireless communications, radio telegraphy, radio waves, short wave technology, electromagnetic radiation, microwave radio beacons, radar, beam communication systems, beam stations, microwaves, and radiotelephony, revolutionized the fields of electronics and communications, laid the groundwork for the development of TV, and helped usher in the modern era of wireless technologies. One of the thousands of modern products that resulted from Marconi's research and innovations, the cell phone, for example, now numbers around 8 billion worldwide. Marconi is known as the "father of the radio." Nobel Prize winner.

MARINESCU, GHEORGHE. 1863-1938. Romanian scientist, neurologist, educator, and filmmaker: Created and produced the world's first science films (1898 onward), opening the way for the use of the art of cinematography as an aid in scientific study, education, and research.

MARKHAM, WILLIAM F. Circa 1852-1930. American businessman, entrepreneur, gunsmith, woodworker, and inventor: Invented the world's first BB gun (1886), an ideal weapon for children, beginners, plinkers, target practice, pest control, and small game hunting. Founder of the Markham Air Rifle Company (1886).

MARLET, RENE LOUIS. Dates unknown. Nationality unknown. Credited with patenting the first toroidal aircraft propeller (date unknown), a type of propeller with enclosed, curved, ring-shaped blades. The toroidal design improves fuel efficiency, reduces noise, reduces vibration, improves air and water flow, improves maneuverability, increases propeller rotation speed (which increases vehicle speed), cuts fuel consumption, increases water- and air-gripping capabilities, increases engine life, increases range, decreases emissions, increases load capacity, and reduces cutting damage (e.g., to underwater plant and animal life), all while increasing the strength and durability of the propeller. Toroidal props, which have revolutionized the science of fluid dynamics, have numerous modern applications, from helicopters, drones, and boats, to ships, wind turbines, submarines, and household and industrial fans.

MARSH, SYLVESTER. 1803-1884. American businessman, entrepreneur, engineer, and inventor: Among numerous other innovations, he is best known for inventing the world's first steep incline-climbing cog railway (1861). Known as the Marsh Rack Railway System, it featured a toothed cog wheel attached to the bottom frame of a steam-powered locomotive (largely designed by Marsh). The cog wheel locked into a single cog rail that was laid down along the center of the railroad track, safely fastening the train (and its passengers) to the railway. It was built for, and first used on, Mount Washington (the tallest natural elevation in the northeast), located in New Hampshire's White Mountains. Marsh's cogway track and train, which continue in operation to this day, led to the further development and refinement of cog wheel train technology, and is now used around the world.

MARTEL, CHARLES. 1860-1945. Swiss (later American) librarian: Co-invented the U.S. Library of Congress classification system (1899), or LC, which allows for the logical organization of its current catalogue of some 40 million books. The LC is still used, not only by the Library of Congress, but also by numerous other libraries and academic institutions around the world.[12]

MARTIN, BILLY. Dates unknown. Nationality unknown. Occupation(s) unknown: Credited by some with having co-invented the electronic or digital thermometer (1970).

MARTIN, GEORGE RAYMOND RICHARD. Born 1948. American novelist: Author of the fantasy novel series, *A Song of Ice and Fire* (1991), which was later turned into the hit TV series *Game of Thrones* (2011).

MARTIN, ROBERT. Dates unknown. Nationality unknown. Occupation(s) unknown: Credited by some with having co-invented the electronic or digital thermometer (1970).

MA'RUF, TAQI AL-DIN MUHAMMAD IBN. 1526-1585. Syrian polymath, scientist, physicist, physician, scholar, mathematician, astronomer, astrologer, theologian, zoologist, engineer, optics researcher, botanist, pharmacist, philosopher, clockmaker, educator, author, and inventor: Proposed the first steam turbine and steam engine (1546), and invented the six-cylinder Monobloc Pump (1559), an alarm clock, an astronomical clock, an early telescope, an earth globe, a framed sextant, a quadrant, and an astronomical observatory (the Istanbul Observatory), as well as a host of other scientific instruments.

MASON, JOHN LANDIS. 1832-1902. American metalworker and inventor: Invented the Mason jar (1858), whose rubber-ringed, self-sealing lid, screw-on metal band, and clear tempered glass jar revolutionized both canning technology and the food preservation industry.

MASSACHUSETTS GENERAL COURT. Established in the 17th Century (1629), this all male legislature is responsible for founding Harvard University (1636).

MASUOKA, FUJIO. Born 1943. Japanese engineer and educator: Invented flash memory (1980) a type of low-cost, efficient, robust, non-volatile memory on which data can quickly by recorded, stored, or erased. Flash memory has thus found widespread usage in numerous applications, from computers, scientific instruments, medical devices, and cameras, to cell phones, video games, robotics, and USB drives.

MATSUSHITA, KONOSUKE. 1894-1989. Japanese industrialist, businessman, entrepreneur, and inventor: Invented, among other unique items, battery-powered bicycle lighting (1923), an electric space heater, an advanced thermostat, and the Matsushita Radio (1931). Most notably he was the founder of Panasonic (1918); however, he was also responsible for the launching of numerous other businesses, such as the Matsushita Electric Industrial Company (1918), the Kyushu Matsushita Electric Company (1955)—later renamed Panasonic, the Matsushita Communication Industrial Group (1958), the Matsushita Electric Corporation of America (1959), and Matsushita Seiko (1962).

MATTEUCCI, FELICE. 1808-1887. Italian scientist, engineer, businessman, and inventor: Credited with co-inventing the first internal combustion engine (1853). Known as the Barsanti-Matteucci engine, it was constructed on a two-cycle, free piston design as an alternative to the steam engine.

MATTHEWS, GORDON. 1936-2002. American businessman, entrepreneur, aviator, and inventor: Credited by some with inventing, or at least improving on and commercializing, voicemail technology (circa 1979)—an electronic system that can record, store, receive, forward, playback, and delete messages. Founder of VMX Inc.

MAUCHLY, JOHN W. 1907-1980. American physicist, engineer, and inventor: Co-invented the Electronic Numerical Integrator and Computer (1946), the first electronic, programmable, multipurpose digital computer. More popularly known by its acronym, the ENIAC, Mauchly's automated device greatly advanced computer technology, laying the foundation for modern computing as we know it today.

MAUDSLAY, HENRY. 1771-1831. English mechanical engineer, blacksmith, businessman, entrepreneur, author, and inventor: A key contributor to the then newly emerging Industrial Revolution (1760-1840), he is best known for his improvements on the metal-cutting lathe (1800). He also worked with, constructed, or improved various measuring instruments, desalination devices, tunneling techniques, cloth printing systems, and engine technology.

Maudslay's first screw-cutting lathe, circa 1797.

MAXIM, HIRAM STEVENS. 1840-1916. American businessman, weapons maker, and inventor: Best known for his invention of the first automatic machine gun (1884). Known as the Maxim Gun, the continuously firing, self-powered weapon was used with deadly efficiency throughout numerous wars, including the Boar War, the Spanish War, and World War I. Led to advancements in machine gun science.

MAXWELL, JAMES CLERK. 1831-1879. Scottish scientist, physicist, engineer, mathematician, photographer, educator, and inventor: Credited with being the first to both propose color photography (1855) and develop the first color photograph (1861). Maxwell also suggested the idea of electromagnetic radiation (1865), which influenced the theories and research of one of his many enthusiastic admirers, Albert Einstein—a venerable scientist with whom Maxwell is often compared.

MAZOR, STANLEY. Born 1941. American computer scientist, computer programmer, computer designer, mathematician, engineer, educator, author, and inventor: Helped develop the world's first CPU (1971), or microprocessor. Known as the Intel 4004, it transformed electronic technology, laying the foundation for the start of the age of personal computing.

MCADAM, JOHN LOUDON. 1756-1836. Scottish civil engineer, road builder, author, and inventor: Invented the Macadam Road surface (1815), a road building system that entails placing a convex layer of cement or asphalt over a compacted substructure of gravel for the purposes of strength, durability, and water drainage. The crushed stone foundation is poured in graduated sizes, with the largest stones at the bottom, the smallest at the top, thereby increasing road stability. The essence of McAdam's process, known as "Macadamization," is still used to construct roadways today—with the usual modern improvements added to his techniques and recipe.

MCCORMICK, CYRUS HALL. 1809-1884. American businessman and inventor: One of the early inventors of the mechanical reaping machine, the business he co-founded, the McCormick Harvesting Machine Company (1847), was later combined with the International Harvester Company (1902). McCormack is known as a "father of modern day agriculture."

MCCOY, ELIJAH. 1843-1929. Canadian mechanical engineer, fireman, railroadman, and inventor: Invented the displacement lubricator (1872), which allowed steam engines to self-lubricate while in operation—a first at the time. His invention caught on quickly, heavily impacting the auto, shipping, train, and factory industries. The phrase "the real McCoy" entered English vocabulary at this time in order to separate McCoy's lubricator from the many inferior quality imitations that were beginning to flood the market. Founder of the Elijah McCoy Manufacturing Company (1920). Though his invention was eventually replaced by modern automatic lubricators, due to its aesthetic appeal the McCoy Displacement Lubricator is still used in such items as model steam engines.

Elijah McCoy.

MCCULLOCH, ERNEST A. 1926-2011. Canadian scientist, cellular biologist, researcher, educator, and inventor: Co-described stem cells (circa early 1960s), an important medical breakthrough that allows for the regeneration of damaged cells and tissue, treatment of disease, drug testing and development, and pain reduction, as well as reducing recovery time and lowering the need for surgery.

MCINTIRE, RAY. 1918-1996. American engineer and inventor: Invented polystyrene foam (1944). Now more popularly known by its brand name "Styrofoam," this lightweight, energy-saving material (which is 98 percent air) is used primarily in the construction industry to insulate homes and buildings.

MCKAY, SR., NICHOLAS. 1920-2014. American businessman, entrepreneur, and inventor: Invented the modern lint roller (1956). Born out of necessity, his product, named the "Lint-Pick-up," began as a simple homemade device made from tape, wire, and a cardboard tube. Co-founder of the company Helmac (1956).

MCLEOD, JUSTIN. Born circa 1985. American businessman and entrepreneur: Founder and CEO of Hinge (2012), a popular dating application.

MCLURKIN, JAMES. Born 1972. American computer scientist, robotics engineer, electrical engineer, educator, and inventor: Invented "swarm bots" (circa 1991), a multi-robot system that uses algorithmic computations to control, guide, and program swarms of autonomous robots. Swarming technology is used primarily for conditions and situations considered too hazardous, dreary, or unpleasant for humans, such as search and rescue, reconnaissance missions, guard duty, perimeter patrol, intelligence gathering, protection, problem solving, medical diagnostics, product assembly, mapping, aerial reconnaissance, and surveillance operations, to name but a few applications.

MECHNIKOV, ILYA ILYICH. 1845-1916. Russian scientist, biologist, zoologist, comparative embryologist, evolutionary embryologist, immunologist, physiologist, comparative pathologist, thanatologist, educator, docent, and inventor: Discovered phagocytosis (1883), the process whereby a phagocyte consumes harmful bacteria, rendering it innocuous. Though disputed by some, among many other innovations he is also said to have discovered the probiotic (i.e., healthful) effect of milk fermented by bacilli (1907)—that is, by lactic acid bacteria, which later helped launch microbiome research and the founding of the science of probiotics. Mechnikov is known as the "father of inherent immunity." Nobel Prize winner.

MÈGE-MOURIÈS, HIPPOLYTE. 1817-1880. French chemist, pharmacist, and inventor: Invented margarine (1869), a butter alternative which he named "Oleomargarine." Mège-Mouriès' original Victorian version was made from beef tallow. In order to lower costs and lengthen shelf life, however, subsequent margarine producers altered and refined his recipe, using skimmed milk, food colorings, vegetable oils, flavorings, emulsifiers, and preservatives.

MEIROWITZ, MORDECAI. Born 1930. Romanian telecommunications engineer, Israeli postmaster, author, and board game designer: Invented the popular board game Mastermind (1971).
MELLON, ANDREW. 1855-1937. American businessman, banker, entrepreneur, politician, and philanthropist: Co-founded (with his brother below) the Mellon Institute of Industrial Research (1913), which later merged with the Carnegie Institute of Technology, becoming Carnegie Mellon University (1967).
MELLON, RICHARD B. 1858-1933. American businessman, banker, industrialist, entrepreneur, and philanthropist: Co-founded (with his brother above) the Mellon Institute of Industrial Research (1913), which later merged with the Carnegie Institute of Technology, becoming Carnegie Mellon University (1967).
MENDELEEV, DMITRI. 1834-1907. Russian chemist and inventor: Founded the scientific principle known as Periodic Law (1869)—which states that the properties of elements reiterate in an expected manner when laid out by increasing atomic number. This concept led to the invention of the Periodic Table (1869)—in which the elements are arranged by their atomic weight. Mendeleev also invented pyrocollodion (1892), a smokeless gunpowder alternative, and co-invented the pycnometer (1859), a liquid density-measuring glass flask.
MENDELL, HARRY. Dates unknown. American computer scientist and inventor: Invented the first digital sampling synthesizer (1976), a musical instrument that can record and store any sound, then play it back as "music" from a keyboard.[13]
MENSINGA, WILHELM. 1836-1910. German scientist and physician: Pursuing more reliable methods of birth control, he invented a cervical cap for women, later known as the diaphragm (1880).
MERLIN, JOHN JOSEPH. 1735-1803. Belgian engineer, musical instrument maker, clock maker, musician, and inventor: Though roller skates were first chronicled in London, England (1743), the name of the inventor was left unmentioned. Thus the honor would later go to Merlin (1760), who is credited with inventing not only the world's first roller skates, but the world's first inline skates—the forerunner of what is now known by the brand name "Rollerblades."
MERRILL, RICHARD B. 1949-2008. American engineer, photographer, and inventor: Invented vertical color filter technology (1997), which is used in various camera models sold by the Sigma Corporation. Founded the company Foveon (1997).
MESMER, FRANZ ANTON. 1734-1815. German scientist, physician, astronomer, astrologer, author, and researcher: Though hypnosis-like therapies have been in use for thousands of years, Mesmer is credited with what might be termed the 18th-Century version, a healing modality he referred to as "animal magnetism" (1774). Later medically known as Mesmerism, it is considered to be the precursor of modern hypnotism. Also called hypnotherapy, hypnosis is a relaxed state in which an

individual can be psychologically influenced. As such, hypnosis is today used to treat a myriad of ailments and issues, including depression, emotional regulation, pain management, phobias, negative habits, memory recovery, and stress management. The word mesmerize derives from his name.

Franz A. Mesmer.

MESTRAL, GEORGE DE. 1907-1990. Swiss electrical engineer, businessman, entrepreneur, and inventor: Invented Velcro (inspired in 1941; patented in 1954), a unique polymer fastening system developed by way of biomimicry: One day while enjoying the great outdoors, De Mestral was inspired to create Velcro after an unpleasant experience with the cocklebur plant (*Xanthium strumarium*), which produces fruit with tiny incurved (hooked) spines—burs that are nearly impossible to remove from clothing. Founded the company Velcro S.A. (circa 1952). Today Velcro is widely used in a myriad of industries, from clothing, aerospace, and medicine, to automotive, household, and construction, among many others. The word Velcro (i.e., vel-cro) derives from the French words *velours* ("velvet") and *crochet* ("hook").

METCALFE, ROBERT M. Born 1946. American computer scientist, electrical engineer, venture capitalist, publisher, author, businessman, entrepreneur, educator, and inventor: Co-invented the Ethernet (1973), a networking technology that utilizes a wired connection to permit communication between electronic devices like computers, TVs, printers, and game consoles. While wireless (Wi-Fi) has become quite popular for a host of applications, the physicality of wired LAN technology (Ethernet) continues to be preferred by most businesses for its robustness, stability, consistency, dependability, and speed. Co-founded the 3Com Corporation (1979).

MEUCCI, ANTONIO. 1808-1889. Italian engineer, chemist, businessman, entrepreneur, and inventor: Though he developed, discovered, or worked on a host of innovations throughout his life (among them smokeless candles, effervescent drinks, oil lamps, paper making, and the physiophonic phenomenon), he is best known for his invention of a working precursor to the telephone (1857), a "sound telegraph" that he called the Teletrofono. To this day Meucci is regarded by many as the true inventor of the telephone, which he had designed nearly 20 years before Alexander Graham Bell was granted his telephone patent (1876). (Note: Meucci was a close friend and business partner of Giuseppe Garibaldi, the famed Italian socialist revolutionary who was once offered a military command in the Union army by Abraham Lincoln—president of the then Left-wing Republican Party.[14] Lincoln's offer fell apart, and, after it was turned down by then U.S. Colonel Robert E. Lee, the position ultimately went to Ulysses S. Grant.)[15]

MEYER, STANLEY ALLEN. 1940-1998. American inventor: Invented the Meyer Water Fuel Cell (circa 1980), a perpetual motion device said to be capable of generating limitless energy from water. While this would violate the Laws of Thermodynamics, according to Meyer, after being retrofitted to a conventional engine, his cell could power the motor using water as a fuel—no gasoline or spark plugs needed. As has occurred with other avant-garde inventors, however, mainstream science labeled him a "crank," his invention was suppressed as "fraudulent," and his ideas were rejected as "pseudoscience."

MICHAELS, LORNE. Born 1944. Canadian producer and screenwriter: Created the hit TV series *Saturday Night Live* (1975).

MICHELIN, ÉDOUARD E. 1859-1940. French industrialist, businessman, artist, and inventor: Invented the first pneumatic tire, as well as the first detachable pneumatic tire (1891). Due to its improved traction, shock-absorbing qualities, and adaptability, Michelin's inflatable rubber tire caught on quickly in both the bicycle and the automotive industries. Co-founder of the Michelin Tire Company (1888). Édouard's brother, André Michelin, was the publisher of the first Michelin Guide (1900). Both the company and the guide continue to flourish to this day.

Édouard E. Michelin.

MICHELL, ANTHONY. 1870-1959. Australian engineer and inventor: Invented the crankless engine using his tilting pad principle (1917). While superior in efficiency, due to manufacturing costs and mechanical complications, Michell's thrust-bearing engine was never completely accepted by the automotive industry. Today crankshaft-powered engines continue to reign supreme.

MIDDLEMARK, MARVIN P. 1919-1989. American inventor: Invented the dipole TV antenna (1953), also known as the "rabbit ears" television antenna. By increasing TV reception—thereby making TV more widely available across the country—it radically altered the television industry, leading to the explosive popularity of the medium beginning in the 1950s.

MIDGLEY, JR., THOMAS. 1889-1944. American mechanical engineer and inventor: Invented tetraethyl lead fuel (circa 1916). Known today as Ethyl gasoline, it greatly reduced oil consumption while revolutionizing aviation technology. Midgley was also responsible for inventing Freon (1928), a nonflammable, nontoxic refrigerant that has become a staple in the air-conditioning and refrigerator industries.

MIESCHER, FRIEDRICH. 1844-1895. Swiss scientist, physician, biologist, educator, and researcher: Discovered the molecule known now as DNA (1869)—a word coined by German biochemist Albrecht Kossel (1881). Miescher's research laid the foundation for the subsequent work of James Watson and Francis Crick, who positively identified DNA 84 years later (1953).

MIKOYAN, ARTEM. 1905-1970. Armenian aircraft designer and inventor: In partnership with Mikhail Gurevich he co-invented the MiG-series Russian fighter aircraft (beginning in 1939). Notable models included the MiG 1, the MiG 3, the MiG-15, the MiG-21, and the MiG-25. Various models are still in use to this day, including Mikoyan's Mig-15 and Mig-21. MiG is a Russian acronym for "Mikoyan and Gurevich."

MIKULIN, ALEXANDER A. 1895-1985. Russian mechanical engineer, aeronautical engineer, aerospace engineer, and inventor: Invented the Mikulin AM-34, AM-35, and AM-38 engines (1928), and the AM-38F engine (1942), among many others. Co-inventor (or co-developer) of the AMBS-1 two-stroke engine, as well as the Tsar Tank (circa 1916). The latter, also known as the Lebedenko Tank, was an imposing but largely impractical combat machine that holds the record for being the largest armored gun transport ever created up to that time.

MIL, MIKHAIL L. 1909-1970. Russian aerospace engineer, helicopter designer and inventor: Beginning in the 1940s, he invented and launched the famous and popular Mil Mi-series helicopter aircraft, still in production today. Founder of Moscow's Mil Helicopters company (1947).

MILES, ALEXANDER. 1838-1918. American mechanical engineer, businessman, entrepreneur, barber, and inventor: Invented an elevator with an automatic, electrically-powered door opener and closer (1887). His invention made elevators safer and paved the way for advancements in elevator technology. His automated door system is still used—with the usual modern improvements and modifications—in elevators today.

MILLAY, GEORGE. 1929-2006. American businessman and entrepreneur: Besides opening over 100 restaurants across the U.S., he is best remembered as the man who invented the idea of the water park (1977), an amusement area typically equipped with slides, artificial rivers, swimming pools, and surfing pools. Millay, also a co-founder of SeaWorld (1964), is known as the "father of water parks."

MILLS, DAVID L. 1938-2024. American computer scientist, educator, and inventor: Co-developed ARPANET (1969), or Advanced Research Projects Agency Network, a forerunner of the modern Internet. More notably he was the inventor of Network Time Protocol (1981), or NTP, an Internet time standard that synchronizes computer clocks, and the first network router (circa late 1970s). Mills was known as the "Father Time of the Internet."

MINSKY, MARVIN. 1927-2016. American cognitive scientist, computer scientist, AI pioneer, mathematician, educator, and inventor: Invented the first neural network simulator (1951), the confocal scanning microscopy (1955)—a high-resolution, 3D optical imager, and the idea of "frames" (1975)—a data structure that gives artificial intelligence systems the ability to think, comprehend, decide, and logically process information. Co-founded the Artificial Intelligence Project (1959), now known as the MIT Computer Science and Artificial Intelligence Laboratory.

MISHIMA, TOKUSHICHI. 1893-1975. Japanese alloy scientist, metallurgist, educator, and inventor: Invented M.K. Steel (1931), a durable, stable, vibration-resistant, and cost effective magnetic steel that retains its magnetic properties in whatever form it is used. To this day his product is employed in a variety of industries, from automotive and aeronautics to electronics and communications.

MITARAI, TAKESHI. 1901-1984. Japanese businessman, entrepreneur, obstetrician, and gynecologist: Co-founder of Canon Inc. (1937), more popularly known as Canon. The company's name is an anglicization of the name of the Buddhist compassion-goddess Kwannon.

MIYAMOTO, SHIGERU. Born 1952. Japanese video game producer and designer, industrial designer, and artist: Created the popular video games Donkey Kong, Mario, the Legend of Zelda, and Star Fox, among others.

MOLCHANOV, PAVEL. 1893-1941. Russian meteorologist and inventor: While he did not invent the first radiosonde, he did invent the first Russian radiosonde (1930)—a telemetric device that when sent aloft (typically attached to weather balloons) measures and records a host of atmospheric data, such as temperature, humidity, and air pressure.

MONET, CLAUDE. 1840-1926. French painter and artistic innovator: Responsible for initiating the French-based Impressionistic Period in art (1867-1886), an era characterized by bright colors, obvious brush strokes, outdoor painting, subjective viewpoints, and above all impressionistic (i.e., often nebulous) portrayals of everyday life.

Claude Monet.

MONIER, JOSEPH. 1823-1906. French gardener and inventor: Credited by some with developing and commercializing the concept of reinforced concrete (patented 1867). Better known as rebar (short for "reinforcing bar"), it was a mesh of iron wire (today steel rods) that is embedded in wet concrete to fortify its tensile strength, thereby greatly enhancing its load-bearing capacity. The rebar system, which is used primarily to strengthen and stabilize concrete foundations, columns, arches, pipes, walls, girders, floors, and beams, greatly advanced the concrete construction industry, permitting the erection of heavier, wider, taller, and more durable structures than ever before (especially beneficial in earthquake-prone and high wind areas). Today's massive modern metropolises, with their huge skyscrapers and expansive bridges, owe their very existence largely to the invention of rebar.

MONTENIER, JULES B. 1895-1962. American cosmetic chemist and inventor: Based on earlier formulas, he improved, and thus invented, the first modern commercialized antiperspirant deodorant (circa 1937). Sold under the brand name "Stopette," the popular product was made from aluminum chloride, was housed in a squeezable polyethylene bottle, and contained a spray nozzle at the tip.

MONTGOLFIER, JACQUES-ÉTIENNE. 1745-1799. French scientist, businessman, papermaker, hot air balloon pioneer, author, and inventor: Although he and his equally famous brother (below) did not invent the hot air balloon (its precursors date back to 3rd Century China), they were the first to carry out an earth-free unmanned flight (June 1783) and an earth-free manned flight (November 1783) in large scale hot air balloons. Their work and discoveries led to advancements in both hot air balloon technology and the meteorological sciences.

Montgolfier brothers balloon flight, 1783.

MONTGOLFIER, JOSEPH-MICHEL. 1740-1810. French scientist, businessman, papermaker, aeronautics researcher, mechanical engineer, hot air balloon pioneer, author, and inventor: Although he and his equally famous brother (above) did not invent the hot air balloon (its precursors date back to 3rd Century China), they were the first to carry out an earth-free unmanned flight (June 1783) and an earth-free manned flight (November 1783) in large scale hot air balloons. Their work and discoveries led to advancements in both hot air balloon technology and the meteorological sciences.

MONTGOMERY, JOHN J. 1858-1911. American physicist, engineer, aeronautical researcher, mathematician, lecturer, educator, and inventor: Basing his theories on ornithological observations, he proposed, designed, and flew the first manned, controlled, heavier-than-air glider (1883-1884) in the West, laying the foundation

for modern glider science and the commercialization of glider recreation.
MONTURIOL, NARCÍS E. 1819-1885. Spanish businessman, artist, and inventor: A self-taught engineer, he was responsible for inventing the first controlled, workable, untethered submarines. His first attempt (1859), the hand-powered submarine *Ictíneo I* (Greek for "fish-ship"), was a 23-foot, double-hulled wooden vessel capable of carrying six individuals and diving to a depth of about 65 feet. Cumbersome and slow, he set out to design a second version. The *Ictíneo II* (1864), also constructed of wood, was steam-powered and could attain a depth of about 100 feet. Though Monturiol's subs drew little attention at the time, his work led to future advancements in submarine technology.
MOOG, ROBERT. 1934-2005. American engineer, businessman, entrepreneur, musician, musical instrument designer and maker, educator, and inventor: Though he designed and invented a host of musical products, he is best known for his invention of the Moog synthesizer (1964), which he developed with Herb Deutsch. The world's first commercial synthesizer, it revolutionized music and helped launch the electronic music genre. Moog was the co-founder of the R.A. Moog Company (1953), later renamed Moog Music (1972).
MOONEY, JOHN J. 1930-2020. American chemical engineer and inventor: Co-invented the three-way automotive catalytic converter (1973), a type of filter that reduces the pollution produced by internal combustion engines. By cutting down on toxic emissions it is estimated that the device has saved hundreds of thousands of lives, prevented millions of cases of throat and lung diseases, and improved the quality of life for billions of people around the world.
MOORE, GORDON E. 1929-2023. American businessman, entrepreneur, computer scientist, and engineer: Co-founder of the Intel Corporation (1968), more popularly known as Intel.
MORAY, THOMAS HENRY. 1892-1974. American radio engineer, author, and inventor: Discoverer of a type of mysterious crystal in Sweden, which, he asserted, could be used to tap into a free and endless "radiant energy" source that is perpetually present in the cosmos. In response, he invented the "Cosray Receiver" (1925) around the stone, an energy-receiving device with which he hoped to generate unlimited power (i.e., electricity). Articles and booklets were penned at the time by scientific eyewitnesses to Moray's demonstrations; many seemed convinced. Although he himself wrote a book on the subject (*A Radiant Sea of Energy*), he never fully revealed the secrets of his cryptic stone to the public (though some believe it may have been spodumene). In 1941 his energy receiver seems to have either disappeared or was intentionally destroyed, an event mired in intrigue, only adding to the mystery. (After Thomas' passing, his son John E. Moray, 1928-2021, carried on his father's work, recording the family's history in his book, *The Sea of Energy in Which the Earth Floats*.) Though Moray's ideas paralleled those of Nikola Tesla (and his "cosmic energy" theory), in the end, with pieces

of the puzzle missing, and with his claims violating the first law of thermodynamics, mainstream science lost interest in Moray's discovery of free unlimited energy, and since then—for obvious reasons—his ideas have been largely ignored and suppressed.

MORENO, ROLAND. 1945-2012. French engineer, journalist, editor, artist, humorist, author, and inventor: Invented the smart card (1974), examples of which today are banks cards, debit cards, identity cards, phone payment cards, driver's licences, and shopping cards. By increasing the security, efficiency, and convenience of money transactions, Moreno's globally used innovation revolutionized the financial world.

MOREY, SAMUEL. 1762-1843. American inventor: Though he is credited with a number of innovations (most notably early steam-powered boat designs), he is best known for his invention of the internal combustion engine, and was the first person to patent the idea (1826).

MORGAN, SR., GARRETT A. 1877-1963. American mechanic, businessman, entrepreneur, philanthropist, and inventor: A son of a former servant of famed Confederate General John Hunt Morgan, he made improvements in industries ranging from sewing machines, cigarettes, and hats, to automotive parts, combs, and belts, and is credited with inventing a hair-straightening cream (circa 1909) and a safety hood (1914)—a forerunner of the gas mask. He opened several stores and shops (1907, 1908), launched a newspaper (1920), and founded his own company, G. A. Morgan Hair Refining Company (1913). He is best known, however, for his invention of what would later become the modern three-way traffic light (1923).

MORGAN, JOHN PIERPONT. 1837-1913. American businessman, entrepreneur, investor, financier, and banker: Co-founder of General Electric (1892) and U.S. Steel (1901), he helped finance the launch of the company International Harvester (1902).

MORITA, AKIO. 1921-1999. Japanese businessman and entrepreneur: Co-founder of the Sony Group Corporation (1946), better known as Sony.

MORLEY, EBENEZER COBB. 1831-1924. English sportsman, solicitor, politician, and justice of the peace: Founder of the FA or Football Association (1863)—a sports game known in the U.S.A., Canada, and Australia, as soccer—and the creator of its first laws. Morley was also the founder of the Barnes and Mortlake Regatta (circa 1853) and the Barnes Football Club (1862). Because the origins of football/soccer date back to the mists of prehistory, its original inventor can never be identified. However, due to the Victorian development and standardization of the game's rules under Morley's leadership, he must be considered the "father of modern football," or, the "father of modern soccer."

MOROZOV, ALEXANDER. 1904-1979. Russian scientist, engineer, military officer, tank designer, and inventor: Made significant improvements to tank design and was the co-developer of a number of notable tank models, including the T-24 (1930), the T-34 (1940), the T-54 (1945), and the T-55 (1957).

MORRIS, THOMAS MITCHELL. 1821-1908. Scottish golfer, golf course designer, golf club maker, golf ball maker, golf educator, businessman, and entrepreneur: The highly disputed origins of golf date back to antiquity, and are therefore unknown. Early forms are recorded, for example, in 10th-Century China and, long before that, pre-Christian Rome. Nonetheless, due to his vital role in standardizing its rules, Morris—more commonly known in his lifetime as "Old Tom"—is credited with being the "father of modern golf." It is due to this fact that bonnie Scotland herself is considered the "original home of modern golf." (Golf was already in existence in Alba before 1457, the year it is first mentioned in historic Scottish records.) Morris' contributions to the game are many, and include unique course layouts and designs, the inclusion of course hazards, regular maintenance of course greens, and perhaps above all, the legitimization and popularization of the sport.

Thomas M. Morris.

MORRISON, WALTER FREDERICK. 1920-2010. American businessman, entrepreneur, and inventor: Conceived (1946) and produced (1948) the "Flyin-Saucer," the first flying disk toy—which Morrison subsequently named the "Pluto Platter" (1955). The sports gliding product was later marketed by Wham-O under the brand name "Frisbee" (1958). This immensely fun, popular, easy-to-use toy has played a large role in expanding both the health and fitness industry and the sports industry.

MORRISON, WILLIAM. 1860-1926. American dentist, attorney, author, and inventor: Besides other innovations, he is most notable for his co-invention of the cotton candy machine (1897).

MORSE, SAMUEL F. B. 1791-1872. American artist, painter, educator, philanthropist, and inventor: Invented an electric telegraph (1832), which led to partnerships with other individuals who assisted in the co-invention of Morse Code (1838)—a electrical communication system that transmits letters and numbers as dots and dashes. Though Morse Code has been replaced by, among other things, telephone, Internet, and satellite technologies, its eponymous inventor, who was also a founding trustee of Vassar College, is now as famous for his historical paintings as for his innovations in telegraphy.

MOSIN, SERGEI IVANOVICH. 1849-1902. Russian engineer, weapons designer, and inventor: Head designer of the Three-Line Rifle, Model 1891 (1891). Known more commonly as the Mosin-Nagant rifle, it saw service throughout World War I and World War II and is one of the most popular military weapons ever created.

MOTORINS, IVAN F. Circa 1665-1735. Russian craftsman and bellmaker: Designed the celebrated bell known as the "Tsar Bell" (1735), housed today on the grounds of the Kremlin, Moscow, Russia. Some 20 feet tall and 22 feet in diameter, it weighs nearly 223 tons, making it the largest bell in the world. Ivan passed away before work on the bell could be completed, and the task of casting it was turned over to his son Mikhail.

MUELLER, FRITZ. Dates unknown. Nationality unknown. Occupation(s) unknown: Credited by some with having co-invented the electronic or digital thermometer (1970).

MUIR, JOHN. 1838-1914. Scottish naturalist, environmentalist, conservationist, explorer, wildlife preservationist, wilderness preservationist, traveler, and author: His writings and work inspired the founding of numerous national parks, including Yosemite National Park (1890), Grand Canyon National Park (1908), Sequoia National Park (1890), and Mount Rainier National Park (1899). Muir, who might rightly called the "father of Yosemite National Park," is also known as the "father of our national park system." He was the co-founder and first president of the Sierra Club (1892).

John Muir.

MULLIS, KARY B. 1944-2019. American molecular biologist and biochemist: Invented the polymerase chain reaction method (1985). Best known by its acronym PCR, the process allows small quantities of DNA to be replicated in large amounts quickly and efficiently, greatly facilitating countless scientific fields, from medicine and forensics, to anthropology and biology. Nobel Prize winner.

MUNTERS, CARL GEORG. 1897-1989. Swiss engineer, businessman, entrepreneur, and inventor: Co-invented the gas absorption refrigerator (1922). Using a process known as "absorption refrigeration," it relies on heat to produce the power that operates the cooling process. Munters' invention was an alternative to refrigerators that depended on mechanical compressors. The modern gas absorption refrigerator, powered by propane gas, can still be found in various applications, mainly those where electricity is unavailable—such as in recreational vehicles or RVs. The owner of over 1,000 patents, Munters is also credited with inventing extruded polystyrene foam (1930s), which later became an integral component in the production of the popular material known by the brand name "Styrofoam."

MURDOCH, COLIN. 1929-2008. New Zealand veterinarian, pharmacist, pharmacologist, and inventor: Invented the disposable hypodermic syringe (1956), the tranquillizer gun (1959), the silent intruder alarm (1966), and an improved version of the child-proof bottle cap (circa 1976). His innovations have saved untold lives and helped reduce stress, trauma, and pain for both humans and animals.

MURDOCH, WILLIAM. 1754-1839. Scottish engineer and inventor: Invented gas lighting (1792) using purified gas from distilled coal. Also credited with innovations involving steam engines, steam locomotives (1786), compressed air steam guns (1803), oscillating engines (circa 1784), and gas production. Murdoch's gas lighting system completely altered light technology and helped usher in the modern world. Among its many benefits: It aided the world economy as stores and factories could remain open longer, it helped lead to the growth of new towns and cities, and it increased safety by reducing crime.

MURGAS, JOZEF. 1864-1929. Slovakian scientist, botanist, clergyman, artist, botanist, architect, newspaper publisher, educator, and inventor: Invented a "wireless-telegraph apparatus" (1903), which aided in the development of radio technology. Founder of the Universal Aether Telegraph Co. (1905).

MURPHEY, ABRAHAM D. 1863-1935. American physician and inventor: Invented the first known antiperspirant (1909). The liquid deodorant's original intended use was as a safety formula: He had created it to wick away sweat from the hands of surgeons in order prevent surgical instruments from slipping during operations. Subsequent inventors refined and improved Murphey's innovation, known by the brand name "Odor-o-no," laying the groundwork for the modern deodorant industry.

MURPHY, WILLIAM LAWRENCE. 1876-1957. American inventor: Invented the "Murphy Bed" (circa 1900), a hinged bed that can be folded up against a wall or into a closet when not being used.

MURZIN, EVGENY. 1914-1970. Russian audio engineer and inventor: Invented the ANS synthesizer (1937). Named for the Russian composer Alexander Nikolayevich Scriabin, this complex photoelectric musical instrument uses rotating glass disks to create sound. The ANS synthesizer continues to be used to this day, though primarily by avant-garde musicians, composers, and musical groups.

MUSK, ELON. Born 1971. South African businessman, entrepreneur, investor, and innovator: Co-founded X Corp (1999), an online bank that later became PayPal (2000). (X Corp is a subsidiary of the X Holdings Corp, a technology company that also owns the social media platform X—formerly known as Twitter.) Additionally, Musk is the founder of Space Exploration Technologies Corp (2001)—abbreviated Space X—a space technology company that owns the satellite communications service Starlink. He is also the founder of the Boring Company (2017), a tunnel construction service, and a co-founder of Zip2 Corp (1995), an online city guide. Most recently Musk invented the idea of creating a temporary U.S. "Department of Government Efficiency" (2024)—which went into effect at the start of President Donald J. Trump's second term

(2025-2029). The primary mission statement of DOGE, as the organization is better known, is to decrease the size of the federal government, uncover fraud and waste, and slash federal spending.[16]

MUSSCHENBROEK, PIETER VAN. 1692-1761. Dutch scientist, physicist, astronomer, astrologer, mathematician, educator, and inventor: Inadvertently discovered, and therefore invented, the Leyden Jar (1746), an early appliance that stores electricity and a forerunner of the capacitor. Musschenbroek is also credited with coining the word pyrometer.

MUSSER, CLARENCE WALTON. 1909-1998. American mechanical engineer and inventor: Invented the harmonic drive gear (1955), a speed-changing device that increases output torque. Also known as a strain wave gear, it is used today in a myriad of industries, from automotive, medicine, and aerospace, to machine tooling, manufacturing, and robotic technology. Musser was also involved in the development and invention of numerous other innovations involving, among other things, the recoilless rifle, jet engines, parachutes, missiles, and oscilloscopes.

MUYBRIDGE, EADWEARD. 1830-1904. English photographer, animator, filmmaker, actor, cinematographer, artist, lecturer, author, and inventor: Through his invention of the zoopraxiscope (1879), an early type of film projector, as well as his detailed photographic studies of human and animal locomotion, Muybridge made numerous contributions to photography science and the film industry.

Eadweard Muybridge.

MYERS, FREDERICK. Dates uncertain. Nationality unknown. Whoever he was (his patent was filed in New York City, March 2, 1869), he is credited with procuring the first patent for a unicycle, or "velocipede," as he called it. As descriptions of the unicycle by other individuals preceded the date of Myers' patent, however, the question as to who actually invented the first unicycle remains open.

Joseph Barrow's Screwing Machine, circa 1860s.

NADIRADZE, ALEXANDER D. 1914-1987. Georgian engineer, rocket designer, and inventor: Invented the RT-21 Temp 2S (mid 1960s), the first mobile intercontinental ballistic missile, or ICBM.

NADJAKOV, GEORGI. 1896-1981. Bulgarian scientist and physicist: Developed and prepared the first photoelectret (1937), an electric equivalent of a magnet. Nadjakov's discovery led to greater insight into the nature of photoelectric solids, finding multiple uses in various applications. These would include computers, medicine, optics, supercapacitors, xerography, prosthetics, electric displays, scientific research, mechanics, and photography.

NAGAYA, TAKAYUKI. Dates unknown. Japanese inventor: Credited by some with co-inventing the Quick Response code (1994). Better known as a QR code, it is a type of gridded, square barcode that, when scanned, accesses whatever data or information its creator attaches to it. Examples: Websites, product information, contact information, payment processing, messaging, sales offers and discounts, etc.

NAGAYOSHI, NAGAI. 1844-1929. Japanese scientist, chemist, pharmacist, and educator: Created the first synthesized form of methamphetamine from ephedrine (1893), a central nervous system stimulant that is used to treat ADHD, or attention-deficit hyperactivity disorder.

NAISMITH, JAMES. 1861-1939. Canadian physician, clergyman, educator, author, sports coach, basketball player, and inventor: Invented basketball (1891), the original game which was played with soccer balls and peach baskets. He is also credited with inventing the protective football helmet (1891), which has saved thousands from serious head injuries and even death. These two innovations alone, as well as his many other contributions to the world of sports, have greatly benefitted society in such areas as physical health, mental health, social interaction, and weight management. The Naismith Memorial Basketball Hall of Fame, at Springfield, Massachusetts, is named after him. Naismith is known as the "father of basketball."

James Naismith.

NAKAMATSU, YOSHIRO. Born 1928. Japanese author, media personality, and inventor: According to the innovator himself, he has been awarded some 3,300 patents, his best known invention, among thousands, being the floppy disk. However, both claims have been disputed.

NAKAMURA, SHUJI. Born 1954. Japanese physicist, electrical engineer, researcher, educator, and inventor: Developed the world's first blue semiconductor LED (1992). This led to his invention of the first low-power blue laser (1996), an instrument used in numerous applications, including manufacturing, laser engraving, 3D printing, medical imaging, textiles, surgical operations, welding, laser microscopy, lighting, laser shows, underwater communications, dental equipment, laser printing, copper production, spectroscopy, and environmental science, among others. Nobel Prize winner.

NAPIER, JOHN. 1550-1617. Scottish physicist, mathematician, astronomer, astrologer, theologian, weapons designer, author, and inventor: Invented logarithms (circa 1594), an important mathematical innovation that simplifies complex calculations, such as those used in astronomy. Along with his Naperian Logarithm system, Napier was also responsible for advancements in trigonometry, including what is known as "Napier's Analogies," as well as a mathematical calculation device called "Napier's Bones"—a precursor to the slide rule.

NARTOV, ANDREY. 1683-1756. Russian scientist, military engineer, author, artist, and inventor: Among numerous other innovations, he is best known for inventing the world's first screw-cutting lathe (1738). The mechanical device, which was used to cut precise screw threads, came with a set of interchangeable gears that added to its efficiency and adaptability.

NASMYTH, JAMES. 1808-1890. Scottish engineer, machine tool maker, astronomer, and inventor: Along with his innovative work on high-pressure steam engines, steam locomotives, and various types of pumps, he is primarily known for his invention of the steam hammer (1839). Also known as a drop hammer, it was used mainly for forging metal, pile driving, and metal strengthening, and helped expand the Industrial Revolution (1760-1840). Though the steam hammer is still sporadically used, it is inevitably being replaced by more modern and efficient technologies, such as mechanical and hydraulic presses.

NATTA, GIULIO. 1903-1979. Italian scientist, engineer, chemist, educator, and inventor: Co-invented the Ziegler-Natta catalyst (1955), a substance that permits alpha-olefins, such as ethylene and propylene, to be polymerized. The resulting product, high-density polyethylene (HDPE), has numerous applications in a wide variety of fields. These include: plumbing pipes, boxes, boat parts, food products, cutting boards, toys, gasoline tanks, plastic bottles and containers, motor oil, film wrapping, outdoor furniture, crates, detergents, marine buoys, housewares, chemical storage tanks, cosmetics, drainage systems, and shoes. Nobel Prize winner.

NEADE, WILLIAM. Lived early 1600s. English inventor: Invented a double-armed weapon he named the "pike and bow" (circa 1624), which combined the lethality of the pike (a long spear) with that of the longbow (a tall, large, accurate, deadly bow that can shoot long distances). Neade's innovation may have been used in the English Civil Wars, which included such battles as Edgehill, Dunbar, and Marston Moor.

NEGISHI, SHIGEICHI. 1923-2024. Japanese engineer and inventor: Invented the very first prototype of the karaoke machine (1967), a device that allows amateur singers to sing along to taped music. The popular music system, which removes the original vocals, includes a microphone and speakers. The word karaoke derives from the Japanese words *kara*, "empty," and *oke*, "orchestra."

NEHER, ERWIN. Born 1944. German physicist and educator: Co-invented the patch clamp technique (1970), which, according to the U.S. government, is an electrophysiological method for "directly measuring the membrane potential and/or the amount of current passing across the cell membrane." The patch clamp technique aids researchers in the fields of medicine and pharmacology. Nobel Prize winner.

NEHER, ROBERT VICTOR. 1886-1918. Swedish scientist, economist, and inventor: Invented aluminum foil (1907), a type of thin rolled aluminum used chiefly for food preservation, with applications ranging from wrappers and containers to lids and cartons. Neher's innovation revolutionized the consumer industry and changed the world as we know it.

NELSON, TED. Born 1937. American sociologist, filmmaker, photographer, educator, lecturer, author, and inventor: Among other innovations Nelson is perhaps best known for both coining the word hypertext (1963) and for inventing hypertext (1965), a type of text that contains electronic links to other items, such as information, graphics, and Websites. Founder of Project Xanadu (1960).

NESSLER, KARL L. 1872-1951. German hairdresser, barber, businessman, entrepreneur, author, and inventor: Invented the permanent wave machine (1909), artificial eyebrows (1902), and the hair curler (1919). He was involved in creating numerous other hair and beautician-related products as well, including artificial eyelashes and a hair growth measuring device.

NESTLÉ, HENRI. 1814-1890. German pharmacist, businessman, and inventor: Upon moving to Switzerland in the 1830s, he changed his birth name, Heinrich Nestle, to Henri Nestlé. Rose to fame as the inventor of a healthy breast-milk alternative for infants called *Kindermehl* ("Children's Flour"), consisting of condensed milk, sugar, and bread (1867). Eventually partnering with chocolatier Daniel Peter (1870s), the two founded the Nestlé Company (1879), which continues to produce chocolate products to this day. Note: Nestle, being a German word for (a bird's) "nest," Nestlé chose an illustration of a mother bird feeding her chicks in a nest as the company trademark—still in use.

NEUMANN, JOHN VON. 1903-1957. Hungarian physicist, mathematician, computer scientist, educator, and inventor: Invented the Von Neumann computer architecture (1945), a basic computer design model that includes a CPU, memory storage, and I/O devices. This schematic forms the basis of all modern computers. Von Neumann, also the inventor of game theory (1944), made mathematical contributions that impacted a wide variety of fields, from economics to quantum theory.

NEWCOMEN, THOMAS. 1664-1729. English engineer, iron maker, and inventor: Credited with inventing the "Newcomen engine" (1712). The popular but rather inefficient atmospheric motor was sold and used throughout Europe, and particularly in England, for nearly a century before it was eventually replaced by James Watt's steam engine.

NEWMAN, LEE. Dates unknown. Nationality unknown. Occupation(s) unknown: Invented the world's first known felt-tipped marking pen (1910), more commonly known after 1953 by the brand name "Magic Marker."

NEWMAN, SYDNEY. 1917-1997. Canadian producer and screenwriter: Created the hit TV series *The Avengers* (1961) and co-created the hit TV series *Doctor Who* (1963).

NEWTON, ISAAC. 1642-1727. English polymath, scientist, physicist, mathematician, chemist, theologian, astronomer, philosopher, historian, and inventor: Invented, most importantly, the science of calculus (circa 1665), a branch of mathematics that calculates rates of change. He was also the first to build a reflecting telescope (circa 1668), a double-parabolic mirrored device that, to this day, is known as a Newtonian telescope. Newton made numerous other vital contributions to science, including his original ideas concerning, and research into, universal gravitation, the laws of motion, and optics.

Isaac Newton.

NICOLELIS, MIGUEL. Born 1961. Brazilian scientist, neuroscientist, biologist, neurologist, neuroengineer, neurobiologist, physician, educator, and inventor: Credited with spearheading research into (or further developing and improving) the concept of chronic, multi-site, multi-electrode recordings (1990s)—an essential component of brain-machine interfaces (BMI). This, in turn, has led to further discoveries of how mammalian brain circuits operate. Nicolelis' work has helped advance treatments for paralysis, epilepsy, and Parkinson's disease, while making important contributions to the fields of neuroscience, computer science, biomedicine, communications, prosthetics, rehabilitation therapies, and robotics.

NIÉPCE, JOSEPH NICEPHORE. 1765-1833. French physicist, engineer, photographer, lithographer, and inventor: While some credit him with "inventing photography," it would be more historically accurate to say that he improved on ideas that had already been developed. This he accomplished by using a camera to take the world's first known *permanent* photographic image (1826), which he created on a pewter plate dusted in light-sensitive compounds. He thus became the inventor of the first photographic reproduction process using mechanical means. (Using knowledge Niépce had acquired, his business associate, Louis Daguerre, later became famous for inventing the process known as daguerreotype photography.) Niépce was also a co-inventor (with his brother Claude Niépce) of an early type of internal combustion engine (1807). The "Pyréolophore," as they called it, was powered by lycopodium powder, predating the invention of the first confirmed commercialized internal combustion engine by 46 years.

NIPKOW, PAUL GOTTLIEB. 1860-1940. Polish/German physicist, engineer, inventor: Invented the Nipkow Disk (1884). Based on scanning technology, a disk, constructed with a series of spiral-arranged holes, is placed in front of a scene and rotated. The reflected light from the scene passes through the spinning holes. These scanned light beams are then picked up by a photoelectric sensor that transforms them into electronic signals. Nipkow's innovation was earth-shaking: It allowed for the development and invention of the television, and was the fundamental component in all early TV sets.

NISHIZAWA, JUN-ICHI. 1926-2018. Japanese scientist, physicist, engineer, educator, author, and inventor: Invented electroepitaxy (1955), photoepitaxy (1961), and GaAs molecular layer epitaxy (MLE) (1984). Additionally, he invented (some say led teams that invented) the static induction transistor (1950), the pin diode (1950), semiconductor inductance (1957), avalanche photodiodes (1952), semiconductor injection laser (1957), solid-state focusing optical fibers (1964), the pnip transistor (1950), hyperabrupt variable capacitance diodes (1959), the pin photo diode (1950), transit time negative resistance diodes (1954), and the ion implantation method (1950). He made numerous other important contributions to LED, optical fiber, electricity distribution, optical communications, and broadcasting technologies, all which helped advance science into the modern age. Nishizawa is known as the "father of Japanese microelectronics."

NOBEL, ALFRED B. 1833-1896. Swedish industrialist, engineer, chemist, armaments manufacturer, photographer, businessman, entrepreneur, poet, and inventor: The founder of numerous companies, he is best known for inventing dynamite (1866), a now common explosive that has greatly facilitated such industries as mining, road building, dam construction, pipeline construction, and oil and gas exploration. The international Nobel Foundation (founded 1900) and the annual science awards it issues, the Nobel Prizes, were both inspired by and named after him. These prestigious intellectual achievement awards cover the following six categories: physics, chemistry, physiology or medicine, literature, peace, and economic sciences. Nobel, who never won a Nobel Prize himself (the first was conferred in 1901, five years after his death), also made contributions in the fields of leather, artificial silk, and synthetic rubber.

Alfred B. Nobel.

NOBEL, LUDVIG I. 1831-1888. Swedish industrialist, oil tycoon, engineer, businessman, entrepreneur, philanthropist, and inventor: A brother of Alfred Nobel (above), he invented the first commercially viable oil tanker (1878) and made improvements to oil pipelines and oil refinery technology. Founder of numerous businesses and companies; responsible for developing the Russian oil industry; co-founder of the oil company Branobel (1876) in what is now Azerbaijan.

NOLLET, JEAN-ANTOINE. 1700-1770. French experimental physicist, clergyman, educator, lecturer, author, and inventor: Of the countless scientific instruments he designed and built, he is best known for his improvements on the electroscope (1748), a device used for detecting and measuring electrical charges. A member of the Royal Society of London, he coined the term "Leyden Jar," and was the discoverer of the natural diffusion process known as osmosis (1748).

NORMANN, WILHELM. 1870-1939. German scientist, chemist, biologist, and inventor: Invented the process of hydrogenation of fats (1901). His partially hydrogenated product, semisolid fat (known today as "trans fat"), increased the shelf life of foods, revolutionizing the food industry. Crisco Vegetable Shortening was the first trans fat-containing food to become commercially available to the public (1911). Common foods that may contain trans fat include doughnuts, cookies, peanut butter, crackers, pastries, chips, nondairy creamers, French fries, pizza, pies, cakes, cookies, and most pre-prepared doughs.

NOVOSELOV, KONSTANTIN S. Born 1974. Russian physicist, researcher, artist, and educator: Co-discovered graphene (2004), the thinnest material currently known to science. Measuring a mere one atom thick, this two-dimensional, pure carbon product is 200 times stronger than steel, which also makes it the strongest material in the world. Additionally, graphene is not only clear, light, strong, and pliable, it is also an excellent conductor of heat and electricity. This makes it suitable for a wide spectrum of uses in industries ranging from science, engineering, lighting, medicine, sports, apparel, and optics, to food, health, energy, nanotechnology, automotive, construction, and electronics. Nobel Prize winner.

NOYCE, ROBERT N. 1927-1990. American scientist, physicist, businessman, and entrepreneur: Co-founder of the Intel Corporation (1968), more popularly known as Intel.

NYBERG, CARL RICHARD. 1858-1939. Swedish industrialist, mechanical engineer, businessman, entrepreneur, aviation leader, and inventor: Best known for inventing the blowtorch (1882), a high temperature, flame-producing device used to heat a wide variety of materials. Nyberg is also known for his research and work in such fields as wind tunnels, aircraft design and construction, boat propellers, and steam-powered engines.

Two 200 horse power engines generating electricity for a large store, 1913.

European book printing in the early 1500s.

O'CONNELL, AARON D. Born 1981. American experimental quantum physicist: Invented the first quantum machine (2009), a manmade machine that uses small particles to execute complicated calculations. Rather than the binary bits system (0 or 1) used in conventional computers, O'Connell's quantum mechanically driven device uses qubits, which can simultaneously exist as both 0 and 1. This makes quantum machines faster and more efficient than conventional computational machines, holding great promise for future technologies.
O'CONNELL, JOSEPH JOHN. 1861-1959. American electrical engineer and inventor: Among the many important inventions credited to him are a signaling circuit (1897), a telephone exchange apparatus (1889), a party telephone line apparatus (1901), and a toll coin collection device (1902). His innovations allowed for numerous advances in electronic and telephonic technologies.
ODHNER, THEOPHIL WILGODT. 1845-1905. Swedish engineer, businessman, entrepreneur, and inventor: Though his innovations include a number of scientific instruments, his best known invention was the Odhner arithmometer (1890), a portable pinwheel calculating machine.
OFFIT, PAUL. Born 1951. American pediatrician: Co-invented the rotavirus vaccine (1998), a pentavalent vaccine for the highly contagious pathogen that causes gastrointestinal infections and inflammation of the stomach and intestines.
OGLE, TOM. Circa 1955-1981. American inventor: Invented the Ogle Carburetor (circa 1971), which the innovator claimed would allow a car to travel an average of 100-150 miles on a gallon of gas. His design

eliminates the carburetor, and instead employs a collection of hoses that deliver an air-gas vapor mixture directly into a conventional engine—increasing fuel economy while cutting both carbon buildup and carbon emissions. Though Ogle had (and still has) his supporters, skeptics eventually outnumbered them. To this day his revolutionary creation remains virtually unknown outside a small circle of engine enthusiasts—another potentially important but suppressed invention.

OHAIN, HANS VON. 1911-1998. German scientist, physicist, aerospace engineer, and inventor: Credited with co-inventing the turbojet engine (1937). Some maintain that it might be more correct to state that his turbojet engine was the first to perform a successful flight (1939).

OIKARINEN, JARKKO. Born 1967. Finnish computer scientist and inventor: Invented the first Internet chat network (1988). Technically known as Internet Relay Chat (IRC), his innovation opened the door to the development of communication between individuals on a single computer network. Though still in use, IRC technology is gradually being replaced by more sophisticated instant messaging technologies that allow communication via text, voice, and video.

OKAMOTO, MIYOSHI. Dates unknown. Japanese chemist, engineer, computer scientist, material scientist, and inventor: Invented microfiber (1970), a synthetic material chemically made from plastics, such as polyester and nylon. One hundred times thinner than a human hair, microfiber is lightweight, soft, wrinkle-resistant, durable, breathable, shrink resistant, water resistant, and stain resistant, making it ideal for clothing, insulation, upholstery, bedding, filters, and cleaning products (e.g., mops, towels, etc.). Okamoto is also the inventor of ultrasuede (1970s), an artificial, wear and tear resistant, microfiber leather known commercially by various brand names, such as Ecsaine and Alcantara.

OLDS, RANSOM ELI. 1864-1950. American automobile manufacturer, engineer, businessman, entrepreneur, and inventor: Designed the first Oldsmobile (1896) and founded Olds Motor Works (1899), which began selling Oldsmobiles two years later (1901). Olds' invention of the world's first assembly line (1901) revolutionized not only the auto industry, but the furniture, appliance, electronic, food, and clothing industries as well. Lastly, Olds was also responsible for

Fanny Brice and Oldsmobile, 1922.

founding the REO Motor Car Company (1904), which began marketing the first REO automobile that same year.

OLIVIER, LUCIEN. 1838-1883. Russian (of French heritage) chef and businessman: Invented one of the world's most famous dishes, the Russian Salad (circa 1860s). Also known as the Olivier Salad, it is made with carrots, peas, mustard, capers, eggs, peppers, mayonnaise, and salt and pepper. Additionally, some recipes call for ham, potatoes, pickles, and dill, as well.

O'NEILL, GERARD K. 1927-1992. American scientist, physicist, astronomer, author, and inventor: Best known for inventing, among other innovations, the colliding-beam storage ring (1956), a scientific instrument that increases the energy output of particle accelerators, and is used in high-energy circular colliders to conduct particle physics research.

ONNES, HEIKE KAMERLINGH. 1853-1926. Dutch physicist, educator, and inventor: The first to liquefy helium (1908). Liquid helium is a strongly correlated quantum fluid that has become a critical component in such fields as medicine (e.g., MRI machines), industry (e.g., cryogenics), and science (e.g., electronics manufacturing). An intellectually wide-ranging scholar, Onnes' work covered a myriad of areas, including radioactivity, optics, thermodynamics, persisting currents, magnetics, superconductivity, the Hall effect, dielectric constants, electricity, and pure metals. Nobel Prize winner.

OPPENHEIM, JOSEPH. 1859-1901. German (later American) scientist, scholar, businessman, entrepreneur, educator, musician, and inventor: Improved on the manure spreader (1899) by incorporating a paddle device that threw the manure in a wide arc over, thereby eliminating the arduous and time-consuming task of hand-spreading.

OPPENHEIMER, J. ROBERT. 1904-1967. American theoretical physicist, engineer, educator, and inventor: The director of the Los Alamos Laboratory (1942-1945) and a crucial figure in the Manhattan Project (1942-1945), he is credited with inventing the first atomic bomb (1945)—an innovation that forever changed the world as we know it. Oppenheimer is called the "father of the atomic bomb."

O'REGAN, BRIAN C. Dates unknown. American scientist, chemist, researcher, author, and lecturer: Co-invented the dye-sensitized solar cell (1988), or DSSC. Known as the Grätzel Cell, the flexible, cost-efficient, biodegradable, easy to produce, extra thin-film cell turns sunlight into electricity using titanium dioxide ($TiO2$) particles coated with dye. O'Regan is also known for his innovative research on meso-porous oxides and colloidal solutions.

ORR, HUGH. 1715-1798. American businessman, entrepreneur, politician, toolmaker, gunsmith, weapons manufacturer, and inventor: Manufactured the first American muskets (1748); produced cannon balls for the Continental Army during the Revolutionary War (1775-1783); co-invented the first American textile producing machinery (1786); invented a flax-cleaning machine.

ØRSTED, HANS CHRISTIAN. 1777-1851. Danish scientist,

physicist, engineer, chemist, pharmacist, researcher, educator, and inventor: Discovered the electromagnetic theory (1820), which led to the science of electromagnetism: the study of the relationship between electricity and magnetism. From this evolved what is called "Ørsted's Law," which states that when an electric current is delivered through a wire it produces a magnetic field. He is also credited with the discovery of piperine (1820), a compound (extracted from the black peppercorn plant) that has numerous medicinal and health benefits. Additionally, he was the discover of aluminum (1825), a durable, lightweight chemical element that is used today in everything from electronics, construction, electrical wiring, consumer goods, power lines, and packaging, to transportation, beverage cans, boat hulls, appliances, wrapping materials, and automobile and aircraft parts. In honor of his work, a physical unit of the magnetic field has been named the "orsted."

ORTELIUS, ABRAHAM. 1527-1598. Dutch geographer, cartographer, cosmographer, author, and inventor: Designed, wrote, compiled, and edited the world's first atlas (1570). Entitled *Theatrum Orbis Terrarum*, it paved the way for the burgeoning atlas industry that was to follow.

OSHCHEPOV, VASILI S. 1893-1938. Russian sports innovator, martial arts researcher, and coach: Credited with co-developing Sambo (early 1920s), a unique martial art form that blends elements of wrestling, judo, and jujutsu, as well as various folk wrestling techniques and styles. The profession, whose name is an acronym for the Russian term *samozashchita bez oruzhiya* ("self-defense without weapons"), is generally divided into two categories: Sport Sambo (intended mainly for competition) and Combat Sambo (primarily intended for self-defense).

O'SULLIVAN, JOHN. Dates unknown. Australian scientist, electrical engineer, and inventor: Invented wireless LAN, or WLAN, a wireless computer network of communication devices, which led to improvements in a type of WLAN known as wi-fi. Thus O'Sullivan is credited with co-developing the modern wi-fi system (1996), which in turn revolutionized the electronic communications industry. Wi-fi is used today in countless devices, from cell phones and watches to tablets and speakers.

OTIS, ELISHA G. 1811-1861. American mechanical engineer, businessman, entrepreneur, and inventor: Among other innovations he created, such as special truck brakes, a bake oven, and a steam plow, he is most familiar as the inventor of an automatic elevator safety system, a life-saving device that prevents an elevator from falling in the event that its lift cable breaks (1852). He was the founder of the Otis Elevator Company (1853), which manufactured this and other related products. The Otis elevator safety brake system is still in use today, though greatly modified with modern technologies using computers and electronic sensors.

OTTENS, LODEWIJK FREDERIK. 1926-2021. Dutch engineer and inventor: Invented the compact cassette (1962) and co-invented the compact cassette recorder (1962), both which revolutionized the music and consumer industries. Though the humble analog cassette has waned in popularity over the decades due to the rise of the compact disk (CD), cassette technology is still being used to record, store, and playback audio and video by various musicians and consumer outlets.

OTTO, GUSTAV. 1883-1926. German engineer, aircraft pilot, businessman, entrepreneur, and inventor: Founder of the Otto Flugmaschinenfabrik company (1910), renamed four years later to Bayerische Flugzeugwerke (1916). It was later renamed once again, this time becoming Bayerische Motoren Werke (1917)—that is, the "Bavarian Motor Works." Today the company is most commonly known simply as BMW, the noted luxury vehicle manufacturer.

OUGHTRED, WILLIAM. 1575-1660. English mathematician, clergyman, astronomer, theologian, author, tutor, and inventor: Invented the slide rule (circa 1620s), a mechanical device that calculates, among other operations, multiplication, division, square roots, and logarithms. Oughtred is also credited with inventing the familiar mathematical symbol, "x," which stands for multiplication.

DC generators (or dynamos), like this one from the early 1900s, convert mechanical energy into electrical energy, and are still used today.

Ford Motor Company's famous Highland Park (Michigan) plant. After opening in 1910, it became known as the "Crystal Palace" due to its many windows and glass roof. It was the site of the world's first assembly line, and at its zenith housed 70,000 workers and produced 1,000 cars a day. In 1924, when this photo was taken, it was the largest car production plant in the world and covered some 278 acres of land.

PACINOTTI, ANTONIO. 1841-1912. Italian scientist, physicist, astronomer, politician, educator, and inventor: Invented the Pacinotti Armature (circa early 1860s), the world's first DC dynamo. His unique ring armature design created a more stable, dependable, and efficient electric current than earlier dynamos, helping open the way for new developments in electric motor technologies.
PACKARD, DAVID. 1912-1996. American businessman, entrepreneur, and electrical engineer: Co-founded the Hewlett-Packard Company (1939), more popularly known as HP.
PAGE, HILARY. 1904-1957. English businessman, entrepreneur, toy maker, and inventor: Invented Kiddicraft Self-Locking Building Bricks (1947), a plastic interlocking toy that became the forerunner of Lego.
PAGE, LARRY E. Born 1973. American computer scientist, engineer, businessman, entrepreneur, educator, and inventor: Co-invented the Google Web Search Engine (1998), a globally popular Internet browser that is accessed 100,000 times a second, 8.5 billion times a day, 2 trillion times a year.
PAINTER, WILLIAM. 1838-1906. American businessman, entrepreneur, mechanical engineer, and inventor: Among other innovations, such as a counterfeit money detecting machine, he is best known for inventing the universal bottle neck (circa 1890), the disposable bottle cap (1892), and the now common bottle opener (1894). Founder of the Crown Cork and Seal Company of Baltimore (1892).
PAJEAU, CHARLES H. 1914-1952. American stonemason, toy designer, and inventor: Invented Tinkertoys (1914), a still popular children's construction toy set. (Some submit that Pajeau co-invented

the toy with Robert Pettit.)

PAJITNOV, ALEXEY L. Born 1956. Russian computer scientist, businessman, mathematician, software developer, and video game designer: Invented Tetris (1985), the popular puzzle video game. Co-founder of the Tetris Company (1996).

PALIN, MICHAEL. Born 1943. English actor, writer, comedian, and presenter: Co-created and acted in the hit TV series and comedy troupe *Monty Python's Flying Circus* (1969).

PALMAZ, JULIO. Born 1945. Argentinian physician, radiologist, educator, and inventor: Invented the world's first commercially viable balloon-expandable intravascular stent (1988), a life-saving device that opened the door to numerous medical advances while saving over 1 million lives.

PALMCRANTZ, HELGE. 1842-1880. Swedish industrialist, engineer, businessman, firearms manufacturer, machine manufacturer, entrepreneur, and inventor: Invented the multi-barrel, lever-actuated, machine gun (1873), later called the "Nordenfelt machine-gun."

PALMER, DANIEL DAVID. 1845-1913. Canadian chiropractor, magnetic healer, spiritualist, businessman, apiarist, educator, and inventor: Invented the medical field known as chiropractic (1895), a type of massage therapy that uses manual manipulation of the spine and joints to improve health. He was also an enthusiastic believer in and practitioner of the scientific art of magnetic healing. Founder of the profession of chiropractic, as well as what is now called the Palmer College of Chiropractic (1897), Palmer is known as the "father of chiropractic."

Daniel D. Palmer.

PALMIERI, LUIGI. 1807-1896. Italian scientist, physicist, meteorologist, volcanologist, and inventor: Invented the first seismograph (1855), which employed tubes, mercury, a recording drum, and a compass to register the approximate duration, time, and intensity of earthquakes. The modern computer-driven seismograph, which uses a sophisticated seismometer to detect seismic waves, and which was developed from Palmieri's work, has saved millions of lives.

PANTONE, PAUL. 1950-2015. American inventor: Invented the "Geet Fuel Processor" (1987), a plasma-based generator that can be fitted to any type of combustion engine, and is, so it is claimed, capable of producing green energy from various fuels (gas, diesel, etc.). GEET, an acronym for the "Global Environmental Energy Technology" fuel reactor, is also said to be able to turn water into a (near) zero emissions fuel source. Pantone's machine, also known as the "GEET Plasma Motor," was his attempt to find a way to produce inexpensive clean energy—a so-called "radical" idea that will always have both its devotees and its detractors. Like nearly all technological visionaries who work in the field of energy production, Pantone was widely censured (presumably for posing a threat to the fossil fuel industry), one of thousands of examples of a suppressed inventor, and perhaps, as many believe, yet another victim of the 1951 National Security Act.

PANTRIDGE, FRANK. 1916-2004. Northern Irish physician, cardiologist, military doctor, educator, and inventor: Invented both the automated external defibrillator or AED (1957) and the portable defibrillator (1965), devices whose wired pads, when applied to the chest of an individual who has suffered a sudden cardiac arrest (SCA), analyze the heart and deliver an electric shock if deemed necessary. These two instruments have saved untold millions of lives, and it is for good reason that Pantridge is called the "father of emergency medicine."

PAPANIKOLAOU, GEORGIOS. 1883-1962. Greek scientist, physician, research biologist, and inventor: Invented the Papanicolaou Test (1928). More commonly called the Pap test or Pap smear, this medical procedure involves swabbing cells from a female's cervix, then checking them for abnormalities; in particular, cellular signs that might indicate cervical cancer. Papanikolaou, whose invention has helped save countless women's lives, is known as the "father of cytopathology."

PARKER, BOB. Dates unknown. Nationality unknown. Credited by some with having invented the liquid crystal thermometer, also known as a temperature strip, liquid crystal display thermometer, or forehead strip thermometer (circa 1970s).

PARKER, THOMAS. 1843-1915. English electrical engineer, chemist, industrialist, molder, and inventor: Among his many innovations, such as the Kyrle Grate (1880), a steam pump (1867), a dynamo, electric plants (1883), electric lighting (1882), Coalite (1904), and electric tramways, he is best known by many for being the inventor of the world's first commercialized electric car (1884). Parker's EV, the "Elwell-Parker car," came with rechargeable batteries (recharged via a steam-powered battery charger), hydraulic brakes on all four wheels, four-wheel steering, and a top speed of 12 mph. Co-founded, with Paul B. Elwell, the Elwell-Parker Company (1882).

PARKER, TREY. Born 1969. Producer, director, writer, actor, animator, and musician: Co-created the hit TV series *South Park* (1997).

PARKES, ALEXANDER. 1813-1890. English chemist, metallurgist, and inventor: Invented the "Parkes Process" (1850), a metallurgical method involving zinc and silver, and the cold vulcanization process (1841), a technique for making water-impermeable textiles. However, he is best remembered for inventing "Parkesine" (1862), or what is now called celluloid. This, the first synthetic plastic, is made from nitrocellulose and camphor. It was first used for photographic film, but later improved upon and re-utilized as a substitute for hard rubber. A thermoplastic (meaning it can be reheated and remolded repeatedly), celluloid is durable, versatile, water-resistant, stain-resistant, inexpensive, and pliable, originally making it ideal for a wide variety of applications, such as office equipment, piano keys, hair brush handles, toys, table tennis balls, baby rattles, musical instruments, board game pieces, linens, plastic pen bodies, hand mirrors, buttons, knife handles, trading cards, eyeglass frames, hair combs, billiard balls, ornaments, and guitar picks. Due to its flammable nature, however, celluloid is gradually being replaced by plastics made from manmade polymers.

PARKINSON, BRADFORD. Born 1935. American engineer and inventor: Along with Ivan A. Getting, Parkinson is the co-inventor (some say the lead inventor and developer) of the Global Positioning System (1973), better known as GPS. Owned by the U.S. government and operated by the United States Air Force, GPS is a space-based radio-navigation system that supplies precise data concerning one's location, speed, and time. GPS has been key in furthering the development of technologies surrounding national security, the military, electronics, agriculture, emergency services, commercial products, meteorology, mapping, environmental protection, surveying, biology, package and food delivery, transportation, mining, seismology, sports, marine concerns, home appliances, public safety, railways, banking, space, recreation, travel, power grids, automobiles, construction, and aviation, among others.

PARRY, FORREST C. 1921-2005. American engineer and inventor: Invented the magnetic stripe card (1960), used for credit cards, debit cards, gift cards, driver's licenses, hotel room keys, store loyalty cards, and identification badges.

PARSONS, CHARLES ALGERNON. 1854-1931. English mechanical engineer, businessman, entrepreneur, author, and inventor: Invented the steam turbine (1884), which was later installed in the marine vessel *Turbinia* (1897)—making her both the world's first turbine-powered steamship and the world's fastest ship that year. Parson's turbine not only revolutionized marine propulsion, it led to numerous advancements in electricity generation technology that are still being used today. Indeed, according to the U.S. Energy Information Administration, "steam turbines are used to generate most of the world's electricity, and they accounted for about 42 percent of U.S. electricity generation in 2022."[17]

Charles A. Parsons.

PASCAL, BLAISE. 1623-1662. French polymath, scientist, physicist, mathematician, theologian, philosopher, researcher, experimenter, author, and inventor: A self-taught prodigy, he is perhaps best known for developing "Pascal's Principle" (1653). Also called "Pascal's Law," according to NASA, it states that "when there is an increase in pressure at any point in a confined fluid, there is an equal increase at every other point in the container."[18] Additionally, Pascal invented the world's first commercially successful mechanical calculator (1642). Known as the "Pascaline," it could add and subtract, but was unable to divide or multiply. Also credited to Pascal is the invention of the syringe, the hydraulic press, the adding machine, various calculus-related concepts, and "Pascal's Theorem" (1639).

PASCH, GUSTAF ERIK. 1788-1862. Swedish chemist, educator, and inventor: Improved on the safety match (1844) by replacing the white phosphorus originally used in match heads with red phosphorus. White phosphorus, a dangerous chemical, was found to be both toxic (causing serious ailments like "phossy jaw") and highly flammable when used in match heads. Pasch realized that when a more stable form, red phosphorus, was placed on the striking surface rather than on the match head itself, it made for a safer and healthier alternative.

PASKOV, DIMITAR. 1914-1986. Bulgarian scientist and chemist: Spearheaded research into potential treatments for memory impairment. He and his team discovered that galantamine, an extract from the Amaryllidaceae plant family (e.g., daffodils and snowdrops), acts as an acetylcholinesterase inhibitor that augments the level of acetylcholine in the brain—aiding in improved reasoning, memory, and other cognitive functions (1959). This chemical extract, once commonly obtained from the snowdrop plant (*Galanthus nivalis*), can now be made naturally or synthetically. Under the brand name Nivalin it is used to treat and manage health conditions such as Alzheimer's disease and dementia.

PATEL, C. KUMAR N. Born 1938. Indian scientist, physicist, electrical engineer, and inventor: Among other innovations (such as the carbon monoxide laser and the spin-flip Raman laser), he was the inventor of the carbon dioxide laser (1964), a high powered, highly efficient CO_2 instrument that emits infrared light. Among the world's most commonly used lasers, it has greatly impacted the industrial and medical fields, where it has found applications ranging from medical surgeries to welding.

PAUL, LES. 1915-2009. American musician, guitarist, radio personality, songwriter, musical instrument maker, educator, and inventor: Among numerous other innovations he is best known for inventing the solid-body electric guitar (1941), multi-track recording (circa 1945), over-dubbing (circa 1930), the eight-track tape recorder (1953), reverb (early 1940s), and digital delay (circa 1940s). Paul is known as the "father of modern music."

PAULRAJ, AROGYASWAMI. Born 1944. Indian electrical engineer, educator, and inventor: Invented MIMO (1992). An acronym for "Multiple Input Multiple Output," it is a wireless system that employs multiple antennas to transmit and receive data simultaneously. By increasing the speed, reliability, coverage, signal quality, and capacity of wireless technology, including wi-fi, 4G LTE ("Long Term Evolution"), and especially 5G cellular networks, MIMO has revolutionized modern communications and forms the core of modern high-speed connectivity.

PAVEL, ANDREAS. Born 1945. German TV programmer, editor, and inventor: Invented the "Stereobelt" (1977), a forerunner of the personal portable stereo listening device.

PAVLOV, IVAN P. 1849-1936. Russian scientist, physiologist, neurologist, chemist, researcher, clinical psychologist, educator, author, and inventor: Developed the concept of classical conditioning (1897). Also known as the conditioned reflex, he is perhaps best remembered for his "Pavlov's dog" experiment, which involved conditioning a hungry canine to associate food with the sound of a bell. Over time, even when the food was removed, the dog would continue to salivate when the bell was rung. The phenomenon of Pavlovian conditioning opened the door to a deeper psychological understanding of how we consciously and unconsciously form connections between various stimuli. Today the concept of classical conditioning is not only used in advertising, but more importantly as a behavioral psychological treatment for mental health issues (e.g., depression, anxiety, phobias, obsessive-compulsive disorder, etc.) and substance abuse—and, of course, as a part of pet training. Nobel Prize winner.

Ivan P. Pavlov.

PAXTON, FLOYD. 1918-1975. American engineer, businessman, entrepreneur, newspaper founder, and inventor: Invented the bread tag (1952). Also known variously as a bread tab, bread buckle, bread clip, or bread tie, this small, removable, plastic notched device is used by bread manufacturers to seal plastic bags efficiently, simply, and securely, thereby helping to maintain freshness and extend shelf life. Additionally, food producers color-code bread tags as a way to track the date a product was made and packaged: blue tags, Monday; green tags, Tuesday; red tags, Thursday; white tags, Friday; and yellow tags, Saturday.

PECK, RICHARD. Dates unknown. American salesman and inventor: Invented the pet fence (1973), also known as "Stay-Put," an invisible containment barrier ("fence") that operates via radio signals. A radio-transmitting wire is buried underground, which sends a radio signal to a radio-collar-receiver worn by the pet. If and when the pet breaches the boundaries laid out by the underground wire, it receives a beeping tone, or in some cases a mild but persuasive electrical shock.

PECQUEUR, ONÉSIPHORE. 1792-1852. French mechanical engineer, automotive engineer, watchmaker, author, and inventor: Invented the first practical automobile differential (1827), a gear mechanism that is used in a myriad of vehicular, horological, and industrial applications to this day. Beside his work in the field of steam power, Pecqueur is also known for his invention of a clock that could display both mean time and sidereal time (1819). His differential gear train, now known as a "mechanical differential," revolutionized both the car and truck industries and the modern era.

PEMBERTON, JOHN STITH. 1831-1888. American chemist, pharmacist, physician, surgeon, businessman, entrepreneur, and inventor: This Confederate States army veteran (still revered as a hero by educated American patriots)[19] is best known for his invention of the soft drink Coca-Cola (1886). First called "Pemberton's French Wine Coca," the Victorian alcoholic beverage was advertised at the time as a "tonic" to aid in digestion, lung and muscle function, and nervous system strengthening. With the commencement of Prohibition in the city of Atlanta, Georgia (1886), Pemberton removed wine from his recipe and replaced it with sugar syrup. He named his new non-alcoholic formula "Coca-Cola," promoting it as a "temperance" drink. Today Pemberton's sugary libation is the number one selling soft drink in the world and is available in over 200 countries. A nephew of famed Confederate General John Clifford Pemberton, Lieutenant-Colonel Pemberton was the founder of the J. S. Pemberton and Company (1860).

John S. Pemberton.

PÉNAUD, ALPHONSE. 1850-1880. French engineer, aviator, and inventor: Invented the *Planophore*, the first rubber band-powered model airplane (1871), as well as a rubber band-powered toy helicopter (1870) and a rubber band-powered toy ornithopter (1872). These were not mere children's toys. They were scientifically designed aircraft that utilized sophisticated, never-before-seen aeronautical principles, many which were later incorporated into full scale flying machines that are still in use today. Although a contributor of numerous advances in aviation technology, Pénaud is best remembered for his invention of the first aircraft possessing a boat hull and retractable landing gear (1876). This later led to the development of the concept of the sea plane and its life-saving offspring: the aerial firefighting "water scooper" or "super scooper"—a lifesaving amphibious plane that skims over the surface of a lake or sea in order to fill its tanks with water, which it then disperses over wildfires.

PENKALA, SLAVOLJUB EDUARD. 1871-1922. Hungarian (Slovakian-Croatian) engineer, chemist, businessman, entrepreneur, and inventor: Among many other innovations he is best known for inventing both the mechanical pencil (1906), which uses an extendable, replaceable graphite core (used primarily by engineers, architects, artists, etc.), and what may have been the first solid ink fountain pen (1907).

PEO, RALPH F. 1897-1966. American engineer, businessman, automobile parts pioneer, and inventor: Among a myriad of innovations he is best remembered for inventing the thermostatically controlled shock absorber (1930s), as well as what seems to have been the first air conditioning system for automobiles (1935).

PERKIN, WILLIAM HENRY. 1838-1907. English scientist, industrialist, businessman, entrepreneur, chemist, and inventor: During his research into synthesizing quinine, he developed a dye subsequently named "mauve" or "aniline purple" (mid 1850s). This was followed by a quick succession of new discoveries and chemical processes involving amino acids, tartaric acid (1860), alizarin (a red dye), unsaturated acids (1867), seasonings, salicyl alcohol, and various dyes. He is best remembered, however, for his invention of the "Perkin Reaction," which was found to aid in synthesizing various compounds, such as courmarin, an organic substance (extracted from Guyana's tonka tree) that was used in the world's first manmade perfume (1868). The Perkin Reaction continues to be used today to produce flavorings, fragrances, industrial products, and pharmaceuticals.

PERKY, HENRY D. 1843-1906. American attorney, businessman, entrepreneur, promoter, author, and inventor: Invented the natural food cereal "Shredded Wheat" (1893), which he marketed as a "digestive aid." Humorously referred to as "little whole wheat mattresses" by its vegetarian creator, Perky is credited by some with having launched the tradition of the "cookless breakfast food"—that is, ready-to-eat breakfast cereal. Founder of The Boston Shredded Cereal Food Company (1893).

Henry D. Perky.

PÉROT, ALFRED. 1863-1925. French physicist, educator, and inventor: Co-invented the Fabry-Pérot Interferometer (1896). This high-resolution spectroscope is used to measure light waves and has found numerous applications, particularly in astronomy and telecommunications.

PERRY, STEPHEN. Victorian Period. English businessman and inventor: Invented the rubber band (circa 1845) using the vulcanization process—which greatly improved the strength, elasticity, and durability of natural rubber. Offices have been better organized ever since.

PERSU, AUREL. 1890-1977. Romanian scientist, dynamics engineer, automobile designer, educator, musician, and inventor: Invented the first aerodynamic car (1923), his motivation being to build an automobile with the lowest drag coefficient possible. Being the first to place the wheels inside the car body (among other aerodynamic innovations), he was able to increase fuel efficiency, safety, stability, road grip, handling, speed, and overall mechanical performance. Though Persu' invention is over 100 years old it continues to benefit drivers of all types of vehicles into the present day.

PETER, DANIEL. 1836-1919. Swiss chocolatier, businessman, entrepreneur, and inventor: Invented milk chocolate, the milk chocolate bar, and chocolate milk (circa 1875)—the latter which he created by combining powdered milk with chocolate. After founding the company

Peter's Chocolate (1867) Peter went on to partner with fellow Swiss chocolatier Henri Nestlé (1879). The innovative pair developed a host of improved as well as new chocolate products, eventually merging their businesses into what is still known today as the Nestlé Company—which continues to produce such as items as chocolate milk, chocolates, cookies, and confectionary.

PETLYAKOV, VLADIMIR M. 1891-1942. Russian aeronautical scientist, aircraft designer, and inventor: Led development on the Soviet ANT bomber aircraft series, as well as a host fighter, dive bomber, training aircraft, heavy fighter, escort fighter, speed bomber, high altitude fighter, long-range heavy bomber, transport, and reconnaissance aircraft (1920s-1930s). The Russian heavy bomber Petlyakov Pe-8 was named after him, and like his other aircraft, was used during World War II to advance the Soviet war effort.

PETRI, JULIUS RICHARD. 1852-1921. German scientist, physician, surgeon, bacteriologist, microbiologist, and inventor: Invented the Petri Dish (1887), a small, shallow, circular, transparent, covered culture plate that contains a thin layer of agar, and which is used to grow and study bacteria. It replaced the earlier bacteria culturing process that relied on a liquid broth bed. The eponymous Petri Dish can still be found in modern laboratories where medical researchers and microbiologists use it to culture, study, and analyze microorganisms in the ongoing global effort to combat disease and save lives.

PETROFF, PETER. 1919-2003. Bulgarian physicist, engineer, adventurer, and inventor: Assisted in the development of numerous innovations, including weather and pollution monitoring instruments, digital wrist watch technology, and wireless heart monitor technology.

PFLEUMER, FRITZ. 1881-1945. German engineer and inventor: Invented magnetic recording tape (1928), which was made by bonding oxide to film strips. By allowing for the analog recording, playback, editing, and storage of video, audio, and digital data, Pfleumer's invention transformed the entertainment, recording, music, audio, film, TV, radio, communications, and media storage industries. Using cassette, reel-to-reel, and video technologies, magnetic recording tape continues to be used today in various fields, such as medicine, finance, government, and libraries, chiefly for archival and backup storage of valuable information.

PICCARD, AUGUSTE. 1884-1962. Swiss physicist, engineer, balloonist, explorer, and inventor: Among many other innovations (primarily having to do with balloon design and flight) he is best remembered for having invented the bathyscaphe (1948), a free-diving, navigable, pressure-sealed, self-propelled, manned submersible used to engage in deep sea exploration. Having been obsolesed by newer technologies, the original bathyscaphe ("free boat") design is little used

Auguste Piccard.

by today's researchers. However, its historical importance cannot be overlooked, for it not only opened up the dark sea depths (and its heretofore hidden marine life) to human study for the first time, it also laid the groundwork for the development of modern submarine and submersible technologies.

PIERPONT, JAMES. 1659-1714. American clergyman: Co-founder (some say the founder) of the Collegiate School (1701), later renamed Yale University (1718) after its main benefactor, American colonial leader Elihu Yale.

PIERSON, ABRAHAM. 1646-1707. American clergyman: Co-founder of the Collegiate School (1701), later renamed Yale University (1718) after its main benefactor, American colonial leader Elihu Yale.

PINCUS, GREGORY G. 1903-1967. American biologist, researcher, and educator: Co-invented the oral contraceptive pill (circa 1950s). Better known as "the pill," it revolutionized the birth control industry and greatly contributed to the explosion of the "free love" movement of the 1960s and 1970s.

PIROGOV, NIKOLAY IVANOVICH. 1810-1881. Russian scientist, military surgeon, anatomist, researcher, educator, and inventor: Considered by many to be one of the most important and famous Russian physicians in the world, as well as one of the most innovative surgeons in history, he is responsible for a myriad of creative medical developments that are still used to this day. Among them, an improved plaster casting process for broken bones, new amputation techniques, and the use of an ether-based anaesthetic. His fame and reputation have been honored, in part, by naming various anatomical features after him, such as "Pirogov's Triangle," "Pirogoff's Aponeurosis," and the "Pirogoff Angle." He is known as the "father of Russian battlefield surgery."

PIROTSKY, FYODOR. 1845-1898. Russian engineer, military officer, and inventor: Invented the first electrified railway system and the first electric street car (circa 1874)—also known as a trolley car or tram.

PITNEY, ARTHUR. 1871-1933. American businessman and inventor: Invented the postage meter (1902), a government authorized postage printing machine that replaces the need for stamps by printing the postmark information (city, state, Zip Code, and date) directly onto one's mail or onto a meter tape. Founded the Pitney Postal Machine Company (1902), and co-founded the Pitney-Bowes Postage Meter Company (1920)—with industrialist-entrepreneur Walter Bowes—today known as Pitney-Bowes. Pitney is known as the "father of the postage meter."

PIXII, HIPPOLYTE. 1808-1835. French scientist, physicist, engineer, scientific instrument maker, and inventor: Created one of the earliest electrical generators: a hand-cranked magneto that produced alternating-current (1832). Pixii's research, like Michael Faraday's work in the same field, helped pioneer new advancements in the electricity producing industry.

PLANCK, MAX. 1858-1947. German physicist, theoretician, educator, and musician: Discovered quanta energy, leading to his development of quantum theory (1900), work that influenced Albert Einstein. Nobel Prize winner.

Max Planck.

PLANTÉ, GASTON. 1834-1889. French physicist and inventor: Invented the first rechargeable electric battery (1859), a type of lead-acid battery that is today widely used in vehicles (cars, trucks, etc.), along with thousands of other applications. Additionally, a large, flightless prehistoric bird genus, *Gastornis*, was named after him in honor of his discovery of its fossil remains near Paris, France (circa 1855).

PLATEAU, JOSEPH. 1801-1883. Belgian scientist, physicist, researcher, educator, and inventor: Invented the phenakistiscope (1832), an early type of stroboscope that used a rotating cardboard disk to produce, when held up to a mirror, the appearance of movement. Plateau's invention broke ground for the motion picture industry explosion that lay on the horizon. Like Simon von Stampfer (who, independently and simultaneously, also invented a stroboscope), Plateau is known as the "father of cinema."

PLATEN, BALTZAR VON. 1898-1984. Swiss engineer and inventor: Co-invented the gas absorption refrigerator (1922). Using a process known as "absorption refrigeration," it relies on heat to produce the power that operates the cooling process. Von Platen's invention was an alternative to refrigerators that depended on mechanical compressors. The modern gas absorption refrigerator, powered by propane gas, can still be found in various applications, mainly those where electricity is unavailable, such as in recreational vehicles or RVs.

PLIMPTON, JAMES LEONARD. 1828-1911. American inventor: Credited with improving on earlier roller skate designs (1863), he invented the quad or rocker skate, which possessed two wheels in front and two in back. As one could now steer merely by leaning, Plimpton's more stable skate design greatly enhanced maneuverability and safety. He is also credited with opening both the world's first skating rink and the world's first skating club.

PLOTKIN, STANLEY ALAN. Born 1932. American physician and inventor: Co-invented the rotavirus vaccine (1998), a pentavalent vaccine for the highly contagious pathogen that causes gastrointestinal infections and inflammation of the stomach and intestines. Plotkin also worked on vaccines for rabies and rubella, among other diseases.

PLOTNIKOV, IVAN. 1902-1995. Russian engineer and inventor: Invented the synthetic fabric known as kirza leather (1935). Intended as a less expensive alternative to authentic leather, kirza is still being manufactured and sold today, and is used mainly in the production of military footwear.

PLUNKETT, ROY. 1910-1994. American scientist, chemist, engineer, and inventor: During his research on tetrafluoroethylene gas (TFE), Plunkett became the accidental discoverer of a heat-resistant, chemically-inert compound that would later become known as "Teflon" (1938). First used as a corrosion-resistant material in the Manhattan Project (1942-1947), it was only years later that Teflon would be applied to cookware (1954).

POENARU, PETRACHE. 1799-1875. Romanian scientist, physicist, engineer, mathematician, agronomist, politician, educator, and inventor: Invented the first fountain pen with an ink tank (circa 1827), an innovation that has had a lasting global impact. According to some studies the fountain pen is making a resurgence and is today more popular than ever.

POITEVIN, ALPHONSE LOUIS. 1819-1882. French chemical engineer, chemist, photographer, artist, daguerreotypist, printer, lithographer, civil engineer, and inventor: An avid researcher in the fields of photographic chemistry and photomechanical printing, he invented the processes of chromolithography, dichromate relief, carbon pigment printing, and collotype (1855). His innovations, collectively known as the "Poitevin process," sparked numerous advancements in both printing technology and photography science.

POLHEM, CHRISTOPHER. 1661-1751. Swiss scientist, physicist, mathematician, industrialist, businessman, entrepreneur, mechanical engineer, dam engineer, canal lock engineer, dry dock engineer, mining engineer, tool maker, clock maker, knife maker, lock maker, author, and inventor: Invented "Polhem's Lock," the world's first padlock (circa early 1700s), and foresaw the mass production system by constructing automated, water-powered factories that relied on conveyor belts, hoists, and the division of mass labor (1704).

POLIKARPOV, NIKOLAI. 1892-1944. Russian aerospace engineer, aeronautics engineer, flight engineer, aircraft designer, aircraft pilot, educator, and inventor: Invented the famous Po-series aircraft, including the Polikarpov Po-2—nicknamed the *Kukuruznik* (the "crop duster"). Though production ceased in 1952, according to some sources, up until 1978 this versatile, reliable, low-cost aircraft was the most widely manufactured biplane in history.

POLLEY, EUGENE. 1915-2012. American electrical engineer, wireless remote pioneer, and inventor: Invented the "Flash-Matic," the world's first wireless TV remote control. His innovation revolutionized both television viewing and the entire electronics industry. Polley is known as the "father of the wireless remote control."

POLZUNOV, IVAN. 1728-1766. Russian mechanical engineer and self-taught inventor: Invented the world's first two-cylinder steam engine (1763), paving the way for later advancements in both steam powered

engines and gasoline powered engines.
POMORTSEV, MIKHAIL. 1851-1916. Russian aerospace engineer, meteorologist, geographer, aeronautics pioneer, rocket designer, military officer, educator, and inventor: Invented the nephoscope (1894), an elementary Victorian scientific instrument that uses the concept of transit time to analyze and measure cloud metrics—that is, their motion (altitude, velocity, and direction). While the nephoscope still has its basic uses, it has largely been replaced by modern meteorological technology.
POPPER, JOSEF. 1838-1921. Austrian engineer, philosopher, political economist, intellectual, lecturer, social systems designer, tutor, author, and inventor: An uncle of the noted philosopher Karl Popper, he is credited with being the first to conceptualize the transmission of power by electricity (circa 1862). Also known by the pseudonym Josef Popper-Lynkeus, his corresponding work in such fields as mass energy (1883), psychology (circa 1889), and quantum energy (1884), greatly impacted both the sciences and other scientists, among them Ernst Mach, Theodor Baer, Sigmund Freud, and Albert Einstein.
POROKHOVSCHIKOV, ALEKSANDR. 1892-1941. Russian military engineer, tank designer, aircraft designer, aircraft pilot, businessman, and inventor: Invented the Vezdekhod (circa 1914), both the first true military tank and the first all-terrain vehicle or ATV.
PORRO, IGNAZIO. 1801-1875. Italian engineer, optician, topographer, geodetist, optics researcher, author, educator, and inventor: Invented the stereogonic telescope (circa 1850), the Porro prism (1851), the strip camera (1853), the cleps or tachymeter (circa 1850), the telemeter (1835), and the telescopic objective (1848), a precursor to the modern range-finder.
PORTA, GIAMBATTISTA DELLA. Circa 1535-1615. Italian polymath, academic, cryptographer, physiognomist, botanist, optics scientist, agriculturist, mycologist, hydrologist, military engineer, philosopher, author, playwright, and inventor: A man of wide-ranging talents and interests, he credited himself (as did his supporters) with being the inventor of the first telescope (circa 1590s)—though this claim continues to be highly debated. Founder of the secret society: *Academia Secretorum Naturae* (1560).
POULSEN, VALDEMAR. 1869-1942. Danish physicist, engineer, and inventor: Invented the Telegraphone (1898), the world's first answering machine. The device worked using a brass cylinder wrapped in wire, an electromagnet, a battery, and a microphone. His public demonstration of the Telegraphone in Paris, France (1900), left us with a rare artifact from the developmental days of communication technology: the earliest known magnetic recording to survive into the 21st Century. He is also credited with inventing an arc converter (1903) that could convey sound over 2,000 miles. Founder of the American Telegraphone Company (1903). Poulsen's magnetic sound recording machine laid the

groundwork for the technology we now know as voicemail.

POUMARÈDE, JEAN-ANDRÉ. Dates unknown. French scientist and inventor: Though the origins of parchment paper, also known today as baking paper, date back to ancient Egypt (and possibly earlier), Poumarède is credited with being the co-inventor of modern parchment paper (1847), which he created by immersing ordinary paper in a mixture of sulfuric acid and water. The result of this chemical experiment was a versatile, grease-resistant material that can now be found in most kitchens around the world.

PRÉVOST, JEAN-LOUIS. 1838-1927. Swiss physiologist, neurologist, and researcher: Credited with co-discovering the process of heart defibrillation (1899), a concept that paved the way for the development of modern defibrillators, as well as the saving of thousands of lives.

PRIESTLEY, JOSEPH. 1733-1804. English scientist, chemist, philosopher, theologian, clergyman, political theorist, author, educator, and inventor: His work in the field of chemistry led to the discovery of ten new gases: oxygen, nitrogen, carbon monoxide, nitrous oxide ("laughing gas"), sulfur dioxide, ammonia, hydrogen chloride, silicon tetrafluoride, nitrogen dioxide, and nitric oxide. Among numerous other discoveries and innovations Priestly is perhaps best remembered for his invention of carbonated water (1767), which would open the way for the development of the massive soda water industry—thriving to this day.

PRIEM, CURTIS R. Born circa 1959. American electrical engineer and businessman: Co-founded the Nvidia Corporation (1993), an electronic graphics company that, among other products, is chiefly known for being a designer, producer, and seller of CPUs and graphics cards for applications ranging from computers, AI, and gaming, to robotics, video editing, and vehicles.

PRIEUR, YVES LE. 1885-1963. French naval officer, aviator, translator, scuba diver, military attache, and inventor: Invented the first SCUBA ("Self Contained Underwater Breathing Apparatus") device (1926), as well as the first full face dive mask (1933). His SCUBA gear outfit, known as the the Fernez-Le Prieur Diving Apparatus, included his dive mask, a regulator, and a compressed air tank that was worn on the back. He was also the inventor of the Le Prieur Rocket (1916), World War I air-to-air missiles that were launched through steel tubes via electricity. Founder of, and first dive instructor at, the Parisian *Club des Sous l'Eau*, or "Underwater Club" (1935). His research and work led to new advancements in underwater exploration, in particular those involving the fields of underwater speleology and archaeology.

PRITCHARD, JOEL M. 1925-1997. American politician, military officer, businessman, and inventor: Along with two of his friends, Barney McCallum and Bill Bell, Pritchard is credited with inventing pickleball (1965). Currently the fastest growing sport in the U.S., this highly popular paddle game combines elements of tennis, badminton, and ping pong.

PRITCHETT, ROBERT TAYLOR. 1828-1907. English businessman, gun designer, gun manufacturer, painter, illustrator, sketch artist, draftsman, lecturer, world traveler, and inventor: Co-invented the Pritchett bullet (1853). A British variant of the French Minié bullet, it possessed a hollow base and a round elongated tip that enlarged after firing, improving stability and accuracy. Pritchett is best remembered by American Southerners as a supplier of Enfield rifles to the Confederate military during America's War for the Constitution—incorrectly (and intentionally) misnamed the "Civil War" (1861-1865).[20]

PROKHOROV, ALEXANDER. 1916-2002. Russian scientist, physicist, educator, editor, author, and inventor: His research in the field of quantum electronics laid the foundation for the invention of both the laser and the maser (circa 1952). Nobel Prize winner.

PROKOPOVYCH, PETRO. 1775-1850. Ukrainian apiarist, apiculturist, commercial beekeeper, writer, and inventor: Invented the first portable frame beehive (1814), the first queen bee excluder, and a method for preventing European foulbrood (EFB)—a bee brood disease caused by the bacterium *Melissococcus plutonius*. Founder of the School of Beekeeping (1828). Prokopovych's apiarian innovations helped lead to the modern commercialization of beekeeping and commercial honey production.

PROKUDIN-GORSKY, SERGEY. 1863-1944. Russian scientist, chemist, photographer, color photography pioneer, educator, and inventor: Invented an early method for making color film slides, as well as a color motion picture projection technique. His work aided in the development of both color photography and color films.

PUBLICOVER, MARK. Born 1958. American entrepreneur and inventor: Invented the first cost-effective trampoline safety net enclosure (circa 1995). He is also credited with making improvements to the shock-absorption aspects of trampoline beds. Co-founder of JumpSport, Inc. (1997).

PULLMAN, GEORGE M. 1831-1897. American industrialist, businessman, entrepreneur, and inventor: Invented the "Pullman Sleeper" (circa 1859), a extra-comfortable railroad car created for luxurious overnight passenger travel. Though expensive, it was an instant success, particularly among the upper classes. Operations of the Pullman Sleeper were halted in 1968. Founder of the Pullman Palace Car Company (1867), as well as the city of Pullman, Illinois (1880).

George M. Pullman.

PUPIN, MICHAEL I. 1858-1935. Serbian scientist, physicist, electrical engineer, educator, author, and inventor: Invented "Pupinization" (1899), a technique for placing wire coils (known as loading coils or "Pupin's Coils") at various distances along telephone lines. Pupin's innovation led to major improvements in the field of long-distance telephony.

PUSKÁS, TIVADAR. 1844-1893. Hungarian scientist, physicist, engineer, telephony pioneer, and inventor: Invented the telephone exchange (1877), an early telecommunications system that employed a telephone operator who manually connected calls on a telephone network via a switchboard of cords, plugs, and switches. The telephone exchange will be familiar to those who lived before 1965, after which the intricate device was replaced by all-digit dialing (1968). Additionally, Puskás is credited with inventing the multiplex switchboard (1887). A system that also required a human telephone operator, it allowed for the transmission of multiple streams of data over a single telephone line. Although the concept of "multiplexing" is still used by the telecommunications industry, the inventor's Victorian multiplex switchboard technology was eventually obsolesed by modern, automated, digital switching systems.

Telephones, early 1900s.

The Elias Ring Armature Motor, 1882.

QUATE, CALVIN FORREST. 1923-2019. American electrical engineer, educator, and inventor: Co-invented the atomic force microscope (1986), or AFM, an instrument that employs a high-resolution imaging technique allowing it to scan almost any type of surface down to the atomic level. This microscope, which uses a technology known as scanning probe microscopy, visualizes the surface topography of materials at the nanoscale by which it can determine nearly any measurable force interaction. Due to its high resolution capabilities, versatility, and non-destructive imaging, the AFM has a broad degree of applications in such industries as molecular research, biotechnology, pharmacology, physical science, vision science, microelectronics, medicine, live cell imaging, mechanical property analysis, biology, and electrical property analysis, among others.

QUETELET, ADOLPHE. 1796-1874. Belgian scientist, physicist, naturalist, astronomer, sociologist, author, statistician, botanist, criminologist, and inventor: One of the first to use statistics, based on the *homme moyen* ("average man"), to understand and analyze social phenomenon. He is best known today, however, as the inventor of the Body Mass Index (1830s). More commonly known as BMI, this medical screening tool estimates body fat percentage by taking a person's body weight (in kilograms) and dividing it by the square of their height (in meters). Using Quetelet's formula, also

Adolphe Quetelet.

known as the "Quetelet Index," modern health authorities have determined that a person of normal weight should have a BMI of between 18.5 and 24.9. A BMI below 18.5 is considered underweight, while a BMI over 25.0 is considered overweight. A BMI of 30 or over is considered obese.

Ford Motor Company workers, 1924.

RABENS, NEIL W. 1929-2020. American artist, cartoonist, toy and game designer, and inventor: Co-invented (with Charles Foley) the popular game Twister (1966).

RABINOW, JACOB. 1910-1999. American mechanical engineer, researcher, lecturer, and inventor: Among the many innovations credited to his name he is most familiar as the inventor of the first magnetic-disk computer memory storage device (1954), the first magnetic automobile clutch (1956), the first self-regulating clock (1960), and the forerunner of modern processing and sorting machines, the "Rabinow Reading Machine" (1960). Founder of RABCO (1968).

RAD, SEAN. Born 1986. American businessman, entrepreneur, philanthropist, and inventor: Founder of Tinder (2012), the globally popular dating application. Also founder of Ad.ly (2009) and co-founder of AllVoices (2018).

RAND, JOHN GOFFE. 1801-1873. American businessman, painter, and inventor: Invented the collapsible metal paint tube (1841), which, for the first time, allowed for the convenient storage and transport of art paints. Thanks to Rand's cleverly designed tin paint tubes, artists could now paint anywhere, even outdoors. This, according to some, paved the way for the birth of the Impressionistic Period (roughly 1867-1886).

RANSOME, ROBERT. 1753-1830. English businessman, entrepreneur, ironmonger, and inventor: Made numerous improvements to the plow (circa early 1800s), which helped advance farming methods, equipment efficiency, agricultural economics, and crop yields.

REBER, GROTE. 1911-2002. American astronomer and radio

engineer: Credited by many with inventing the first radio telescope (1937), thereby launching the field of radio astronomy.

REEVES, ALEC. 1902-1971. English scientist, electrical engineer, and inventor: Invented pulse-code modulation (1938), or PCM. This versatile communications system is capable of transmitting multiple "pulses" of analog data, such as telephone calls, as digital signals that can be transferred on the same radio frequency. Reeves' innovation is still in use, and is the veritable scaffolding upon which today's digital communications technology is built. Industries include sound recording, film, and telephony.

REGNAULT, HENRI VICTOR. 1810-1878. French physicist, thermodynamicist, organic chemist, educator, photographer, and inventor: Developed and discovered polyvinyl chloride (1835), or PVC, a manmade resin created by polymerizing vinyl chloride. Today one of the most common plastics in the world, this lightweight, general-purpose, tough material is now used in an enormous variety of applications, from construction (siding, plumbing, roofing, insulation, window frames, door frames, etc.), garden hoses, vinyl records, home appliances, and camera bodies, to credit cards, medical equipment, synthetic leather, camping equipment, packaging, footwear, electrical equipment, telecommunications, and shower curtains—among many others. Regnault is known as the "father of PVC."

REICH, WILHELM. 1897-1957. Austrian (later American) psychoanalyst, psychiatrist, author, and inventor: An associate of Sigmund Freud and one of the most famous and influential figures in the world of psychology, for our purposes he is best known as the inventor of the "Cloudbuster" (1953), a large device consisting of long, parallel, water-filled tubes pointed skyward. According to Reich, his "orgone accumulator" not only provided mental and physical health benefits, it could also influence the weather by atmospherically controlling "orgone energy," a power that he equated with "cosmic energy"—or what would now be known as the universal life force, life energy, or chi. His invention, his experiments with ("unlimited free") energy, and his health claims, all three which mainstream science labeled "pseudoscience," soon attracted the attention of the U.S. government, in particular the FDA. Numerous investigations, problems, and conflicts ensued, all which finally resulted in Reich being arrested and charged for contempt of court. Many surmise, however, that the real reason behind his troubles was the attempt to halt the selling and distribution of his "fraudulent" invention—to a *very interested* public, it should be noted. By order of a U.S. Federal Court, his machines were destroyed and his books censored, suppressed, and burned. Some 8 months later the inventor died in prison (1957). Reich is known as the "father of orgone therapy."

REICHENBACH, KARL VON. 1788-1869. German scientist, chemist, industrialist, businessman, geologist, naturalist, metallurgist, botanist, astrologer, philosopher, author, and inventor: Among numerous other chemicals and compounds, he is best known for his discovery of 1) paraffin wax—used in lubricants, candles, insulation, fuels, cosmetics, matches, etc.; 2) creosote oil—used in pesticides, shampoos, wood treatments, antiseptics, etc.; 3) phenol—used in mouth washes, shore throat lozenges, cleaning products, etc.; and 4) kerosene—used in aviation fuel, lubricants, heating oil, solvents, lighting fuel, etc. (1830s). Reichenbach was also a keen scientific researcher of the paranormal and of vitalist phenomenon, which included experiments in such areas as mesmerism, the human aura, dowsing, crystals, table-tipping, magnetic healing, Spiritualism, and the Odic Force (named for the Anglo-Saxon/Norse god Odin or Woden, after whom our weekday Wednesday is named: "Woden's Day."

REICHSTEIN, TADEUS. 1897-1996. Polish scientist, chemist, lecturer, researcher, educator, and inventor: Among his numerous scientific achievements (including his award-winning work on hormones), he is best known to many as the inventor of the Reichstein Process (1933), a technique for synthesizing Vitamin C (ascorbic acid)—an essential nutrient that is vital to wound healing, disease prevention, protein synthesis, iron homeostasis, antioxidant protection, and immune system regulation. The importance of Reichstein's discovery becomes manifest when we realize that, due to an absence of the enzyme gulonolactone oxidase, the human body cannot produce Vitamin C. And though we can procure Vitamin C from foods (mainly fruit and vegetables), this is not always possible. Thus the need for a manmade version of Vitamin C that can be used as a dietary supplement. Nobel Prize winner.[21]

REMINGTON, ELIPHALET. 1793-1861. American businessman, entrepreneur, engineer, gunsmith, and inventor: Founder of Remington Arms (1816), noted firearms manufacturer, which supplied arms to the U.S. government during both World Wars. Remington's original gun designs, as well as subsequent models, were extremely popular, some which became standard weaponry for a number of law enforcement agencies and armies. The company is perhaps best known for its Remington Model 700 bolt-action rifle (1962).

Eliphalet Remington.

REMSEN, IRA. 1846-1927. American chemist, educator, author, magazine founder and editor, and inventor: Credited by some with the co-discovery of the manmade sweetener saccharin (1879). (Others, however, credit only his research associate Constantin Fahlberg with the finding.) President of Johns Hopkins University (1901-1912).

RESKI, RALF. Born 1958. German biotechnologist, geneticist, botanist, educator, and inventor: Invented the moss bioreactor (circa 1999), a device used to grow genetically engineered moss, which in turn is used to produce pharmaceuticals as well as proteins for medical applications.

RESSEL, JOSEF L. F. 1793-1857. Czechoslovakian forester, marine researcher, and inventor: Among many other innovations credited to him, he is perhaps best known as the (disputed) inventor of the ship propeller (1820s).

REYNOLDS, WILLIAM. 1758-1803. English ironmonger, steam engine designer, canal builder, and inventor: Though the inclined plane, or something like it, dates back to ancient Egypt thousands of years ago, Reynolds is credited with inventing the modern inclined canal plane (1792). A true inclined plane is nothing more than a sloping ramp that allows for the efficient up-and-down movement of objects, such as boxes. Though Reynolds' design was similar, it was created specifically for canal-traveling boats. His tilted ramp, which involved physically lifting boats up and down through different water levels, greatly improved both canal navigation and the marine economy.

RICHTER, CHARLES FRANCIS. 1900-1985. American scientist, physicist, seismologist, mathematician, geologist, and inventor: Inventor (some say co-inventor) of the eponymous Richter Magnitude Scale (1935), which records ground movement in order to analyze the size and intensity of earthquakes.

RICHTER, RICHARD. Dates unknown. German physician and gynecologist. Invented the intrauterine device or IUD (1909), a T-shaped contraceptive device that is placed in a woman's uterus in order to protect her against pregnancy.

RICKENBACKER, ADOLPH. 1886-1976. Swiss (later American) engineer, musical instrument designer, and inventor: Co-invented the first electric guitar (1931). The influence and popularity of Rickenbacker guitars was greatly enhanced throughout the 1960s and 1970s by bands such as the Beatles, the Who, Jefferson Airplane, and Creedence Clearwater Revival, who performed and recorded with Rickenbacker's unique stringed instruments. Rickenbacker was a co-founder of the Rickenbacher Guitar Company (1931), which continues to manufacture various types of guitars and guitar accessories.[22]

RICKOVER, HYMAN GEORGE. 1900-1986. Polish (later American) electrical engineer and U.S. military officer: Invented the world's first nuclear submarine (1954), the USS *Nautilus* (SSN-571).

RIFE, ROYAL RAYMOND. 1888-1971. American alternative medicine advocate and inventor: Invented the Rife Machine, a "beam ray" device he claimed could destroy pathogenic organisms, including cancer cells. Rife's invention, along with his "radical" ideas concerning the healing properties of electromagnetic frequencies, have been relegated by mainstream medicine to the trash bin of "pseudoscience," labeled "worthless," "fraudulent," and even "harmful." Rife, however, emphatically maintained that his inventions worked, and that his findings had been suppressed by an "establishment conspiracy" conducted by conventional health organizations. The controversy continues.

RIGGENBACH, NIKLAUS. 1817-1899. Alsatian mechanical engineer, railwayman, machinist, rail locomotive builder, and inventor: Though he did not invent the first rack railway, he did invent the Riggenbach Rack Railway System (1862)—which was outfitted with another one of his inventions, the Riggenbach Counter-Pressure Brake (1862). The rack railway design features a toothed rail that is laid between the two outer running rails. Cog wheels located beneath the train cars lock into the "teeth" on this center rail, providing extra rail grip, allowing the train to navigate steep gradients that would be impossible for ordinary locomotives.

RITCHIE, DENNIS. 1941-2011. American computer scientist, mathematician, software developer, author, and inventor: Inventor of the C programming language (1972); co-inventor of the UNIX operating system (1969), as well as the B programming language (circa 1969).

ROBBINS, MERLE. 1911-1984. American barber and inventor: Invented the popular card game Uno (1971).

ROBERVAL, GILLES DE. 1602-1675. French scientist, physicist, mathematician, educator, and inventor: Made significant contributions to the field of geometry, including the development of what were later called "Robervallian Lines." Best known for his invention of the Roberval Balance (1669), a commercial instrument that uses balances and scale pans to accurately calculate weight. Roberval's idea is still in use—though it has been incorporated into the software and hardware of modern electronic balances and scales.

ROCK, JOHN. 1890-1984. American scientist, obstetrician, gynecologist, author, and researcher. Co-invented the oral contraceptive pill (circa 1950s). Better known as "the pill," it revolutionized the birth control industry and greatly contributed to the explosion of the bohemian "free love" movement of the 1960s and 1970s.

ROCKEFELLER, SR., JOHN D. 1839-1937. American businessman, entrepreneur, oil tycoon, and philanthropist: Founded the Standard Oil Company (1870), the University of Chicago (1890), Rockefeller University (1901), and the Rockefeller Foundation (1913), among numerous other institutions.

RODDENBERRY, GENE. 1921-1991. American producer and screenwriter: Created the hit TV series *Star Trek* (1966).

ROEBUCK, JOHN. 1718-1794. English scientist, chemist, physician, industrialist, businessman, mechanical engineer, and inventor:

John D. Rockefeller, Sr.

His background in chemistry helped lead to a number of important contributions in the field of metal production. Arguably his most significant concept was a technique for producing sulfuric acid using lead condensing chambers (1746). He also provided financial aid to James Watts during the latter's research into the steam engine. Roebuck's ideas helped lay the groundwork for the upcoming Industrial Revolution (1760-1840).

ROGERS, WILLIAM BARTON. 1804-1882. American scientist, physicist, geologist, educator, and lecturer: Founder and first president of the Massachusetts Institute of Technology (1861), better known today as MIT.

ROHRER, HEINRICH. 1933-2013. Swiss physicist and inventor: Co-invented the scanning tunneling microscope (1981), or STM, which allows the viewer to observe atoms in 3D. Nobel Prize winner.

RÖNTGEN, WILHELM CONRAD. 1845-1923. German scientist, physicist, engineer, researcher, educator, lecturer, and inventor: Discovered X-rays (1895). A type of electromagnetic radiation used in medical imaging, X-rays allow internal structures to be observed, aiding in the diagnosis and screening of disease as well as various types of surgery. Sometimes known as "Röntgenograms," Röntgen called them "X" rays because at the time he did not know he had discovered a form of ionizing radiation. Nobel Prize winner.

RORSCHACH, HERMANN. 1884-1922. Swiss psychoanalyst and psychiatrist: Best known for inventing the Rorschach Test (1921), a projective examination that uses cryptic inkblots to help identify, diagnose, and study a variety of cognition-related conditions, from mental health issues to intelligence level.

ROSENTHAL, SIDNEY. 1907-1979. America engineer and inventor: Improved upon earlier felt-tipped marking pen designs, eventually inventing what is popularly known by the brand name "Magic Marker" (1953).

Hermann Rorschach.

ROSHAL, EUGENE. Born 1972. Russian software developer: Invented RAR file format (1993), WinRAR file archiver (1995), and the FAR file manager (1996), all which have led to advancements in the fields of data archiving, compression, and management.

ROSING, BORIS L. 1869-1933. Russian scientist, physicist, engineer, television pioneer, author, and inventor: Invented what seems to be the world's first TV to utilize a cathode-ray tube as a receiver (1907). Better known as CRT television, Rosing's innovation helped open the door to new developments during the beginning stages of the TV age.

ROSSUM, GUIDO VAN. Born 1956. Dutch computer engineer, software developer, and systems programmer: Invented Python (1991), a comprehensive programming language.

ROTHMAN, MICHAEL A. Dates unknown. American engineer, author, novelist, and inventor: Among his over 1,000 patents, he is best known by many for being a co-developer of the Unified Extensible Firmware Interface Standard (early 2000s). The UEFI standard, which replaced the BIOS system (2006), specifies how a computer's hardware and operating system interact.

ROUND, HENRY JOSEPH. 1881-1966. English electronics engineer and inventor: Made many improvements in the field of radio technology, which included such instruments as radio receivers, radio transmitters, radio telephones, and radio directional detectors—all which aided Britain's war efforts in the first half of the 20th Century. An associate of famed Italian electrical engineer Guglielmo Marconi, Round is best remembered, however, for being the first to research, describe, and observe the phenomenon of electroluminescence (circa 1907), important work that inevitably opened up the way for the development of the light-emitting diode, or LED—one of the most significant and world-altering inventions of modern times.

ROZIÈR, JEAN-FRANÇOIS PILÂTRE DE. 1754-1785. French scientist, physicist, chemist, aviator, aviation pioneer, balloonist, educator, and inventor: He and Marquis d'Arlandes made history when they became the first to fly in an untethered manned balloon flight (1783)—marking them as the world's first true balloonists. However, Rozièr is also remembered for inventing the Rozière Balloon (1784), which he and his associate Pierre Romain later attempted to fly over the English Channel (1785). Tragically, the balloon was pushed back overland by the wind where it caught fire, taking the lives of both passengers—making this the first recorded incident of death due to an aviation accident.

RUBIK, ERNŐ. Born 1944. Hungarian engineer, architect, mathematician, businessman, entrepreneur, game designer, educator, sculptor, and inventor: Invented the mechanical puzzle toy "Magic Cube" (1974). Better known today as Rubik's Cube, this enormously popular gadget has sold over 100 million units globally, and continues to be used today as an educational device by science teachers.

RUGGLES, KAY LEROY. 1932-2012. American businessman, designer, and inventor: Invented UMBO shelving and furniture (1970), an adjustable, modifiable, plastic shelving system. The holder of numerous patents covering a myriad of industries, Ruggles is also credited with inventing the tubular water slide.

RUSKA, ERNST A. F. 1906-1988. German electrical and optics scientist, physicist, educator, lecturer, author, and inventor: Invented the electron microscope (1933), a type of high-resolution research instrument that replaces the conventional light beam illumination with an electron beam. Nobel Prize winner.

RUSSELL, NOADIAH. 1659-1713. American clergyman: Co-founder of the Collegiate School (1701), later renamed Yale University (1718) after its main benefactor, American colonial leader Elihu Yale.

RUTGERS, HENRY. 1745-1830. American Revolutionary War icon and state legislator: Rutgers University was named after him (1825).

RUTH, WILLIAM CHESTER. 1882-1971. American industrial engineer, businessman, machinist, blacksmith, draftsman, and inventor: A developer of farm machinery, he is best known for his invention of the baling-press feeder (1924) and the Ruth Cinder Spreader. Additionally, he made improvements to already existing technologies, such as the farm elevator, the manure spreader, and the hay baler.

RYSSELBERGHE, FRANÇOIS VAN. 1846-1893. Belgian scientist, mathematician, meteorologist, educator, and inventor: Invented the Telemareograph (1875) and the Telemeteograph (1875), scientific instruments that were used to measure and analyze meteorological phenomena. He is also credited with inventing the Van Rysselberghe System (1884), a telephonic technology that allowed the transmission of both telegraph and telephone signals over the same cable.

A factory roller conveyor, 1925.

SABIN, ALBERT BRUCE. 1906-1993. Polish (later American) biomedical scientist, physician, pathologist, virologist, educator, and inventor: Invented the oral polio vaccine (1961). Also known as OPV, it is credited with saving tens of millions of lives since its development.
SABLUKOV, ALEXANDER. 1783-1857. Russian engineer, military officer, and inventor: Invented the centrifugal fan (1832). Also known as a radial fan, this dependable and long-wearing mechanical instrument is designed to propel air in a specific direction, useful for such applications as cooling and extraction.
SABUNCUOĞLU, ŞERAFEDDIN. 1385-1468. Turkish physician, surgeon, and author: His book *Imperial Surgery* (1465) seems to be the world's first known illustrated surgical atlas (1465). This novel reference book included instructions, illustrations, anatomy descriptions, diagnoses, instruments, and methods related to medical operations and diseases, and contains references to female surgeons.
SACHAR, ABRAM L. 1899-1993. American historian and author: Founding president of Brandeis University (1948), named after Louis Dembitz Brandeis, American lawyer and associate justice of the Supreme Court.
SAKHAROV, ANDREI D. 1921-1989. Russian nuclear scientist, physicist, human rights activist, and inventor: Among numerous other innovations, he is best known for co-developing the Soviet Union's first hydrogen bomb (1946), a response to America's wartime nuclear program. Despite this, he advocated for the beneficent use of nuclear technology, culminating in his invention of the Tokamak (1954), a magnetic component of nuclear fusion reactors—intended to peacefully

produce energy during the Cold War (1947-1991). Sakhoarov is known as the "father of the Soviet hydrogen bomb." Nobel Prize winner.

SAKMANN, BERT. Born 1942. German physician and research scientist: Co-invented the patch clamp technique (1970), which, according to the U.S. government, is an electrophysiological method for "directly measuring the membrane potential and/or the amount of current passing across the cell membrane." The patch clamp technique aids researchers in the fields of medicine and pharmacology. Nobel Prize winner.

SALK, JONAS EDWARD. 1914-1995. American physician, biomedical researcher, bacteriologist, biologist, educator, and inventor: Invented the injection polio vaccine (early 1950s), saving tens of millions of lives while preventing countless cases of paralytic polio.

SALMON, ROBERT. 1763-1821. English mechanical engineer, carpenter, mechanic, architect, construction builder, musical instrument maker, and inventor: Invented or improved upon numerous pieces of farming equipment, among them haymaking machines, cultivators, seed drills, reaping machines, tree-pruners, and plows. He is best known, however, for his invention of the chaffcutter, an agricultural implement that chops plant material (such as hay) into small sections to make it easier for animals to consume.

SALÒ, GASPARO DA. 1542-1609. Italian stringed instrument maker, musician, and inventor: Although the originator of the first violin cannot be known with certitude, many credit Salò, also known as Gasparo Bertolotti, with being one of the first, if not the first, to invent it (mid 1500s). The debate continues.

SAN-GALLI, FRANZ FRIEDRICH WILHELM. 1824-1908. Polish businessman, entrepreneur, engineer, and inventor: Invented the first cast-iron heating radiator (1855), making him the inventor of central heating.

SANGER, FREDERICK. 1918-2013. English biochemist, educator, and inventor: Besides being the discoverer of the structure of insulin (circa 1943), he is best remembered for his invention of the Dideoxy Technique (1977), a method used to sequence DNA molecules. Also known as Sanger Sequencing, his efficient, economical system is able to analyze genes at the nucleotide level, making it an integral medical aspect of molecular biology. Nobel Prize winner.

SANKAI, YOSHIYUKI. Born 1958. Japanese businessman, educator, and inventor: Though he did not invent the idea of the exoskeleton (the earliest known conception dates back to the Victorian Period), he did co-develop a modern version of the robotic exoskeleton (1989), a machine-powered body covering that provides additional support and protection for muscles and limbs. His innovation has numerous applications, among them the military, medicine, and general industry. Founder and CEO of Cyberdyne (2004), a manufacturer of cyborg robots.

SANTORIO, SANTORIO. 1561-1636. Slovenian (Italian) physician, physiologist, educator, author, and inventor: An associate of Galileo, Santorio is chiefly credited with inventing the thermoscope (1612), what seems to have been the first truly accurate thermometer. Along with numerous other scientific instruments, such as a pulse clock, a wind gauge, and a water meter, he is also known for inventing the pulsilogium (1602). The world's first known reliable pulse-measuring device, it laid the foundation for the development of the modern oximeter.

SANTOS-DUMONT, ALBERTO. 1873-1932. Brazilian aviator, aeronaut, balloonist, engineer, balloon and airship designer, author, and inventor: Made many contributions to airship and airplane technology. Was the first to fly a heavier-than-air aircraft in Europe (1906). His title as the "first aviator," while heartily defended in his native country Brazil, continues to be debated as this honor generally goes to the Wright Brothers.

SAUSSURE, HORACE BÉNÉDICT DE. 1740-1799. Swiss physicist, geologist, educator, and explorer: Invented the first practical hygrometer (1783), a device used to measure humidity. He also invented the first electrometer (circa 1760s), an instrument for measuring electric potential. Saussure coined the word "geology."

SAVAGE, ARTHUR WILLIAM. 1857-1938. British (born in Jamaica) businessman, entrepreneur, adventurer, explorer, gun and ammunition designer and manufacturer, tire designer, and inventor: Though during his life he racked up a large list of accomplishments (including railroad manager, orange-grower, oil driller, coffee grower, gold miner, brick manufacturer, banana grower, race car driver, etc.), as well as innovations (including torpedoes and hemp cleaners), this restless inventor is best remembered as the founder of Savage Repeating Arms Company (1894), the noted gun manufacturer that supplied the U.S. military with arms through both World War I and World War II. His Savage Model 1895 rifle employed a revolutionary new type of ammo for the time: the 303 Savage .30-cal. cartridge, made especially for the new smokeless gunpowder. During this period Savage invented the concept of the gun magazine, a highly readable publication that featured, among other things, exciting first-person hunting accounts from customers. One of the companies most popular firearms was the Savage Model 99 lever-action rifle (production ended in 1997). Savage was also the founder of the Savage Tire Company (1911) at which time he invented (disputed) the first radial tire (1916).

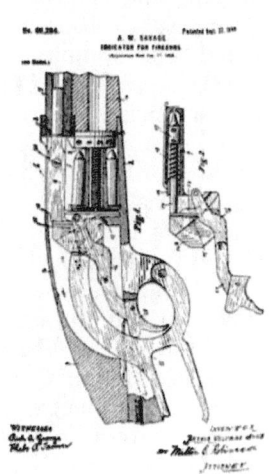

Savage rifle patent drawing, filed Sept. 27, 1898.

SAVERY, THOMAS. 1650-1715. English engineer and inventor: Invented the first commercially viable steam engine (1698), revolutionizing mechanical technology and paving the way for the Industrial Revolution (1760-1840).

SAVILLE-KENT, WILLIAM. 1845-1908. English marine biologist, naturalist, ichthyologist, author, and photographer: Credited with growing the first artificial pearl, opening the door for the development of today's cultured pearl industry.

SAX, ANTOINE-JOSEPH. 1814-1894. Belgian musician, musical instrument maker, educator, and inventor: Invented the saxophone (1846), a popular wind instrument used in a variety of musical genres, including jazz, pop, and classical music.

SCHAEFER, VINCENT JOSEPH. 1906-1993. American meteorologist, chemist, educator, and inventor: Invented cloud seeding using pellets of dry ice (1946). His chance experiment, which artificially generated precipitation (snow and rain), launched the meteorological field of weather control, benefitting farmers individually and humanity generally.

SCHICK, BÉLA. 1877-1967. Hungarian pediatrician, businessman, researcher, educator, author, and inventor: Invented the diphtheria test (1913), better known as the Schick Test. Noted for his work on both the immune system and children's nutrition, he was also the discoverer of what he and his associate Clemens von Pirquet would come to call "allergies." Co-founder of the American Academy of Pediatrics (1936). Nobel Prize nominee.

SCHICKARD, WILHELM. 1592-1635. German polymath, scientist, mathematician, astronomer, educator, cartographer, theologian, photogrammetrist, clergyman, academic, Hebraist, engraver, surveyor, and inventor: Though often disputed, he is credited with inventing the first mechanical calculator (1623). He is also known for aiding in improving astronomical calculations as well as mapmaking technology.

SCHIFF, HUGO. 1834-1915. German chemist: Invented the Schiff Test (1866), a chemical analysis that is used in the creation of antibacterial drugs and antibacterial compounds. Still in use.

SCHILLING, PAVEL. 1786-1837. Estonian electrical engineer, historian, diplomat, and inventor: Developed the idea for both the world's first electromagnetic telegraph (1828), and the first long-distance system for detonating explosives with electrified wires (1834).

SCHJELDAHL, GILMORE T. 1912-2002. American businessman, entrepreneur, plastics engineer, and inventor: Though he had numerous innovative credits to his name, he is perhaps best remembered for inventing the airsickness bag (1949) and for his work on Echo I (1960), the first communications satellite. Founder of the Schjeldahl Company (1955).

SCHLAFLY, HUBERT J. 1919-2011. American electrical engineer, television engineer, and inventor: Invented (some say co-invented) the

Teleprompter (1950). This visual cue device—with a scrolling script screen that aids speakers and actors in delivering their lines more accurately—has revolutionized the media industry, which includes entertainment, print, broadcast, and online media.

SCHLENK, WILHELM JOHANN. 1879-1943. German organic chemist and inventor: Invented the Schlenk Flask (1913). An instrument that connects two flasks via a glass tube, it was the forerunner of today's standard laboratory filtration device, the swivel-frit—commonly used in air-sensitive chemical research. Nobel Prize nominee.

SCHMIDT, BERNHARD V. 1879-1935. Estonian scientist, physicist, optics researcher, astronomer, engineer, technologist, photographer, and inventor: Invented the Schmidt Camera, also known as the Schmidt Telescope (1930), a type of catadioptric telescope with an exceptionally wide field of view. Mainly used in astrophotography, it is still in use.

SCHMIEDL, FRIEDRICH. 1902-1994. Austrian rocket designer and inventor: Invented the concept of rocket mail (1931), a system of delivering mail from town to town using solid-fuel rockets. Schmiedl's plan, which included allowing his mail-filled, remotely-controlled projectiles to fall gently to earth via parachutes, never caught on (probably due to the dangers of the explosive fuels he used). It is interesting to note, however, that, though living in the pre-drone age, he seems to have anticipated modern drone package delivery technology.

SCHMITT, OTTO. 1913-1998. American scientist, biophysicist, biomedical engineer, research engineer, zoologist, biologist, educator, and inventor: Among his many innovations and contributions to science, he is perhaps best known for his co-invention of the Schmitt Trigger (1934), a square-wave generator employed in signal conditioning applications and which is still in use. Schmitt is also credited with founding the field of biomedical engineering.

SCHRADER, AUGUST. 1807-1894. German engineer, mechanic, businessman, entrepreneur, author, and inventor: Among his many innovations he is chiefly known for his co-invention (along with his son George) of the Schrader Valve (circa 1891), a pneumatic tire valve that is still used on a majority of the world's bicycles, motorcycles, cars, trucks, buses, and construction vehicles (forklifts, excavators, etc).

SCHWARZ, DAVID. 1845-1897. Hungarian businessman, timber merchant, airship pioneer, self-taught mechanic, and inventor: Invented the world's first all metal airship (circa 1890). Schwarz did not live to see the success of his innovative aircraft, however. According to some (disputed), his ideas lived on after him in the form of the famous rigid airship known as the Zeppelin, developed by Ferdinand von Zeppelin.

SCHWARZE, BERTHOLD DER. Lived 14th Century. German alchemist, monk, educator, and inventor: Though considered legendary by some, others regard him as a historical figure and credit him with inventing gunpowder (circa 1315). Either way, the true original inventor of gunpowder must be regarded as unknown, for the invention of this world-altering black powder explosive was first recorded four centuries earlier in 9th-Century China. It might be safest then to view Schwarze (assuming he was a real person) as the first European to discover gunpowder.

SCOTT, RAYMOND. 1908-1994. American musician, record producer, sound engineer, film score composer, songwriter, conductor, music arranger, audio engineer, actor, pianist, bandleader, and inventor: A prolific inventor of all manner of devices, instruments, and gadgets, he is probably best remembered for his many musical works, including TV music and film scores, as well as his experiments with, and development of, electronic music.

SEGATO, GIROLAMO. 1792-1836. Italian naturalist, cartographer, Egyptologist, anatomist, and inventor: Considered (falsely) by some of his contemporaries as a ghoulish dabbler in the Black Arts, this eccentric scientist is credited with inventing a secret mummification process known as "petrifaction" (1823). Here, the body parts of deceased humans and animals were mineralized; in other words, "turned to stone," perfectly preserving their color as well as other natural physical properties. Unfortunately for both science and history, Segato never discussed his "recipe" with anyone, taking the still unsolved mystery to his grave. Only a few of the body parts he "petrified" survive.

SEGUIN, MARC (THE ELDER). 1786-1875. French civil engineer, mechanical engineer, entrepreneur, author, and inventor: Invented (some say co-invented with his brother Camille) the wire cable suspension bridge (1824), as well as the multiple-fire-tube boiler (1828)—which replaced the conventional water-tube boiler used in steam engines. Additionally, Seguin made contributions to railroad technology, physics, and navigation. Seguin's highly effective and popular wire cable suspension bridge design is still very much in use, one of the more notable examples being San Francisco's Golden Gate Bridge.

Marc Seguin.

SEINFELD, JERRY. Born 1954. American actor, comedian, producer, and writer: Co-created the hit TV series *Seinfeld* (1989).

SEISHŪ, HANAOKA. 1760-1835. Japanese surgeon, anesthesiologist, pharmacist, surgical innovator, educator, and inventor: The first to use general anaesthetic on a patient during surgery (1804). Seishū's liquid pain-killer, which he named "Tsusensan," consisted of seven different herbs and oils, including wolfsbane, ginseng, and wild celery.

SELKER, TED. Dates unknown. American computer scientist, futurist, educator, coach, and inventor: The creator of a myriad of computer interface technologies, he is perhaps best known by many as the inventor of the "pointing stick" (1984), an isometric analog pointing device, or trackpoint, with a rubber cap, usually located in the center of computer

keyboards.
SERLING, ROD. 1924-1975. American producer and screenwriter: Created the hit TV series *The Twilight Zone* (1959).
SERPOLLET, LÉON E. 1858-1907. French industrialist, businessman, entrepreneur, engineer, race car driver, and inventor: Invented the flash steam boiler (1886), which he installed in his famous Gardner-Serpollet Steam Car (1903). Serpollet's flash boiler was such a success that it was eventually used to power not just cars, but trucks, buses, trolley cars, and even street lights. His early death steered the public's attention away from steam powered engines, however, and internal combustion powered engines quickly took center stage in the auto world.
SERRURIER, IWAN. 1878-1953. Dutch mechanical engineer, electrical engineer, businessman, entrepreneur, draftsman, and inventor: Invented the Moviola (1924), a traditional analog film editing machine. While it was the conventionally accepted film editor in the film industry for some 50 years, it was eventually obsolesed by technology and the introduction of digital film editing.
SERRURIER, MARK U. 1904-1988. American mechanical engineer, businessman, and inventor: Son of Iwan Serrurier (above), he is credited with inventing the Serrurier Truss (1935). A support system originally built for the Hale Telescope, the technology continues to be used in many modern telescopes. Won an Academy Award for his contributions to film editing technology.
SESSLER, GERHARD. Born 1931. German scientist, physicist, electrical engineer, educator, and inventor: Invented the foil electret microphone (1962), designed around the micromachining of silicon. The small, low-cost, highly sensitive foil electret mic revolutionized the sound and recording industries, and has today found many applications, from computers, laptops, hearing aids, smartphones, and diagnostic equipment, to baby monitors, surveillance equipment, audio and recording equipment, and sound measuring devices.
SEVERSKY, ALEXANDER PROCOFIEFF DE. 1894-1974. Russian (later American) aircraft designer, aeronautical pioneer, businessman, entrepreneur, military pilot, author, and inventor: Made hundreds of contributions to aircraft technology, including inventing an automatic stabilized bombsight, an inflight refueling system, a metal monoplane, and an amphibious aircraft (early 1930s). Co-designed the all-metal, single cockpit P-35 fighter aircraft (1935), the first of its kind and a precursor to the famous American P-47 Thunderbolt aircraft. Additionally, Seversky was responsible for the heavier-than-air aircraft known as the Ionocraft (1959), as well as the electrostatic precipitator (1973)—the latter which continues to be used by various industries as a standard environmental air-cleaning device.
SEYMOUR, ED. (inv. c. 1949). American businessman, paint salesman, and inventor: Invented aerosol spray paint (1951), revolutionizing the paint, construction, and art industries.

SHAMSHURENKOV, LEONTY. 1687-1758. Russian inventor: Invented the first self-propelling carriage (1752), a precursor to both the bicycle and the automobile, as well as an improved odometer.

SHARROW, GREGORY. Dates unknown. Nationality unknown. Businessman, entrepreneur, producer, director, and inventor: Credited with patenting a toroidal marine propeller with twisted loops (circa 2010s), a type of propeller with enclosed, curved, ring-shaped blades. The toroidal design improves fuel efficiency, reduces noise, reduces vibration, improves air and water flow, improves maneuverability, increases propeller rotation speed (which increases vehicle speed), cuts fuel consumption, increases water- and air-gripping capabilities, increases engine life, increases range, decreases emissions, increases load capacity, and reduces cutting damage (e.g., to underwater plant and animal life), all while increasing the strength and durability of the propeller. Toroidal props, which have revolutionized the science of fluid dynamics, have numerous modern applications, from helicopters, drones, and boats, to ships, wind turbines, submarines, and household and industrial fans. Sharrow founded Sharrow Engineering (2012) and Sharrow Marine (2012).

SHAW, LOUIS AGASSIZ. 1886-1940. American scientist, physician, and inventor: Co-invented the first iron lung (1928), known as the Drinker Respirator, a biomedical respiratory device that was used to treat polio patients. Replaced by modern mechanical ventilators (1940s).

SHILOVSKY, PYOTR. 1871-1957. Russian politician, jurist, and inventor: Invented (some say produced) the first gyrocar (1912), a motor-powered two-wheeled vehicle. Unlike a bicycle or a motorcycle, the gyrocar maintains its stability and balance through the use of gyroscopes. Although the gyrocar design has spawned several models over the years—most notably the Ford Gyron (1961)—it has never completely caught either the public's imagination or the car industry's attention. As a result, it has never gone into worldwide production and remains largely a mechanical curiosity.

SHIMA, MASATOSHI. Born 1943. Japanese computer scientist, electrical engineer, chemist, educator, and inventor: His research and work helped lead the way to the development of the microprocessor (1960s).

SHIRAZI, FATHULLAH. Born circa 1582. Iranian polymath, physician, mathematician, mechanical engineer, scholar, historian, philosopher, astronomer, diplomat, artist, author, and inventor: His many innovations included a gun-cleaning device, a wooden pen box, a passenger carriage, an early anti-infantry volley gun, an agricultural calendar, and a corn grinder.

SHIVERS, JOSEPH C. 1920-2014. American chemist and inventor: Invented the manmade fiber Lycra (1958). More commonly known as Spandex or Elastane, this artificial, lightweight, soft, durable, highly elastic material revolutionized the clothing industry and is today found in a host of different products, including tech products, undergarments, yoga pants, home furnishings, leggings, sportswear, the car industry, cycling shorts, and swimsuits, to name but a few.

SHOCKLEY, WILLIAM BRADFORD. 1910-1989. American physicist, educator, and inventor: Co-inventor of the transistor (1947), which did away with the more cumbersome vacuum tube. The transistor, a tiny semiconducting device used for generating and controlling electrical signals, is an integral component of the microchip, and today serves as the very foundation of computer-information age.

SHOLES, CHRISTOPHER LATHAM. 1819-1890. American printer, newspaper publisher and editor, politician, and inventor: Invented (some say co-invented) the first usable commercially viable typewriter (1867), later known as the "Remington Typewriter" (circa 1874). His invention transformed communications, greatly accelerated business efficiency, increased commercial productivity, created entirely new occupational categories, advanced the literary world (writers and publishers), regularized writing, and last but not least, paved the way for the modern computer. Indeed, Shole's QWERTY keyboard (which I am now using to type this entry) is standard on every computer, laptop, tablet, and cell phone. Sholes is known as the "father of the typewriter."

Christopher L. Sholes.

SHRAPNEL, HENRY. 1761-1842. English artillery officer and inventor: Invented the eponymously named ammo, "Shrapnel Shell Ammunition" (1784), a hollow metal casing loaded with lead shot. The idea behind this deadly antipersonnel projectile is simple: An explosion-timed cannonball detonates near the enemy, throwing lethal fragments of metal across a wide area, giving an advantage to outmanned military units. His surname (shrapnel) has entered the English language as a term for flying metal fragments that result from bombs, bullets, mines, shells, or any other type of exploding device.

SHUGART, ALAN. 1930-2006. American computer engineer, businessman, entrepreneur, and inventor: Credited with leading the IBM team that invented the floppy disk (1967), originally an 8 inch wide, flexible "memory disk" that was used to backup, store, and retrieve data. Floppy disks were eventually replaced by USB drives (i.e., thumb drives or memory sticks), cloud storage, and optical disks.

SHUKHOV, VLADIMIR G. 1853-1939. Russian polymath, physicist, civil engineer, structural engineer, mathematician, architect, and inventor: Among his many diverse innovations he is best remembered as the inventor of the aesthetically pleasing hyperboloid design (1896), a double-curved, anti-buckling, quadric surface that allows heavy features to be supported at the top of different types of structures. Though curved, these hour-glass shaped architectural wonders, most famously used in towers of various sorts (e.g., nuclear power station cooling towers), are made entirely of straight lines, enabling them to carry extra heavy weight loads using less materials. One of the most famous examples of a Shukhov hyperboloid is the Shukhov Tower in Moscow, Russia (1922).

SHURE, SIDNEY N. 1902-1995. American businessman, entrepreneur, audio-electronic equipment pioneer, photographer, and philatelist: Founder of the Shure Radio Company (1925). Now known as Shure, Inc., it is heralded as the world's largest manufacturer of microphones, and more generally, audio-electronic equipment, such as headphones, mixing boards, and sound monitoring systems. An avid stamp collector, Shure was also the founder of the Israel-Palestine Philatelic Society (1940s).

SIEBE, AUGUSTUS. 1788-1872. German engineer and inventor: Among his many innovations he is best known for various inventions related to underwater diving gear, in particular, the removable, valved dive helmet (circa 1830s).

SIEMENS, ERNST WERNER VON. 1816-1892. German engineer, businessman, entrepreneur, and inventor: Invented the first trolleybus (1882). Known as the "Electromote," it was a trackless, public transport vehicle that draws power from four trolley wheels mounted on the end of trolley poles, that are, in turn, attached to two electrified overhead cables. He was also responsible for inventing the electric elevator (1880), and was the company founder of the technology giant Siemens (circa 1847). Siemens' innovations revolutionized both the transportation and the construction industries, and are still in use today worldwide.

SIEMENS, WILLIAM. 1823-1883. German electrical engineer, industrialist, metallurgist, businessman, entrepreneur, and inventor: Invented a water meter (1851), the open hearth furnace (1861), arc lights, and an electric railway, all which aided in the development of various industries, including factories, heating, metal works, electrical cables, and telegraphy.

SIKORSKY, IGOR. 1889-1972. Russian aerospace scientist, aircraft designer, aircraft pilot, businessman, entrepreneur, and inventor: His many innovations are legendary, but he is perhaps best known for inventing the eponymous Sikorsky helicopter line (1939).

SILVER, BERNARD. 1924-1963. American scientist, physicist, engineer, and inventor: Co-invented the barcode (1949), a two-dimensional, lined, machine-readable square version of the UPC code. Barcodes store information that can be quickly accessed by scanners. With its ability to identify, manage, monitor, inventory, price, and track products (as well as people), barcode technology has revolutionized the commerce industry.

SILVERBROOK, KIA. Born 1958. Australian scientist, businessman, entrepreneur, and inventor: Though he has nearly 10,000 patents (and patent applications) to his credit, this prolific innovator is best known by many for inventing the Memjet Printer (2010), a high speed, full color inkjet printer that has greatly advanced printing technology, as well as the commercial printing industry. Founder of Fairlight Instruments (1977), among numerous other businesses.

SIMJIAN, LUTHER G. 1905-1997. Armenian engineer, businessman, entrepreneur, and inventor: Made important contributions to a variety of fields including photography, postage machinery, musical instruments, air flight, range finders, medicine, and more; however, he is best remembered for his invention of the automated teller machine (1939). More familiar today as the ATM, it revolutionized the banking industry by allowing individuals to withdraw, deposit, and transfer money in an easily accessible and effective manner.

SIMON, DAVID. Born 1960. American journalist, author, and producer: Created the hit TV series *The Wire* (2002).

SIMONOV, VLADIMIR. 1935-2020. Russian engineer, radio technologist, and inventor: Invented the SPP-1 Underwater Pistol (1960s) as well as the APS Underwater Assault Rifle (1970s).

SINA, IBN. 980-1037. Iranian scientist, physician, philosopher, author, and inventor: Also known by the name Avicenna, he made numerous contributions to a wide variety of fields, including physics, meteorology, alchemy, medicine, theology, psychology, music, astrology, botany, philosophy, astronomy, mathematics, zoology, mineralogy, poetry, oneiromancy, logic, philology, geometry, physiognomy, and metaphysics. His views, knowledge, and inventions have greatly impacted the modern world, particularly in his native country.

SINCLAIR, CLIVE. 1940-2021. English engineer, computer scientist, businessman, entrepreneur, and inventor: Made contributions to a variety of fields, including computers, amplifiers, electric cars, calculators, electric bicycles, record players, and radios. Some of his more famous inventions were the System 2000 (1968), the ZX80 (1980), the Sinclair C5 (1985), the X-10 PWM (1960s), the Zike (1992), the ZX Spectrum (1982).

SINGER, ISAAC. 1811-1875. American businessman, entrepreneur, machinist, actor, and inventor: Though he did not invent the sewing machine, he is credited with making modifications and improvements to it, which resulted in the famous Singer Sewing Machine (1851). Co-founder of the Singer Manufacturing Company (1863).

SKINNER, B. F. 1904-1990. American psychologist, behaviorist, educator, author, philosopher, and inventor: Invented the Skinner Box (1930s). More technically known as the

Singer Sewing Machine, 1851.

operant conditioning chamber, it is still used today to observe and analyze animal behavior. Author of numerous heavily influential books, including the social engineering, utopian novel *Walden Two* (1948).
SLAVYANOV, NIKOLAY GAVRILOVICH. 1854-1897. Russian inventor: Invented shielded metal arc welding (1888), an electric arc welding process that fuses metals with one another with the use of a consumable electrode.
SLIWA, CURTIS. Born 1954. American activist, politician, actor, talk show host, political commentator, and radio personality: Founder of the international organization Guardian Angels (1979), a non-profit, volunteer, safety patrol group dedicated to the prevention of crime.
SMAKULA, ALEXANDER. 1900-1983. Ukrainian scientist, physicist, photographer, educator, and inventor: Among numerous other innovations he is best remembered for inventing the Smakula Effect (1935), a method—scientifically known as "illumination of optics"—that layers lenses with a unique film coating, greatly enhancing their optical characteristics. It is used to this day in a myriad of optical industries, including cameras, microscopes, telescopes, binoculars, and periscopes.
SMEDT, EDWARD JOSEPH DE. Dates unknown. Belgian educator: Invented the first practical version of asphalt (1870), a durable and inexpensive roadway material that today covers a large percentage of the land in all developed countries.
SMITH, MICHAEL. 1932-2000. English (later Canadian) scientist, molecular biologist, chemist, DNA research pioneer, businessman, entrepreneur, and inventor: Invented (some say co-invented) Site-Directed Mutagenesis (1978). Also known as SDM, it is a method for producing specific changes in double stranded plasmid DNA, aiding researchers in studying DNA, as well as screening for mutated genes. Nobel Prize winner.
SMITH, ROBERT H. 1879-1950. American physician and surgeon: One of the co-founders of Alcoholics Anonymous (1935)—an addiction recovery organization devoted to conquering alcoholism.
SMITHIES, OLIVER. 1925-2017. English scientist, physical biochemist, geneticist, educator, and inventor: Co-developed gene targeting (early 1980s), which opened up the fields of genetic modification and knockout mice (the latter which are animal models used for research into genetic diseases). Nobel Prize winner.
SMITHSON, JAMES. 1765-1829. French (British) scientist, geologist, chemist, and writer: Founded the Smithsonian Institution (1846), which was named after him, as was the mineral smithsonite (zinc spar).
SMOLIN, YEFIM. Lived late 1600s-early 1700s. Russian glassmaker and inventor: Invented table-glass, an early type of faceted drinkware. Smolin is said to have proudly given one of his newly invented, multi-angled glasses to Peter the Great, asserting that it was unbreakable. Wanting to test the claim, the Russian emperor threw it to the floor in front of his guests, whereupon it shattered to pieces. Everyone had a

hearty laugh. Despite this, the Tsar continued to support Smolin and his invention, and here began the tradition of breaking glass for good luck. Known in Russian as *granyonyi stakan*, Smolin's faceted drinking glass design is still commonly produced, sold, and used around the world.

SNITZER, ELIAS. 1925-2012. American scientist, physicist, electrical engineer, laser pioneer, educator, and inventor: Invented the first fiber amplifier (1961) and the first fiber optical laser (1961), and co-demonstrated the first double clad fiber laser (1988). He was also the first to describe the single-mode optical fiber (1961), or SMF, and was involved in the development of the praseodymium fluoride glass fiber laser amplifier and the mask fabrication of fiber Bragg gratings. His many innovations led to a myriad of advancements in the fields of laser science, photonics, and fiber optics, impacting everything from the military to medicine—eventually spurring the invention of broadband networks. Snitzer is known as the "father of the glass laser."

SOENNECKEN, FRIEDRICH. 1848-1919. German businessman, entrepreneur, graphic artist, graphic designer, and inventor: He re-popularized the calligraphy type known as Round Script (*Rundschrift* in German), but he is best remembered as the inventor of the paper hole punch (1886) and the paper ring binder (1886). Founder of the German office products company Soennecken (1875).

SOMERSET, HENRY. 1847-1924. British peer, aide-de-camp to Queen Victoria, Hereditary Keeper of Raglan Castle, and justice of the peace: Though the origins of badminton appear to date back to early India, the development of the modern sport has been credited to Somerset, who is more commonly known as the 9th Duke of Beaufort. For in 1873 it was he who introduced the game we know today on the lawn of his estate, "Badminton"—from which the sport takes it name.

Badminton House, Gloucestershire, UK; the seat of the Duke of Beaufort, and the estate where the sport of badminton was invented in 1873.

SONG, SU. 1020-1101. Chinese polymath, scientist, mechanical engineer, politician, zoologist, pharmacologist, mineralogist, cartographer, mathematician, astronomer, metallurgist, horologist, hydraulic engineer, philosopher, geologist, botanist, poet, architectural engineer, author, ambassador, and inventor: Contributed to a myriad of scientific fields, but he is best known for his invention of the first power-producing chain drive—which he installed in another one of his many innovations: the hydro-mechanical astronomical clock tower. The cost-effective chain drive (better known now as a roller chain), with its ability to handle high torque and produce superior power, is still in use to this day, and is standard equipment on most bicycles and motorcycles. It is also used in modern forklifts, farming equipment, conveyor belts, cranes, and industrial assembly lines as well.

SOUTHERN, EDWIN. Born 1938. English scientist, molecular biologist, biochemist, educator, and inventor: Invented the Southern Blot (1973), an accurate method for analyzing and measuring DNA. His work has helped advance the field of genomics.

SOUTHWICK, ALFRED P. 1826-1898. American engineer, steam boat pioneer, dentist, educator, and inventor: Credited with inventing the electric chair (1881), touted at the time as a more civilized and ethical form of execution compared to hanging. "Old Sparky" is still used in a number of American states.

SPENCER, PERCY. 1894-1970. American scientist, physicist, electrical engineer, and inventor: Invented the microwave oven (1945), a now globally common kitchen appliance that cooks and heats food using high-frequency electromagnetic waves. Spencer's invention has revolutionized both the cooking and the appliance industries.

SPERRY, ELMER AMBROSE. 1860-1930. American industrialist, engineer, businessman, entrepreneur, and inventor: Made numerous improvements to already existing technologies, such as the arc lamp, the dynamo, streetcars, locomotive engines, searchlights, electric automobiles, metal processing, and electric wire. However, he is best remembered for his contributions to gyroscope design (circa 1910), which were successfully used on ships, torpedoes, and aircraft. The owner of some 400 patents, he was the founder of the Sperry Electric Mining Machine Company (1888), as well as at least seven other companies.

SPINETTI, GIOVANNI. Lived 1500s. Italian musician and luthier: Invented the clavichord (early to mid 1500s), a primitive keyboard instrument and forerunner of the pianoforte or piano. The clavichord is comprised of tangents (metal blades) connected to the keys of the keyboard. When a key is pressed, its tangent strikes a metal string, producing a soft, pleasing, but single-volume tone. Because it lacked the decibels and pressure sensitivity needed for live performance, the clavichord was used mainly for composition and practice The modern day word spinet (a small keyboard) probably derives from his surname.

SPIRIDONOV, VIKTOR A. 1882-1944. Russian sports innovator, martial arts researcher, and coach: Credited with co-developing Sambo (early 1920s), a unique martial art form that blends elements of wrestling, judo, and jujitsu, as well as various folk wrestling techniques and styles. The profession, whose name is an acronym for the Russian term *samozashchita bez oruzhiya* ("self-defense without weapons"), is generally divided into two categories: Sport Sambo (intended mainly for competition) and Combat Sambo (primarily intended for self-defense).
SPITZER, LYMAN. 1914-1997. American scientist, astronomer, physicist, astrophysicist, mountaineer, educator, and inventor: Helped develop sonar technology (mid 1940s); but more importantly he conceived the idea of the space-orbiting telescope (1946), inspiring the construction of the Hubble Space Telescope (1990)—the world's first sophisticated optical space observatory. Spitzer is also known for his invention of the stellarator (1951), an innovative but unsuccessful attempt to harness thermonuclear fusion for the peaceful production of electricity.
SPRAGUE, FRANK J. 1857-1934. American electrical engineer, businessman, military officer, and inventor: A specialist in both vertical and horizontal transport, among his many contributions were his work on electric elevators, electrical light systems, the electric motor, and power distribution systems. He is best remembered though for his development of the world's first commercially viable electric street railway (1887). Founder of the Sprague Electric Railway and Motor Company (1884). Sprague is known as the "father of electric traction."
STALLMAN, RICHARD. Born 1953. American software programmer and developer, author, and inventor: A promoter of the notion of free software, he is credited with inventing the UNIX-like GNU operating system (1983), a free, open-source software collection. Founder of the Free Software Foundation (1985).
STAMPFER, SIMON RITTER VON. 1792-1864. Austrian scientist, physicist, mathematician, astronomer, optics researcher, geodeticist, surveyor, topographer, educator, and inventor: Invented the stroboscope (circa 1832), an early type of animation technology that used a rotating cardboard disk to produce the appearance of movement when held up to a mirror. The word "stroboscopic" derives from Stampfer's invention, a device that led to advancements in what would later become the motion picture industry. Like Joseph Plateau (who, independently and simultaneously, developed a type of stroboscope), Stampfer is known as the "father of cinema."
STANFORD, LELAND. 1824-1893. American businessman, entrepreneur, philanthropist, politician, and California governor: Co-founder of California's Stanford University (1885), which was named after his son Leland Stanford, Jr.
STAREVICH, WŁADYSŁAW. 1882-1965. Polish film director, cinematographer, film director, film producer, film editor, film art

director, animator, illustrator, production designer, screenwriter, author, and inventor: Respected in his day as a "film magician," this independent filmmaker is credited with inventing puppet animation, hence the term for which he is best known: the "father of stop-motion animation."

STARKWEATHER, GARY K. 1938-2019. American physicist, optics specialist, engineer, lecturer, and inventor: Credited with inventing the world's first laser printer (1969), as well as developing color management printing technology. He also aided the film industry with his work on digital effects and color film scanning.

STARLEY, JOHN KEMP. 1855-1901. English industrialist, engineer, designer, businessman, entrepreneur, and inventor: Though he did not invent the bicycle, he is responsible for creating the modern bicycle (1884), which replaced earlier oversized models known as high wheelers. Also called "penny farthings," these attractive but ungainly pedal-powered Victorian bikes had tiny handlebars and a seat located over a massive front wheel—and set at a height that required a ladder to mount. These

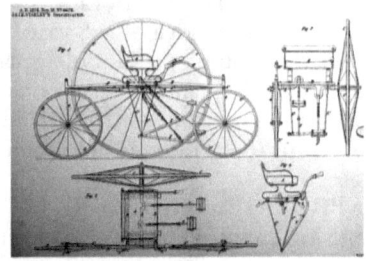

An 1877 patent drawing of one of Starley's "velocipedes," this one of a three-wheel bicycle "suitable for ladies' use."

characteristics made high wheelers difficult to manage, frightening, and even dangerous, for all but the most fit and athletically inclined. By greatly reducing the forward wheel size, enlarging the handlebars, and adding a chain drive to the back wheel and a spring-cushioned seat set further back on the frame, Starley's much more comfortable bike (brand name "Rover") made bicycling safe and accessible for all ages and both sexes. With its solid reputation as a "safety bicycle," the Rover became the first commercially viable bicycle, opening the door to the development of the fully modern bicycle we enjoy today. Starley, a co-founder of the Starley and Sutton Company (1877), is known as the "father of the modern bicycle."

STEPHENSON, GEORGE. 1781-1848. English mechanical engineer, civil engineer, railway pioneer, and inventor: With the aid of his son Robert, he introduced steam power to rail locomotives and was responsible for inaugurating both the first rail transport line (1825) and the first intercity railway line (1830). Stephenson, also credited with inventing the Stephenson Safety Lamp (1816) for the mining industry, is known as the "father of the modern railway."

STEVENS, JOHN. 1749-1838. American attorney, engineer, U.S. Revolutionary War officer, and inventor: Credited with inventing the first American internal combustion engine (1798). His lifelong work in the field of steam engine design also deserves mention, as it helped pave the way for the further development of both locomotive and water-going ferry technologies.

STEVENS, ROBERT LIVINGSTON. 1787-1856. American engineer and ship designer: Invented the inverted-T rail (1830) and the first railroad spike (1832), both which revolutionized the railroad industry.

STIHL, ANDREAS. 1896-1973. Swiss engineer, businessman, and inventor: Invented the world's first electric chain saw (1926). Stihl, founder of Andreas Stihl AG and Company (1926)—known more commonly today as Stihl—is regarded as the "father of the chainsaw."

STIRLING, ROBERT. 1790-1878. Scottish engineer, clergyman, and inventor: Invented the Stirling Engine (1816). The first practical hot-air motor (i.e., powered by heated air), it was more efficient than steam powered engines; yet it never caught on and was eventually outpaced by both steam and later internal combustion technologies.

STODOLA, AUREL. 1859-1942. Slovakian scientist, physicist, mechanical engineer, educator, author, and inventor: Invented the gas turbine (1939), an electricity-generating machine that is still in use today—particularly in the shipping industry, the aircraft industry, and the power plant industry. He is also credited with co-inventing the Stodola Arm (1915), the world's first artificial limb.

STOKES, WILFRED. 1860-1927. English engineer, bridge designer, and inventor: Invented the modern mortar (1857), known as the Stokes Mortar. This short smoothbore military gun was capable of firing shells at sharp high angles, at low speed, and over short distances.

STOLETOV, ALEKSANDR G. 1839-1896. Russian scientist, physicist, engineer, educator, author, and inventor: Invented the first photoelectric device using a photoelement (1888), which helped lead to the development of the modern solar cell.

STONE, MATT. Born 1971. Producer, animator, writer, actor, and musician: Co-created the hit TV series *South Park* (1997).

STRAUB, AMBROSE. Dates unknown. American physician and inventor: Invented the first peanut butter making machine (1903), which allowed peanut butter to be mass-marketed for the first time.

STRAUSS, LEVI. 1829-1902. German (Bavarian) industrialist, businessman, entrepreneur, tailor, designer, and inventor: Invented blue jeans (1871), the world's first durable, riveted, denim clothing. Founder of Levi Strauss and Company (1853) in San Francisco, California (still in business). Strauss' textile innovations transformed our culture, while greatly impacting both the clothing industry and the fashion industry.

STRINGFELLOW, JOHN. 1799-1883. English engineer, businessman, entrepreneur, and inventor: Stringfellow's work on steam engines, hot air balloons, air gliders, and steam powered aircraft—in particular his transport plane or "aerial steam carriage" (1842), helped lead to numerous advancements in modern aeronautics and aviation.

STROUSTRUP, BJARNE. Born 1950. Danish computer scientist and program designer: Invented the C++ programming language (1979), the primary language for nearly all of today's operating systems, Internet browsers, software applications, and game engines.

STROWGER, ALMON B. 1839-1902. American undertaker, educator, and inventor: Also a cavalryman in the American Union army, he is credited with inventing the automatic telephone exchange (circa late 1800s), revolutionizing the telephone industry with advancements that were in use well into the 1970s. Founder of the Strowger Automatic Telephone Exchange (1891).

STRUB, EMIL V. 1858-1909. Swiss railway engineer, mechanical engineer, railway pioneer, railway developer, and inventor: Invented the Strub Rack Railway System (1896). An improvement on earlier rack railway designs, Strub's version incorporated many traditional features, such as a toothed flat bottom rail (the rack), cog wheels, and a train-mounted pinion. Essentially rack and pinion rail systems lock trains to the rails, preventing slippage while aiding them in ascending and descending steep inclines. Used mainly in hilly and mountainous regions.

STUBBS, JOSEPH. Dates unknown. British inventor: Invented the first adjustable wrench (circa 1850), known as a spanner in the United Kingdom. To this day this globally popular implement can be found in most tool boxes and is considered an indispensable piece of equipment in machine shops and on construction sites.

STURGEON, WILLIAM. 1783-1850. English electrical engineer, lecturer, writer, and inventor: Invented the first electromagnet (1824), a device in which electricity is sent through a copper wire-wrapped piece of iron, producing a strong magnetic field. Electromagnets are used in a variety of important applications, from scientific instruments, hard disk drives, and loudspeakers, to engines, generators, and medical equipment, among many others.

SUÁREZ, RENÉ NÚÑEZ. Born circa 1945. El Salvadorian engineer: Invented the Turbococina ("Turbo-cooker"), a stainless steel cookstove equipped with an electric fan that governs fuel and gas levels. Considered more environmentally friendly than traditional wood stoves due to its lower fuel requirements.

SUGDEN, DAVID B. Dates unknown. Nationality unknown. Engineer: Credited with patenting a toroidal marine propeller (date unknown), a type of propeller with enclosed, curved, ring-shaped blades. The toroidal design improves efficiency, reduces noise, increases air flow, cuts fuel consumption, and reduces cutting damage, while increasing the strength and durability of the propeller. It has numerous modern applications, from helicopters, drones, and boats, to wind turbines, submarines, and household fans.

SUKHOI, PAVEL. 1895-1975. Russian aircraft designer, aircraft builder, aircraft engineer, aerospace engineer, and inventor: Invented the Su-series fighter aircraft series, including the noted Sukhoi Su-7 (1955), a supersonic fighter-bomber that was used to protect and defend the Soviet Union for many decades. Founder of the Sukhoi Design Bureau (1939).

SUNDBÄCK, GIDEON. 1880-1954. Swedish electrical engineer and inventor: While he did not invent the zipper (which was originally conceived by Whitcomb L. Judson), he is credited with inventing the modern zipper (1909), the familiar device that is now used in hundreds of modern applications.

SUNG-GI, RI. 1905-1996. North Korean chemist and inventor: Invented Vinylon or Vinalon (1939). A manmade fiber that is used chiefly in Asia as a cotton and nylon replacement, it has countless applications ranging from thread, clothing, fishing nets, cement reinforcement, bedding, and paper, to shoes, films, ropes, batteries, resins, and filters.

SUSHRUTA. Lived circa 600 BC. Indian physician, surgeon, researcher, author, educator, and inventor: An early wholistic doctor, Sushruta is credited with being the first to perform plastic surgery (such as rhinoplasty), as well as cataract surgery, lithotomy, and surgical kidney stone removal, among many other procedures. He was also the first doctor known to recommend exercise to his patients, and held that physical movement could rid one of numerous ailments, including diabetes and obesity. He made important contributions to the fields of pharmacology, diet and nutrition, pediatrics, physiology, toxicology, dissection, and Ayurveda (*ayur*, "life," and *veda*, "knowledge")—a traditional, wholistic, body-healing system that originated in India. Sushruta is known as the "father of surgery."

SUTTON, THOMAS. 1819-1875. English photographer, businessman, architect, editor, author, and inventor: Invented the earliest known wide angle lens and camera (1859) as well as the single-lens reflex camera (1861). He was also involved in creating the first color photograph (1861).

SVEDBERG, THEODOR. 1884-1971. Swedish scientist, chemist, educator, author, lecturer, and inventor: Invented the ultracentrifuge (1924), a machine that separate particles by spinning them at high speeds, allowing them to be studied and analyzed—a process known as ultracentrifugation. Nobel Prize winner.

SWAN, JOSEPH WILSON. 1828-1914. English scientist, physicist, chemist, businessman, and inventor: Invented the first known incandescent light bulb (1860), a design that not only later inspired Thomas Edison (and others), but which helped usher in the era of modern technology, permanently altering the world in countless ways. Swan's other innovations include bromide paper (photograph development) and advancements in the textile industry.

Joseph W. Swan.

SWANSON, ROBERT. 1905-1994. Canadian mechanical engineer, researcher, businessman, railway specialist, horn and whistle developer, poet, and inventor: Invented the first five- and six-note air-horns (1949), which are still used on rail locomotives to this day.

SYCKEL, SAMUEL VAN. Dates unknown. American oil worker and inventor: Invented and constructed the first known commercially viable oil pipeline (1865), transforming the modern world in countless ways.

SYLVESTER, WALTER. 1867-1944. English engineer and inventor: Credited with inventing the "Sylvester" (1895), a device that was used to remove pit props (the lumber support beams in mines) without endangering miners or causing mine collapses. Having been replaced by newer and safer technologies, the Sylvester is no longer used.

SYMINGTON, WILLIAM. 1763-1831. Scottish engineer and inventor: Invented a unique rod and crank engine (1801) which he used to power the first paddlewheel steamboat, the *Charlotte Dundas* (1802). Simultaneously, Symington also became the inventor of the first inboard boat motor.

SYROMYATNIKOV, VLADIMIR. 1933-2006. Russian aerospace engineer: Best known for inventing the Androgynous Peripheral Attach System (1975), a successful docking mechanism used for spacecraft that is still in use.

SZE, SIMON MIN. 1936-2023. Chinese electrical engineer and inventor: Co-invented the metal oxide semiconductor field-effect transistor (1967), or MOSFET, a floating gate transistor (FGT) that serves as the foundation of non-volatile memory cells. Also known as a FGMOS, this vital electronic device is a key digital element used in a wide variety of applications, including flash memory, analog storage, neural networking, signal switching, integrated circuits, thin-film transistors, amplification, and digital potentiometers.

SZILÁRD, LEÓ. 1898-1964. Hungarian scientist, physicist, molecular biologist, educator, author, and inventor: Co-developed research that allowed for the invention of the atomic bomb (late 1930s), which in turn sparked the founding of the Manhattan Project (1942-1945).

Men at work.

TAKÁTSY, GYULA. 1914-1980. Hungarian microbiologist, physician, and inventor: Invented the first microplate (1951). Also known as a well plate or microtiter plate, it operates as an alternative to the test tube and is used primarily by microbiologists, virologists, and serologists to clinically study, test and analyze cells, DNA, antibodies, various compounds, and nucleic acids.

TAN, HENN. Dates unknown. Singaporean businessman, entrepreneur, and inventor: Invented (singly or perhaps as the leader of a research team) the USB drive (circa 2000), or what is variously known as a thumb drive, flash memory, flash stick, or memory stick. His innovation obsolesed the floppy disk and helped advance technologies involving the storage, backup, retrieval, and transfer of computer data.

TANG, CHING W. Born 1947. Hong Kong chemist, chemical engineer, educator, and inventor: Co-invented the organic light emitting diode (1997). Better known as OLED, it is brighter, with less motion blur, faster pixel response time, and more overall power efficiency than LED technology. OLED, which can be made into flexible, foldable, stretchable, and rollable displays, is used in computers, flat-panel displays, cell phones, and televisions.

TARPENNING, MARC. Born 1964. American engineer, businessman, entrepreneur, and technologist: Co-founded Tesla, Inc. (2003), an electric vehicle company. Served as its first CFO.

TAUSCHEK, GUSTAV. 1899-1945. Austrian self-taught computer scientist, engineer, and inventor: Invented the first optical character recognition device (1929), or OCR, which he called the "Reading Machine." Tauschek is also credited with inventing magnetic drum memory (1928), an early type of computer memory. Though drum memory was used through the 1960s, it was eventually replaced by faster, more economical, more efficient semiconductor RAM technology.

TAYLOR, KENYON. 1908-1986. English scientist, electrical engineer, and inventor: An innovator of numerous devices, he is chiefly known as the inventor of the flip-disk display (1960s). Also known as the flip-dot display, it is comprised of small, flat, bicolored disks with a different color on each side (such as black and white). Programmed electromagnets cause the disks to "flip," displaying contrasting images and words. At one time this familiar form of dot matrix display was widely used, for example, on sports scoreboards, train station departure boards, and buses. Though still seen occasionally, the flip-disk display is slowly being obsolesed by LED technology, which offers both a sharper image and lower costs.

TELLEGEN, BERNARD. 1900-1990. Dutch physicist, electrical engineer, scientific advisor, educator, and inventor: Invented the pentode (1926), a type of vacuum tube with five electrodes. While the pentode was commonly used in computers, radios, musical instruments, musical instrument amplifiers, televisions, radar, and microphones through the 1960s, since then it has largely been replaced by transistor technology.

TELLER, EDWARD. 1908-2003. Hungarian scientist, nuclear physicist, theoretical theorist, educator, and inventor: Chief architect of the hydrogen bomb (1951) and a leader in the Manhattan Project (1942-1945). Teller is known as the "father of the H-Bomb."

TER-POGOSSIAN, MICHEL M. 1925-1996. Armenian scientist, medical physicist, nuclear medicine pioneer, educator, and inventor: A pacesetter in the field of cyclotron-produced radionuclides, he is best known for leading the development (some say improving on the already existing technology) of positron emission tomography (1960s). Also known as PET, it is an imaging methodology that creates three-dimensional images that are used primarily for diagnosing anatomical diseases, but also for clinical research and study. Ter-Pogossian is known as the "father of positron emission tomography."

TERRY, ELI. 1772-1852. American businessman, entrepreneur, clockmaker, and inventor: Though he did not invent the clock, he was awarded the first American patent for his clock design (1797), and was, more importantly, the catalyst behind the commercialization of clockmaking in the United States. Terry is known as the "father of the American clock industry."

TESLA, NIKOLA. 1856-1943. Serbian scientist, physicist, electrical engineer, mechanical engineer, electrician, lecturer, futurist, author, and inventor: An innovator of countless new, eccentric, and fabulous technologies, he is most famous for inventing the rotating magnetic field (1882), the induction motor (1883), the three-phase electric power system (1887), alternating-current dynamos and transformers (1883), terrestrial stationary waves (1899), and perhaps most notably, the Tesla Coil (1891). An instrument that produces high voltage electricity using electromagnetic induction, the Tesla Coil is still used in various industries, such as radio, oil, streetlight illumination, education, construction, and gas. Tesla's rivalry with Thomas Edison (who promoted the use of direct current, or DC, as opposed to Tesla's alternating-current, or AC) was scientifically important at the time (1880s-1890s). The pair's infamous "War of the Currents" was eventually settled on sheer practicalities: cost, efficiency, and potential usages. Today Tesla's AC electrical system and Edison's DC electrical system operate side-by-side, with both being used by different industries for different applications. Tesla's conviction that extraterrestrials are real (and that they are both communicating with us and controlling us) continues to cause chagrin in many parts of the scientific community; yet it has won him worldwide devotion among true believers.[23] Other controversies surrounded Tesla as well, including those stemming from such inventions as his earthquake maker, his flying machine, his drone, and his "death ray." Arguably foremost among these, however, was Wardenclyffe Tower, a proposed wireless transmission station located on Long Island, New York (eventually constructed in 1901). Also known as "Tesla Tower," the inventor intended using it to generate limitless energy by tapping into earth's natural electricity—which he then planned to give away for free. Extraordinarily, the grandiose scheme was soon cut short: Much of Wardenclyffe was suddenly torn down (1917), and the property was foreclosed five years later (1922). Derelict and vacated, the buildings were vandalized and looted, finally forcing the now financially incapacitated Tesla to completely abandon the project. What was behind all of this? While mainstream historians insist that it came down to funding and patent issues, conspiracy theorists maintain that Tesla's plan to disperse free unlimited energy to every country in the world posed a major threat to the globe's mega power-producing companies, sealing the tower's fate. This ultimately caused it to be demolished and the inventor's wireless transmission ideas banned and suppressed. Nikola, whose legendary status has been commemorated in the name of the popular electric vehicle manufacturer Tesla, is known as the "father of electricity."

Nikola Tesla.

An exhibit of some of Tesla's many motors.

TEVANIAN, AVIE. Born 1961. American computer scientist, software developer, software engineer, and inventor: Invented the Mach Operating System (beginning in 1985), the NeXTSTEP Operating System (1989), and macOS (1999)—later to become iOS (2007).

THEREMIN, LÉON. 1896-1993. Russian physicist, engineer, and inventor: Among his many innovations (such as the Rhythmicon, the world's first drum machine, 1931), he is most famously credited with inventing the eponymous Theremin (1920). Also known as the aetherphone, this odd electronic instrument consists of a wooden box with an antenna on each end. Skilled musicians (thereminists) can play musical melodies by waving their hands in the air between the two poles, which transforms their motions into sound frequencies (note that the Theremin itself is never physically touched during a performance). The instrument was most famously used in the films *Spellbound* (1945), *The Day the Earth Stood Still* (1951), and *The Lost Weekend* (1945), and by rock groups such as the Rolling Stones and Led Zeppelin.

THOMPSON, KENNETH LANE. Born 1943. American computer scientist and inventor: Credited with inventing (some say co-inventing) the UNIX operating system (1969). He is also the inventor of the B programming language (1969).

THOMSON, ELIHU. 1853-1937. English scientist, electrical engineer, chemist, businessman, entrepreneur, educator, and inventor: The owner of nearly 700 patents and an ardent supporter of the AC electrical system, he is credited with developing and/or improving upon such items as X-rays, generators, transformers, incandescent lights, welding equipment, arc lights, the wattmeter, and dynamos. Co-founder, with Thomas Edison, of what would later become General Electric (1892).

THOMSON, ROBERT WILLIAM. 1822-1873. Scottish civil engineer, businessman, entrepreneur, and inventor: Credited with inventing the world's first pneumatic or air-filled tire (1845)—some 46 years before John B. Dunlop became world famous for his patent on the same invention. Thomson's creation not only replaced solid rubber tires (with all of their many issues), but by greatly increasing traction, speed, cost-effectiveness, and comfort, his pneumatic tire completely transformed the car, truck, bicycle, motorcycle, and aircraft industries. Thomson is also known for his innovative work in electricity, steam cranes, steam traction, explosives, machinery, writing pens, flexible belts, road hauling, and railways.

THOMSON, WILLIAM. 1824-1907. Irish scientist, physicist, engineer, mathematician, astronomer, author, politician, and inventor: Considered one of history's greatest scientific synthesists, he spent his life working in countless fields, inventing, developing, and improving, contributing to humanity's welfare while helping carve out the modern world as we know it. He is best remembered for his invention of the Kelvin absolute temperature scale (1848), a measuring unit system in which "0 Kelvin" represents the absolute zero temperature at which all molecular and/or thermal energy terminates. The designation immortalizes his title of nobility: Lord Kelvin, or more technically, 1st Baron Kelvin. Known to have influenced scientists from nearly every branch of science, and even aiding in the evolution of physics, Thomson has rightfully been elevated to the auspicious ranks of Galileo, Nikola Tesla, Albert Einstein, Niels Bohr, Thomas Edison, Johannes Kepler, Alexander Graham Bell, Stephen Hawking, Charles Darwin, Isaac Newton, and Michael Faraday.

TIGERSTEDT, ERIC. 1887-1925. Finnish electrical engineer, early film and sound pioneer, businessman, entrepreneur, and inventor: Though he developed dozens of audio products, including microphones, amplifiers, hearing aids, and loudspeakers, he is best known for inventing sound-on-film technology (circa 1913), which ended the era of silent movies. Tigerstedt is known as the "Thomas Edison of Finland."

TIHANYI, KÁLMÁN. 1897-1947. Hungarian scientist, physicist, electrical engineer, and inventor: He made countless contributions to the fields of television and electronics, among them the first infrared video camera (1929), drone military aircraft (circa 1929), the electronic TV system (1926), and the plasma display (1936). Most importantly, he co-developed (some say improved) the cathode ray tube (1926). This last invention, which Tihanyi called the "Radioskop," opened the door for further developments in TV technology.

TIKHONRAVOV, MIKHAIL. 1900-1974. Russian aircraft engineer and inventor: Along with his work on interplanetary probes, intercontinental ballistic missiles, and high-altitude rockets, he is best remembered for being a co-developer of the first liquid propellant engine rocket (circa 1933), as well as the world's first artificial satellite (1957), known as Sputnik.

TIKHOV, GAVRIIL ADRIANOVICH. 1875-1960. Belarusian astronomer, astrophysicist, astrobotanist, astrocolorimetrist, photographer, educator, and inventor: Invented the feathering spectrograph (circa early 1900s), a unique photographic lens diaphragm that aids in determining the spectral classification of stars.

TILGHMAN, BENJAMIN CHEW. 1821-1901. American businessman, entrepreneur, chemist and inventor: Among his many innovations he is best known for inventing 1) sandblasting (circa 1870)—used today mainly for cleaning and restoring surfaces, and 2) paper production using sulfite pulping (1866)—a process still utilized to manufacture high quality paper.

TILL, JAMES. Born 1931. Canadian scientist, physicist, researcher, educator, and inventor: Co-described stem cells (circa early 1960s), an important medical breakthrough that allows for the regeneration of damaged cells and tissue, treatment of disease, drug testing and development, and pain reduction, as well as reducing recovery time and lowering the need for surgery.

TOMBAUGH, CLYDE. 1906-1997. American astronomer and

educator: Discovered what was once considered the ninth planet in our solar system: Pluto (1930). It has, however, recently been reclassified as a microplanet or dwarf planet.

TOMLINSON, RAY. 1941-2016. American computer scientist, computer programmer, and inventor: Sent the first computer-to-computer email, invented the at sign (@), and co-invented both Internet protocol, or IP (1974), and Transmission Control Protocol, or TCP (1974)—also known as TCP/IP.

TOMPION, THOMAS. 1639-1713. English mechanical engineer, clockmaker, watchmaker, and inventor: Among his many innovations he is best known for inventing the Tompion Regulator (1674), a type of balance spring regulation system used in pocket watches for some 200 years (up to the late Victorian Era). Tompion is known as the "father of English clockmaking."

TORRICELLI, EVANGELISTA. 1608-1647. Italian scientist, physicist, mathematician, educator, author, and inventor: His mathematical research aided in the development of integral calculus. However, he is mainly remembered for inventing the world's first barometer (1643), a device used to analyze atmospheric pressure.

TORVALDS, LINUS B. Born 1969. Finnish computer scientist, software developer and engineer, businessman, author, and inventor: Invented the Linux operating system (1991), a type of robust, reliable, efficient, open source software that continues to be used in numerous applications.

TRAEGER, ALFRED. 1895-1980. Australian electrical engineer and inventor: Invented the pedal radio (circa 1923), an environmentally friendly radio powered by a generator that is cranked using bicycle-like foot pedals. First developed for the Australian military, it was a boon to soldiers located in areas where electricity was inaccessible. Still in use.

TREVITHICK, RICHARD. 1771-1833. English mechanical engineer, mining engineer, explorer, and inventor: After developing high-pressure steam technology, he went on to invent the world's first steam rail locomotive (1803).

TRKMAN, FRANC. 1903-1978. Slovenian businessman, entrepreneur, and inventor: While credited with developing a host of innovations, he is most widely known for his invention of various types of electrical switches as well as the water-tight window.

TROPSCH, HANS. 1889-1935. German chemist and inventor: His most important contribution was the co-development with (Franz Joseph Emil Fischer) of the Fischer-Tropsch Process (1925), a complex technique for producing fuels (mainly gasoline and diesel) from biomass, hydrogen, natural gas, carbon monoxide, and coal. Still in use.

TROUVÉ, GUSTAVE. 1839-1902. French polymath, electrical engineer, businessman, entrepreneur, and inventor: Although the origins of the electric vehicle (car) are bathed in mystery and controversy, Trouvé is generally credited with inventing the first self-powered EV

Gustave Trouvé.

(1880). Activated by a small electric motor and a rechargeable battery, both mounted on a Victorian tricycle, the world's first electric car could attain a top speed of only 3 mph. As a public demonstration, Trouvé once drove his EV up and down a French avenue; but it aroused little interest at the time. His ideas, however, would have world-altering effects, lasting well into the 21st Century, when Tesla Motors unveiled its first EV: the lithium-ion battery powered "Roadster" (2008). Besides his EV, as well as numerous other innovations, Trouvé is also celebrated for inventing the first outboard motor (1881), which he created by attaching a portable battery-powered engine to a boat.

TRUFFAULT, J. M. M. Dates unknown (Victorian). French bicyclist and inventor: Credited with inventing the world's first shock absorber (1898). Known as *La Fourche Truffault* ("The Truffault Fork"), the fork-shaped device employed coiled springs and vibration dampening devices, and was first installed on bicycles. After partnering with American businessman Edward V. Hartford, the absorbers were outfitted for automobiles and renamed the "Truffault-Hartford Suspension." Cost for four car suspensions in the year 1905: $60.00.

TSIEN, ROGER Y. 1952-2016. American biophysicist, biochemist, educator, and inventor: Co-discovered (with Martin Chalfie and Osamu Shimomura) green fluorescent protein (1962). Also known as GFP, it is a substance taken from jellyfish that glows green under blue light. It is used as a marker protein in the observation, analyzation, monitoring, study, and mapping of various types of cells and cellular processes. Nobel Prize winner.

TSVET, MIKHAIL S. 1872-1919. Russian biochemist, botanist, educator, and inventor: Invented chromatography (1903), a biophysical method for separating, studying, and identifying complex compounds for research purposes.

TUPOLEV, ALEXEI A. 1925-2001. Russian aerospace engineer, aircraft designer, and inventor: Among his many aircraft designs he is best remembered for his invention of the first supersonic passenger jet, the Tupolev Tu-144 Rossiya (1955).

TURING, ALAN. 1912-1954. English computer scientist, logician, statistician, cryptographer, educator, and inventor: Among his many contributions to humanity, he is best known for inventing the Turing Machine (1936), an abstract mathematical device specifically modeled to understand the inner and outer boundaries of computational possibilities.

TWEDDELL, RALPH HART. 1843-1895. English mechanical engineer: Invented the portable hydraulic riveter (1871), a necessity on large worksites. While it is still used, it has been replaced in some industries with the electric, cordless portable riveter.

Jet aircraft production plant, 1957.

UDA, SHINTARO. 1896-1976. Japanese engineer, educator, and inventor: Co-invented the Yagi-Uda Antenna (1926). Also known as the Yagi Antenna, it is a directional, wireless antenna comprised of numerous metal rods arrayed in parallel fashion, an assembly that is attached to a pole. Used primarily for receiving local broadcast (mainly VHF and UHF) TV signals, and also for boosting ham radio signals, it will be most commonly recalled as the wire "fishbone" antenna mounted on millions of rooftops around the world. Although slowly being replaced in many areas by cable and satellite TV (which offer higher resolution and much greater channel selection), the Yagi-Uda Antenna still reigns supreme globally due to three items: it offers free TV, it is easy to install, and it is simple to use (simply point it in the direction of the desired signal).

UPTON, FRANCIS ROBBINS. 1852-1921. American scientist, physicist, mathematician, and inventor: A former assistant to Thomas Edison, Upton is responsible for inventing (some say co-inventing) the world's first electric fire alarm (1890), opening the way for the development of the modern fire alarm, a device that has saved countless thousands of lives.

URRY, LEWIS F. 1927-2004. Canadian scientist, chemical engineer, and inventor: Credited with inventing two devices that transformed the modern age: the alkaline battery (1959) and the lithium battery (1985). Urry's innovations revolutionized the world of electronics and opened the door to the development of the portable electronics industry. Urry is known as the "father of the modern battery."

UZELAC, TOMISLAV. Dates unknown. Croatian engineer,

programmer, video game designer, businessman, entrepreneur, and inventor: Invented the AMP MP3 Playback Engine, the first commercially viable MP3 player (1997).

Machine worker, 1928.

VANDERBILT, CORNELIUS. 1794-1877. American business tycoon, investor, entrepreneur, and philanthropist: Founder of the school that today bears his name: Vanderbilt University (1873).
VASSAR, MATTHEW. 1792-1868. English businessman, entrepreneur, brewer, and philanthropist: Founder of Vassar College (1861).
VEKSLER, VLADIMIR I. 1907-1966. Ukrainian scientist, physicist, and inventor: Invented the synchrotron (1944), a particle accelerator used chiefly by physicists, chemists, and biologists to perform high resolution atomic and molecular research on various materials. Note: The synchrotron was independently invented a year later by American physicist Edwin McMillan (1945).

Cornelius Vanderbilt.

VENN, JOHN. 1834-1923. English mathematician, logician, educator, clergyman, lecturer, author, and inventor: Invented the Venn Diagrams (1880). Based on a branch of logic known as set theory, a Venn Diagram is a graphic mathematical illustration using overlapping circles to visually demonstrate the logical relationships between two or more data sets or categories; often used in computer science, statistics, and linguistics.
VERITY, CLAUDE HAMILTON. 1880-1949. English engineer, film and audio pioneer, businessman, entrepreneur, and inventor: Among his many creations, he is perhaps best known for inventing the Veritiphone (1922), a sound-on-disk system that synchronized sound with film. His work helped break new ground in both the British cinema industry and the global film industry.

VERNEUIL, AUGUSTE VICTOR LOUIS. 1856-1913. French chemist, educator, and inventor: Invented the Verneuil Process (1902), the world's first successful method for producing manmade gemstones, such as synthetic rubies, emeralds, spinel, sapphires, and alexandrite.

VERNIER, PIERRE. 1584-1638. French mathematician, civil servant, and inventor: Invented the Vernier Scale (1631). Also known as the Vernier Caliper, it is a non-electronic, sliding ruler-like instrument that uses two graduated scales to make precise measurements. It is employed primarily by specialists and industries needing highly accurate linear measurements, such as science laboratories, machines shops, engineering firms, and manufacturing companies.

VIDI, LUCIEN. 1805-1866. French physicist and inventor: Invented the barograph (1844), a type of barometer that uses a pen to record changes in barometric pressure on a chart mounted on a slowly turning drum. The mechanically operated barograph is gradually being supplanted by electronic digital meteorological instruments.

VIEILLE, PAUL M. E. 1854-1934. French scientist, physicist, and inventor: Through his research into shock waves as well as the substance nitrocellulose, he invented smokeless gunpowder (circa early 1880s), which in the modern era has largely replaced regular gunpowder, known as "black powder."

VILLCHUR, EDGAR M. 1917-2011. American audio pioneer, audio engineer, educator, author, and inventor: Invented the acoustic suspension loudspeaker (1953). Capable of reproducing distortion-free bass frequencies for the first time, it transformed the audio, recording, and music industries. Villchur also made important contributions to the hearing-impaired and vision-impaired industries.

VIRTANEN, ARTTURI ILMARI. 1895-1973. Finnish biochemist, educator, and inventor: Invented AIV fodder (1929). An eponymous acronym (i.e., it represents the initials of his formal name), AIV fodder is a type of crop silage treated with dilute sulfuric acid or hydrochloric acid in an effort to slow down the deterioration process of fermentation. A great boon for ranchers, cattlemen, stockmen, and farmers, the AIV method not only helps preserve crop nutrients, it has numerous other benefits as well. For example, among such animals as cows, horses, poultry, sheep, pigs, and goats, it has been found to strengthen the immune system, increase fertility, and improve milk yield. Nobel Prize winner.

VOGT, RICHARD. 1894-1979. German aircraft engineer, mechanical engineer, aircraft designer, military pilot, and inventor: Known chiefly for his invention and promotion of the oblique wing design (1942), a variable aircraft wing system in which the entire wing pivots, causing one wing to sweep forward and the opposite one to sweep backward. While this slewed wing arrangement reduces drag at high speeds while increasing fuel efficiency, a number of issues remain unsolved. Thus it has yet to be successfully manufactured on a large scale for either military or commercial use.

VOLTA, ALESSANDRO. 1745-1827. Italian physicist, chemist, educator, author, and inventor: Invented (some say co-invented with Luigi Galvani) the first electric battery (1800). Volta's battery, called the "voltaic pile," made him a sensation. In honor of his achievement Napoleon I conferred upon him the aristocratic title "count" (1810), while later the scientific unit representing electromotive force was named a "volt" (1881). Volta is also remembered for his discovery of methane (1776), paving the way for the development of the natural gas industry and its associated branches: fertilizers, fuel, electricity production, heating, cooking, antifreeze, sanitation, and chemical manufacturing.

Alessandro Volta.

VONNEGUT, BERNARD. 1914-1997. American meteorologist, atmospheric scientist, physicist, chemist, theorist, educator, author, and inventor: Discovered (some say co-discovered with Vincent J. Schaefer and Irving Langmuir) a cloud seeding method using silver oxide (1946). Vonnegut's precipitation producing technique is still used globally to modify the weather, in particular, as a method for increasing both rain and snowpack. Brother of noted novelist Kurt Vonnegut, Jr.

VORHAUER, BRUCE W. 1941-1992. American businessman and entrepreneur. Invented the birth control sponge (1976), a contraceptive device for women.

VUČETIĆ, IVAN. 1858-1925. Croatian anthropologist, criminologist, police specialist, and author: Credited with inventing the science of dactyloscopy (1891). Better known as fingerprinting, ever since the late Victorian Era his system has greatly aided law enforcement agencies around the world in the identification, capture, and punishment of millions of criminals.

Turbo pump, 1909.

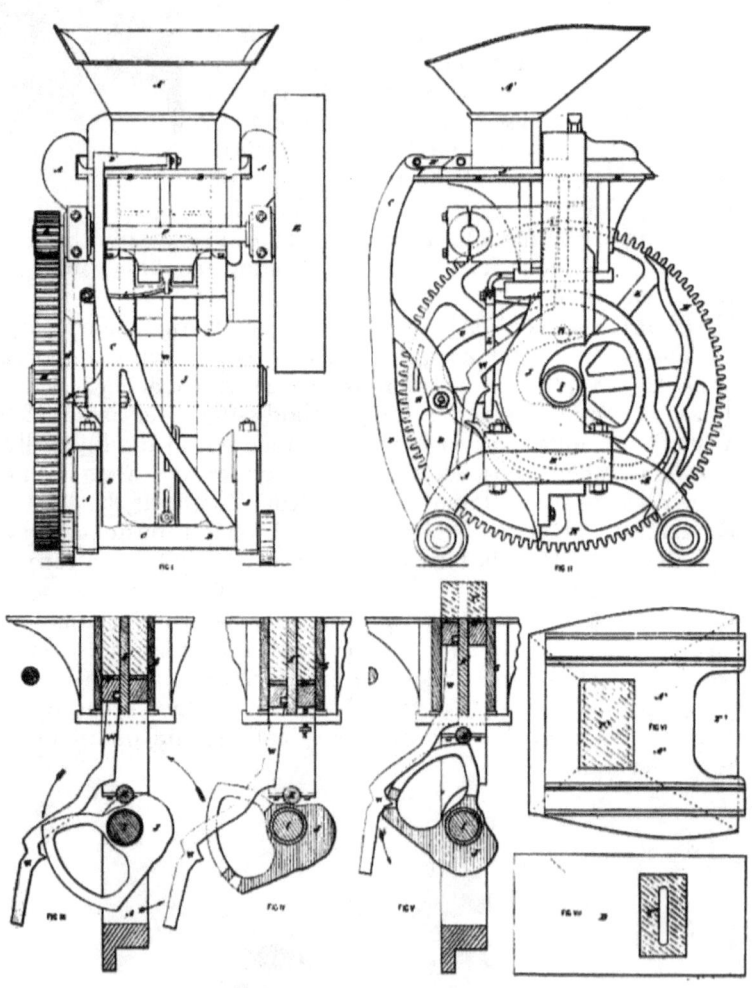

Patent drawing of a Victorian machine, 1857.

WALDEN, PAUL. 1863-1957. Latvian chemist, educator, scholar, and inventor: Invented the Walden Inversion (1896), a complex chemical reaction used by chemists in scientific fields such as pharmacology.
WALKER, ADAM. 1730-1821. English mechanical engineer, natural philosopher, author, lecturer, and inventor: Among his many innovations, which included musical instruments, mills, water control devices, plows, and wind-powered vehicles, he is best remembered for inventing the eidouranion (circa 1785). Walker's 27 foot wide planetarium, a unique type of orrery that featured a central sun orbited by the planets and the 12 astrological sun signs, was employed during his lectures to demonstrate his astronomical views. His shows and equipment were forerunners of the modern planetarium projection hall.
WALLIS, BARNES. 1887-1979. English scientist, physicist, aerospace engineer, and inventor: Among his many scientific innovations, which included contributions to supersonic flight, remote control aircraft, bomber geodetic designs, and swing-wing aircraft, he is most famous for inventing the bouncing bomb (1942). After the deadly depth charge-like device was dropped from a moving aircraft over water, the barrel-shaped explosive skipped along the surface until it reached its target (dam, boat, submarine, etc.), after which it sank and detonated. The bomb, which was specifically designed to "spin" over torpedo nets, was successfully used by Britain during World War II, and is memorialized in the dramatic film *The Dam Busters* (1955), which depicts the real-life heroic raid (using the bouncing bomb) that took place May 16-17, 1943.
WALSON, JOHN. 1915-1983. American businessman, entrepreneur, and inventor: Credited with inventing cable TV (1948), a technology

that revolutionized the television industry. Cable TV is slowly replacing both the antiquated "rabbit ear" antenna TV system and the Yagi-Uda antenna—the latter familiar to many as the "fishbone" antenna mounted on millions of rooftops around the world. At the same time, cable TV is being obsolesed by satellite TV, which itself, with the next generation of inventors, will be replaced by newer technologies.

WALTON, FREDERICK. 1834-1928. English businessman, manufacturer, chemist, entrepreneur, and inventor: Invented linoleum (1860), a floor covering made from natural materials (sawdust, cork, pine resin, linseed oil, minerals, etc.), that possesses numerous benefits: It is easy to clean, durable, contains low VOCs, and is both environmentally friendly and antimicrobial. Many reading this book will have grown up with linoleum, while the current generation is more likely to be familiar with its modern replacement: sheet vinyl flooring.

WARBURG, OTTO HEINRICH. 1883-1970. German physician, physiologist, oncologist, researcher, educator, and military officer: Discovered the cause of—and in consequence, the cure for—cancer (1930), which, in essence, he attributed to an anaerobic (low oxygen) and consequently highly acidic condition in the body.[24] The Otto Warburg Medal was named after him. Nobel Prize winner.

Otto H. Warburg.

WARD, MAURICE. 1933-2011. English hairdresser and inventor: Invented Starlite (1985), a proven, heat-resistant, thermal shielding material that was intended to be used primarily as a flame retardant (fire protection). According to some sources, however, Ward died without passing along the recipe for Starlite, making it impossible to manufacture the product commercially.[25]

WARREN, DAVID. 1925-2010. Australian scientist, engineer, chemist, educator, and inventor: Though he did not invent the idea of either the flight data recorder (FDR) or the cockpit voice recorder (CVR), he is responsible for creating the first data/voice combination recording device: the FDR/CVR (1958). According to the U.S. government an FDR records flight parameters, such as airspeed, heading, and altitude, while a CVR records sounds in the cockpit and radio transmissions. Warren's innovation—originally called the "black box," but now painted safety orange for greater visibility—has proven invaluable in the investigation of aircraft accidents.

WARREN, NEIL CLARK. Born 1934. American psychologist, theologian, businessman, educator, author, and entrepreneur: Founder of eHarmony (2000), a popular dating application.

WARTHIN, ALDRED SCOTT. 1866-1931. American pathologist, researcher, educator, author, and inventor: Among numerous other medical contributions, he is best remembered for co-developing (with

Allen Chronister Starry) the Warthin-Starry Stain (1920). Also called silver staining, it is a stain based in silver nitrate that is used primarily by histologists in the study of microscopic tissue structures.

WASHINGTON, GEORGE. 1732-1799. U.S. president, military commander-in-chief, surveyor, farmer, and inventor: Invented a sixteen-sided threshing barn (1794). Inside, horses were separated from the grain they were threshing by placing them on a circular platform. Because it prevented food grains from being contaminated by horse waste, this idea was a great agricultural advancement at the time. The city of Washington, D.C. (1789), the state of Washington (1889), and Washington and Lee University (founded 1749) were all named after him.[26]

George Washington.

WATERMAN, LEWIS. 1837-1901. American engineer, businessman, entrepreneur, salesman, educator, carpenter, and inventor: Invented the modern fountain pen (1884), which obsolesed dip pens and inkwells and revolutionized the handwriting industry. While less expensive ballpoint pens have now taken over the market, the humble fountain pen remains a cherished item by many, valued for its vintage aesthetic, comfort, smooth flowing ink, reusability, muscle fatigue reduction, improved handwriting, artistic possibilities, colored inks, and pure enjoyableness.

WATSON-WATT, ROBERT. 1892-1973. Scottish scientist, physicist, electrical engineer, educator, author, and inventor: Among his many contributions to science, which included, among other things, his research into cathode ray technology and electromagnetic radiation, he is mainly remembered for his invention of radar (1935). While radar had been described far earlier by various scientists (the first being Heinrich Hertz), Watson-Watt was the original developer of the first complete and commercially viable radar system—coming just in time to aid Britain during World War II.

WATT, JAMES. 1736-1819. Scottish scientist, physicist, mechanical engineer, chemist, scientific instrument maker, mathematician, land surveyor, businessman, entrepreneur, and inventor: An important contributor to a great many fields and industries, particularly those having to do with mechanical engineering. Though he did not invent the steam engine, he is credited with greatly improving upon it with his invention of the Watt Engine (1769), which utilized his two most famous innovations: the separate condenser (1765) and parallel motion (1784). The former of the two, a separate tank that helped reduce heat loss while maximizing fuel and engine efficiency, was an outstanding technological leap forward, one lacking in earlier steam engine

James Watt.

designs. Watt, who was the founder of the company Boulton and Watt (1794), was a key player in the opening up of the Industrial Revolution (1760-1840). Indeed, so profound were his influences that many believe that 1784, the year he invented parallel motion, marks the beginning of the Anthropocene Epoch (the "time period of Man")—the era in which humans began to make significant and irreversible changes to the natural environment.

WAYNE, RONALD G. Born 1934. American businessman, entrepreneur, and electronics industry executive: Cofounder (with Steve Wozniak and Steve Jobs) of the Apple Computer company (1976).

WEBB, JAMES EDWIN. 1906-1992. American government bureaucrat and attorney: Served in numerous official capacities, among them Director of the U.S. Bureau of the Budget (1946-1949), military officer (Marines Corps), and perhaps most notably, second administrator of NASA (1961-1968) during the Kennedy and Johnson administrations. The James Webb Space Telescope was named after him (2002).

WEDGWOOD, JOSIAH. 1730-1795. English potter, businessman, entrepreneur, and inventor: The famed maker of fine European pottery and the grandfather of naturalist Charles Darwin, Wedgwood is credited by some with inventing the pyrometer (circa early 1780s), a contact-free instrument that measures temperature based on the infrared radiation an object releases. Founder of the eponymous Wedgwood Company (1759).

Josiah Wedgwood.

WEDGWOOD, THOMAS. 1771-1805 English chemist, photographer, photogramer, researcher, and inventor: The son of Josiah Wedgwood (above), many claim that he is the inventor of the first photograph (circa 1800). However, this is not technically true, for what Wedgwood created was not the first photograph, but the first photogram: an image produced without a camera using light-sensitive photographic paper. Here is the 19th-Century method he would have used to make a photogram: An object, like a flower, is taken into a darkroom, placed on photographic paper, and exposed to light for a few moments. What remains is a low resolution, black and white negative image of the flower. (Note: Modern color photograms can be produced by using color negative photographic paper.) Thus, as noted, what Wedgwood actually invented was the first (known) photogram, a precursor to the photograph. As there was no way at the time to preserve photograms, we have no evidence for his work outside of various writings in which he documented his experiments.

WELSBACH, CARL AUER VON. 1858-1929. Austrian scientist, chemist, physicist, engineer, and inventor: Chiefly known for three innovations: 1) The Welsbach Mantle (1885), a mesh made from fabric that is soaked in, for example, nitrates of cerium, magnesium, yttrium, and thorium oxides; when heated by flame the device produces a white incandescent light that is much brighter than those created by earlier illumination technologies. Welsbach's nonelectric gas mantle lamp, which he called "Auerlicht," revolutionized street lighting during the Victorian Era, and is still used to this day in camping lanterns and other types of lights. 2) Through his work on the mantle lamp he developed the metal filament bulb (1897), replacing the earlier carbon filament light system. Though once the most common type of electric light, metal filament bulbs are now being replaced by more efficient LED bulb technologies. 3) Welsbach's invention of ferrocerium (1903), a spark-producing, manmade alloy comprised of cerium and iron, also changed the world, as it is used in most lighters today; it is also familiar to campers, hikers, and outdoorsmen as a ferrocerium rod or "ferro rod." Used to create sparks for igniting fires, Welsbach's "spark stick" replaces matches and is particularly useful in survival situations.

WENSTRÖM, JONAS. 1855-1893. Swedish scientist, engineer, and inventor: Independently invented (along with several other individuals) the three-phase electric power system (1890), used to generate and transmit electricity, as well as the slotted armature (1880), a grooved rotating metal core used in generators, motors, and engines. Both innovations are still in common use around the world.

WESTINGHOUSE, GEORGE. 1846-1914. American industrialist, electrical engineer, businessman, entrepreneur, and inventor: Invented, among other products, a rotary steam engine (1865), an electrical railway air signal system, a water meter, the rail air brake (1869), and natural gas piping (1883), as well as making improvements in transformer and generator technologies. Founder of the Westinghouse Air Brake Company (1869) and the Westinghouse Electric and Manufacturing Company (1889). He was instrumental in promoting and establishing the electricity-producing alternating-current (AC) system in the U.S.

George Westinghouse.

WESTON, EDWARD. 1850-1936. English physicist, electrical engineer, electrician, chemist, businessman, entrepreneur, and inventor: Invented the first precise, direct current, portable voltmeter (1888), as well as the first direct-reading light meter for photographers (1932). A holder of nearly 350 patents, he also made improvements on the dynamo and generator, developed a saturated cadmium cell (1893) and a magnetic speedometer, founded the Weston Electric Light Company (1888), and was the inventor of constantan and manganin (1884)—two stable-resistance alloys that were used to increase the accuracy and stability of his large line of electronic measuring devices (e.g., voltmeters, frequency meters, ohmmeters, transformers, ammeters, transducers, and wattmeters).

WHEATSTONE, CHARLES. 1802-1875. English scientist, physicist, educator, and inventor: Invented the concertina (1829), a small accordion, the stereoscope (1838), a three-dimensional image viewing device, the Playfair cipher (1854), and the pseudoscope (1852), used in ophthalmological research.

WHEELOCK, ELEAZAR. 1711-1779. American clergyman and educator: Founded Dartmouth College (1769), becoming its first president. The school was named after one its benefactors, British statesman William Legge, the Second Earl of Dartmouth.

WHITCOMB, RICHARD T. 1921-2009. American aviation pioneer, aeronautical engineer, aerospace engineer, and inventor: Considered by NASA to have invented many of the most important aerodynamic advancements in aviation history. These include the winglet (1969), a small, vertical (turned up), drag-reducing airfoil mounted on the wingtip of aircraft; the supercritical airfoil (1970), a wave-drag reduction wing shape; and the area rule (1952), an aeronautical principle pertaining to wave drag and the longitudinal cross-section of aircraft.

WHITE, ANDREW DICKSON. 1832-1918. American businessman, entrepreneur, diplomat, scholar, lecturer, and educator: Co-founded Cornell University (1865), becoming its first president. The school was named after its other co-founder, American businessman Ezra Cornell.

WHITEHEAD, ROBERT. 1823-1905. English engineer, draftsman, and inventor: Based on designs by Giovanni Luppis, Whitehead invented the first successful self-propelled torpedo (circa 1865). Known as the "Whitehead Torpedo," it operated via a compressed air-powered engine and used a horizontal rudder (to control depth) and a gyroscope (to control direction). His invention, and the modern improvements that inevitably came later, revolutionized naval warfare.

WHITEHOUSE, CORNELIUS. 1795-1883. English scientist, engineer, businessman, and inventor: Invented a welding technique for producing wrought-iron gas pipes that improved on both efficiency and cost. Victorian-style wrought-iron pipes are notorious for corrosion and leaking issues, the main reasons they have been replaced today by steel and plastic pipe materials.

WHITNEY, ELI. 1765-1825. American mechanical engineer, businessman, manufacturer, entrepreneur, educator, and inventor: Best known for inventing the cotton gin (1794), an engine ("gin") that separated cotton seeds from their fibers using hand power, horse power, or steam power. By speeding up cotton production, the cotton gin greatly improved the profitability of the cotton trade. It also transformed farming as a whole, opened the door to developments in other related technologies, launched the idea of mass production, and aided in the technological explosion known as

Eli Whitney.

the Industrial Revolution (1760-1840). Whitney is known as the "father of American technology."

WHITTLE, FRANK. 1907-1996. English engineer, military officer, and inventor: Inventor (some say co-inventor) of the first turbojet engine (1930), which he bench-tested (though did not fly) seven years later (1937). Note: Maxime Guillaume is credited with patenting the first turbojet engine (1921), while Hans von Ohain is credited with flying the first turbojet engine (1939).

WICHTERLE, OTTO. 1913-1998. Czechoslovakian scientist, chemist, educator, and inventor: A specialist in organic synthesis, he is most famous for inventing both soft contact lens and the technique for producing them (1961). His invention, which was made from hydroxy ethyl methacrylate (HEMA), replaced earlier more rigid lenses that were made from polymethyl methacrylate (PMMA). Today soft contact lenses, which are more comfortable and inexpensive than earlier hard plastic designs, are worn by millions of people around the world.

WILKINSON, NORMAN. 1878-1971. English artist, illustrator, camoufleur, etcher, and inventor: Invented dazzle camouflage (1917). Used primarily during World War I and World War II, it was painted on hundreds of both British and U.S. ships as a method of evading detection by submarines. The colors and angles used were intended to break up a ship's silhouette, as well as mask its size, speed, and heading. Ship camouflage has largely been obsolesed by modern electronic surveillance technology. However, outside the marine military industry dazzle camo remains popular, and continues to be used, for instance, on clothing, and in architecture, art, structural interiors, and fashion.

WILSON, CHARLES THOMSON REES. 1869-1959. Scottish scientist, physicist, meteorologist, educator, lecturer, and inventor: Invented the Wilson Cloud Chamber (1911), a particle collector used in physics to detect and visualize ionizing charged particles. Among other advancements, the device was most notably used in the discovery of antimatter (1932), as well as the muon (1936). While the cloud chamber continues to be used by some educators, it has been replaced in modern laboratories by electronic instruments able to observe high-energy particles (a task that was outside the capabilities of Wilson's device). Nobel Prize winner.

WILSON, WILLIAM G. 1895-1971. American military officer: Most familiar to millions of people as "Bill W," one of the co-founders of Alcoholics Anonymous (1935). An addiction recovery organization devoted to conquering alcoholism, it has saved many lives and families.

WINCHELL, PAUL. 1922-2005. American character actor, film actor, voice actor, comedian, ventriloquist, and inventor: Working in conjunction with Henry J. Heimlich (Heimlich Maneuver), Winchell invented an implantable, mechanical, artificial heart (1956). Many of Winchell's design ideas were adopted by Robert K. Jarvik, who later developed the Jarvik 7: the first artificial heart to be successfully implanted in a human being (1982).

WINCHESTER, OLIVER F. 1810-1880. American businessman, entrepreneur, politician, and inventor: Founder of the Winchester Repeating Arms Company (1866). One of its most popular firearms is the Winchester Model 70 (1936), still known to this day as the "rifleman's rifle" and the "bolt-action rifle of the century."

WINGFIELD, WALTER CLOPTON. 1833-1912. Welsh military officer, tennis pioneer, ceremonial bodyguard to Queen Victoria, author, and inventor: Credited by many with inventing the sport of tennis (1874). While this claim continues to be debated, whoever the true inventor may turn out to be, as the owner of the original tennis patent Wingfield must, at the very least, be acknowledged as the founder of *modern* tennis—a game he first demonstrated (to Henry Petty-Fitzmaurice, Lord Lansdowne) as early as 1869. Furthermore, it was Wingfield who formalized the game, set up its rules, and codified regulations concerning rackets, nets, and balls, and who published the first book on the sport. Wingfield is thus rightfully known as the "father of lawn tennis."

WINKEL, DIETRICH NIKOLAUS. 1777-1826. Dutch inventor: Invented the first metronome (circa 1814), which he called the chronometer. The device, mechanical, electronic, and digital, is still in common use in households and music academies around the world.

WINOGRADSKY, SERGEI. 1856-1953. Russian-Ukrainian scientist, biochemist, microbiologist, bacteriologist, botanist, scholar, and inventor: Invented the Winogradsky Column (1880s). A transparent glass or plastic cylinder filled with sediment (e.g., mud) and various supplemental organic compounds (e.g., cellulose), it is an educational tool used to breed stratified microbial communities for study and analysis.

WINTERHALDER, ROBERT. 1866-1932. German innkeeper, farmer, and inventor: Credited with inventing the world's first ski lift (1908), revolutionizing the modern ski industry.

WIRTH, NIKLAUS. 1934-2024. Swiss computer scientist, electronics engineer, educator, and inventor: Invented the Pascal programming language (1970). Though widely employed in computer systems at the time of its development, it is used today primarily as an educational tool. Wirth named the program after French physicist Blaise Pascal.

WITHERSPOON, JOHN. 1723-1794. American Founding Father, clergyman, farmer, and educator: Co-founder of the College of New Jersey (1746), later renamed Princeton University (1896). Witherspoon, a signatory of the Declaration of Independence (1776), served as Princeton's sixth president.

John Witherspoon.

WOOD, ALBERT BALDWIN. 1879-1956. American scientist, engineer, and inventor: Specialized in pumping, sewage systems, and drainage, but is best known for his invention of the Wood Screw Pump (1913), a high volume drainage pump originally intended to solve flooding issues in New Orleans, Louisiana. Wood's efficient, robust technological designs are still in use around the world.

WOODLAND, NORMAN JOSEPH. 1921-2012. American engineer and inventor: Invented (some say co-invented) the barcode (1949), a two-dimensional, lined, machine-readable square version of the UPC code, that stores information that can be quickly accessed by scanners. With its ability to identify, manage, monitor, inventory, price, and track products (as well as people), barcode technology has revolutionized the commerce industry.

WOODMAN, NICK: Born 1975. American businessman and entrepreneur: Founded GoPro, Inc. (2002). More popularly known as GoPro, the company designs and manufactures action cameras, accessories, and software.

WOODS, GRANVILLE. 1856-1910. American self-taught mechanical engineer, electrical engineer, and inventor: The author of some 60 patents, including a trolley wheel and a telephone transmitter, he is best known for his invention of the Synchronous Multiplex Railway Telegraph (1887), a magnetic field system that permitted rail trains to communicate with train stations while in transit—increasing efficiency and safety. While Woods' Synchronous Multiplex Railway Telegraph has been replaced by newer technologies, the basic concept behind his invention has been carried over into modern communications systems.

WOOG, PHILIPPE-GUY E. 1932. Swiss inventor: Invented the world's first electric toothbrush (1954). Called the "Broxodent," it revolutionized the dentistry industry and has helped millions prevent cavities and gum disease.

WOZNIAK, STEVE. Born 1950. American businessman, entrepreneur, computer scientist, software developer, electrical engineer, actor, film producer, educator, investor, and inventor: Cofounder (with Steve Jobs and Ronald Wayne) of the Apple Computer company (1976), he invented the Apple I and II computers, early Macintosh concepts, and the CL 9 CORE universal remote, as well as a number of other devices and applications.

WRIGHT, JAMES HOMER. 1869-1928. American pathologist, microbiologist, bacteriologist, researcher, and inventor: Invented Wright's Stain (1902), used in blood analysis by physicians to identify, diagnose, study, and treat infections.

WRIGHT, JOHN LLOYD. 1892-1972. American architect, businessman, entrepreneur, and toy inventor: A son of preeminent architect Frank Lloyd Wright, as well as a notable architect in his own right, he was also the inventor of Lincoln Logs (1916), a still popular children's construction toy set comprised of miniature notched wood logs.

WRIGHT, ORVILLE. 1871-1948. American aviation pioneer, aircraft designer, aviator, pilot trainer, businessman, self-taught mechanic, publisher, printer, and inventor: While he did not invent either the idea of manned flight or the first aircraft, he is deservedly famous for being the first man, along with his brother Wilbur (below), to successfully fly an engine-powered, heavier-than-air aircraft in a sustained and controlled flight (1903). This particular plane, called the *Wright Flyer*, took off from Kill Devil Hills, near Kitty Hawk, North Carolina. A subsequent aircraft, their *Wright Flyer III*, became the world's first fully functional, fix-winged airplane (1905). To this day debate rages as to who truly owns the title of "the first to pilot a successful manned flight." Even while they were alive, the two lifelong bachelors were plagued with patent issues, personal feuds, public derision, lawsuits, competitors, professional ridicule, and scornful skeptics from around the world, all questioning the legitimacy of their claims. Despite this ongoing global resistance, mainstream history has preserved as official, legal, and authorized, the Wright Brothers' assertion that they were indeed the first to conquer the skies.

Orville Wright.

WRIGHT, WILBUR. 1867-1912. American businessman, entrepreneur, aircraft pilot, self-taught mechanic, writer, and inventor: While he did not invent either the idea of manned flight or the first aircraft, he is deservedly famous for being the first man, along with his brother Orville (above), to successfully fly an engine-powered, heavier-than-air aircraft in a sustained and controlled flight (1903). This particular plane, called the *Wright Flyer*, took off from Kill Devil Hills, near Kitty Hawk, North Carolina. A subsequent aircraft, their *Wright Flyer III*, became the world's first fully functional, fix-winged airplane (1905). To this day debate rages as to who truly owns the title of "the first to pilot a successful manned flight." Even while they were alive, the two lifelong bachelors were plagued with patent issues, personal feuds, public derision, lawsuits, competitors, professional ridicule, and scornful skeptics from around the world, all questioning the legitimacy of their claims. Despite this ongoing global resistance, mainstream history has preserved as official, legal, and authorized, the Wright Brothers' assertion that they were indeed the first to conquer the skies.

Wilbur Wright.

WYNNE, ARTHUR. 1871-1945. English journalist, editor, musician, and inventor: Inventor of the crossword puzzle (1913), a puzzle game in which answers to clues must be written (one letter at a time) across horizontally- and vertically-aligned squares.

XING, YI. 683-727. Chinese Buddhist monk: Invented the world's first mechanical clock (725). The elaborate timing device was powered by water slowly dripping on a wheel, which revolved fully once every 24 hours. It would be another 500 years before the first mechanical clock was invented in Europe, finally supplanting Xing's invention.

European Hughes Printing Telegraph, 1876.

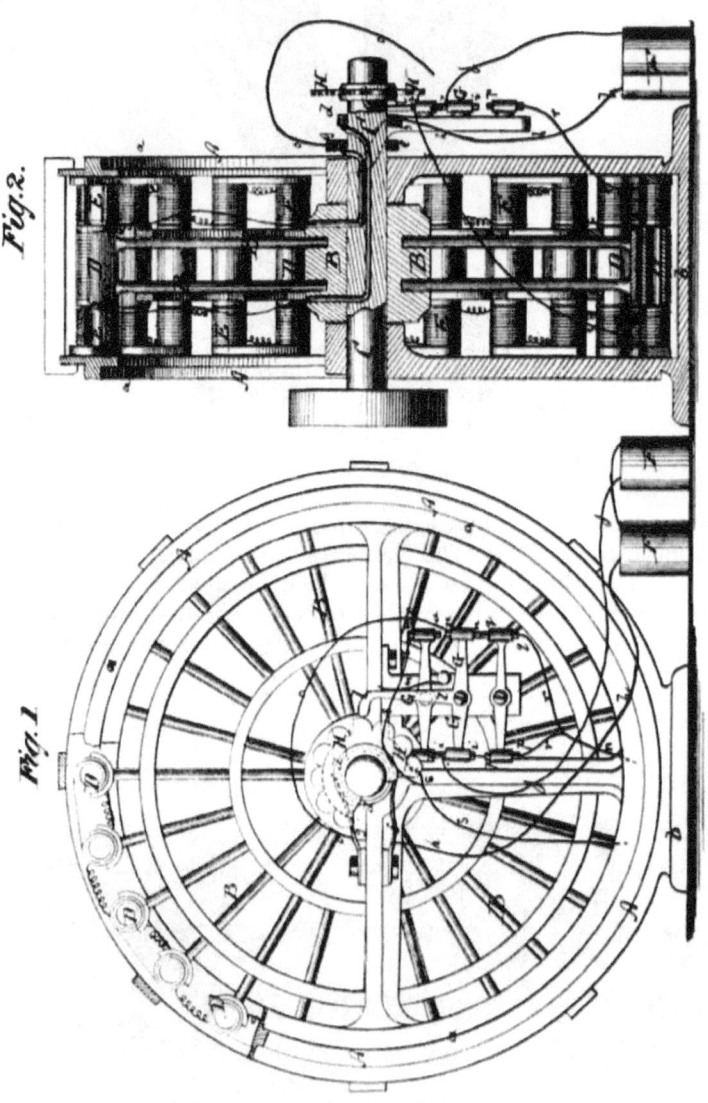

Patent drawing of electro magnetic motors, 1872.

YABLOCHKOV, PAVEL. 1847-1894. Russian electrical engineer, businessman, and inventor: Invented the Yablochkov Candle (1876), the world's first commercially successful arc lamp. Popular at the time throughout Europe's street lighting industry, it was eventually replaced by modern incandescent light technology.

YAGI, HIDETSUGU. 1886-1976. Japanese electrical engineer and inventor: Co-invented the Yagi-Uda Antenna (1926). Also known as the Yagi Antenna, it is a directional, wireless antenna comprised of numerous metal rods arrayed in parallel fashion, an assembly that is attached to a pole. Used primarily for receiving local broadcast TV signals (mainly VHF and UHF), and also for boosting ham radio signals, it will be most commonly recalled as the wire "fishbone" antenna mounted on millions of rooftops around the world. Although slowly being replaced in many areas by cable and satellite TV (which offer higher resolution and greater channel selection), the Yagi-Uda Antenna still reigns supreme globally due to three items: it offers free TV, it is easy to install, and it is simple to use (point it in the direction of the desired signal).

YALE, SR., LINUS. 1797-1858. American businessman, entrepreneur, metalsmith, and inventor: Father of Linus, Jr. (below), he is most noted for his invention of, among other lock designs, the modern pin tumbler lock (1848), still used in homes, doors, buildings, factories, cabinets, game machines, vending machines, loading docks, vehicles, washing machines, bicycle locks, safes,

Linus Yale, Sr.

padlocks, retail displays, and cabinets, to name but a few applications.
YALE, JR., LINUS. 1821-1868. American mechanical engineer, lock making pioneer, lock manufacturer, metalsmith, businessman, entrepreneur, portraitist, and inventor: Son of Linus, Sr. (above), he invented both the Yale Infallible Bank Lock (1851), the world's first cylinder lock (1861), and the world's first keyless combination lock (1862). Co-founder of the Yale Lock Manufacturing Company (1868). Yale's innovations, many of them based on lock designs developed thousands of years ago in ancient Egypt, broke new ground in lock technology, a majority of them still in use today.
YAMAKI, MICHIHIRO. 1933-2012. Japanese engineer, businessman, and entrepreneur: Founded the Sigma Corporation (1961). Although it produces various products (such as food items, HVAC equipment, and wastewater products), the company is best known today as a maker of cameras, lenses, and photography accessories.
YAMAUCHI, FUSAJIRO. 1859-1940. Japanese entrepreneur: Founded the company Nintendo (1889), which produced both the Nintendo Entertainment System (1985) and the popular video game Donkey Kong (1981), as well as many other successful games, systems, and franchises.
YAMAZAKI, SHUNPEI. Born 1942. Japanese physicist, engineer, and inventor: Among his 11,000 patents he is best known for his work involving display devices, electronic devices, liquid crystal display devices, cold fusion systems, and non-volatile memory.
YANAGISAWA, KATSUMI. Dates unknown. Japanese businessman: Founded the Pearl Musical Instrument Company (1946), which specializes in percussion instruments, and in particular, drums.
YAŞARGIL, GAZI. Born 1925. Turkish biomedical surgeon, neurosurgeon, educator, and inventor: Invented microneurosurgery (1960s), a surgical method that employs microscopes and precision medical instruments for operating on fragile nervous system structures, such as the brain and spinal cord, as well as vascular conditions. Yaşargil is known as the "father of microneurosurgery."
YAZU, RYŌICHI. 1878-1908. Japanese engineer and inventor: Invented the Yazu Arithmometer (1903), Japan's first mechanical desktop calculator.
YEONG-SIL, JANG. Circa 1390-1445. South Korean scientist, astronomer, metallurgist, weapons engineer, weaponsmith, technician, and inventor: Invented the first Korean water clock, a printing press, a rain gauge, a sundial, an armillary sphere, and the first known water gauge.
YOKOI, GUNPEI. 1941-1997. Japanese engineer, toy maker, video game designer, video game developer, video game producer, and inventor: Best known for his invention of "Game Boy" (1989), a popular handheld game console. He is also credited with inventing the cross-shaped "Control Pad" (1982) and the "Game and Watch" handheld

electronic game system (1980).
YOSHIDA, GORO. 1900-1993. Japanese businessman and entrepreneur: Co-founder of Canon, Inc. (1937), more popularly known as Canon. The company's name is an anglicization of the name of the Buddhist compassion-goddess Kwannon.
YOUNG, ARTHUR M. 1905-1995. American-French mechanical engineer, helicopter designer, astrologer, author, and inventor: Invented the Bell Helicopter series (1942), including the "Bell 47," one of the most popular military and civilian helicopters in history.
YOURKEVICH, VLADIMIR. 1885-1964. Russian ship building engineer, ship design architect, and inventor: Invented numerous cutting edge ship hull designs (early 1900s), most notably the French ocean liner SS *Normandie* (1935).
YUNUS, MUHAMMAD. Born 1940. Bangladeshi economist, businessman, entrepreneur, banker, and educator: Invented microfinance (1976), which provides financial services, such as small loans (microcredit), for low income earners and small businesses. Nobel Prize winner.
YUZPE, ABRAHAM ALBERT. Born 1938. Canadian obstetrician and gynecologist: Invented the Yuzpe Regimen (circa 1974), a type of emergency contraception that, if used within 72 hours of sexual intercourse, inhibits ovulation.

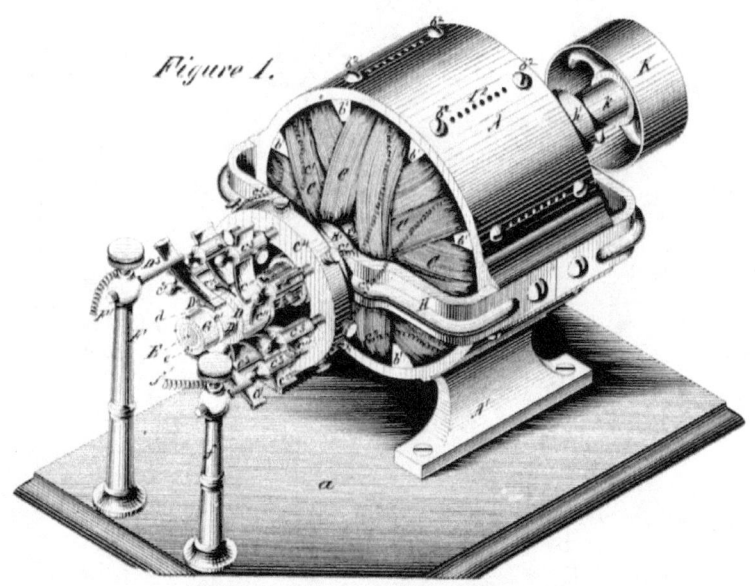

Patent drawing of a dynamo electric machine, 1878.

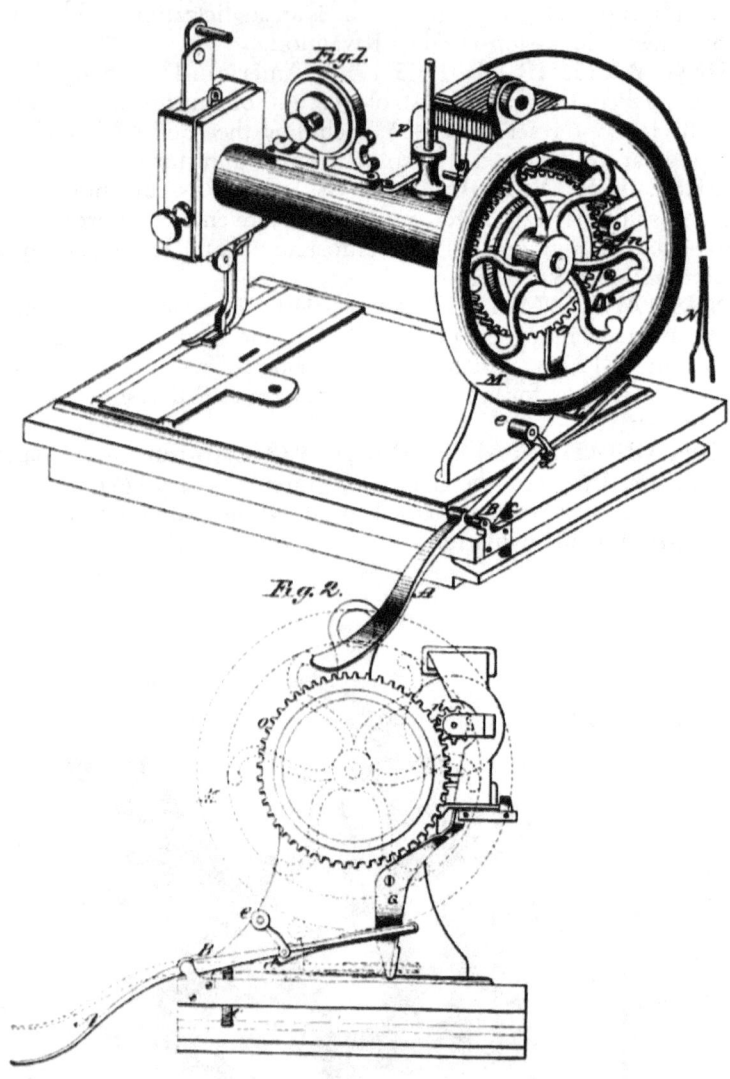

Electro Magnetic Sewing Machine Motor, 1881.

ZAMBONI, FRANK. 1901-1988. American businessman, entrepreneur, and inventor: Invented the Ice Resurfacer (1949). Popularly nicknamed the "Zamboni" after its creator, it is an ice-resurfacing machine that scraps and sweeps ice shavings, washes and mops the ice surface, and sprays a coat of water over the freshly cleaned surface. It is used today on most professional ice rinks and has become an attraction in its own right.

ZAMBONI, GIUSEPPE. 1776-1846. Italian physicist, clergyman, educator, author, and inventor: Invented the Zamboni Pile (1812), an early type of electrostatic battery. Used well into the 20th Century, Zamboni's innovation paved the way for new battery technologies.

ZAMENHOF, LUDWIK ŁAZARZ. 1859-1917. Polish ophthalmologist, poet, and inventor: Invented Esperanto (1887), a manmade language founded on European linguistics. Zamenhof created Esperanto as an international language, one meant to help break down national barriers and foster global unity. Today it is the most commonly used artificial language in the world.

ZAPP, WALTER. 1905-2003. Latvian camera pioneer and inventor: Invented the Minox (1936), a unique subminiature camera. Zapp's camera miniaturization technology, often used in the spy industry, laid new groundwork for small camera design.

ZAVOISKY, YEVGENY. 1907-1976. Russian physicist: Invented electron paramagnetic resonance (1994). Also known as EPR, it is a spectroscopic tool used by scientists in the fields of physics, chemistry, and biology.

ZELINSKY, NIKOLAY. 1861-1953. Russian chemist, educator, and inventor: Invented the world's first operative filtering coal gas mask (1915), a full-face protective mask used to prevent the inhaling of airborne toxins, such as smoke, chemicals, paint spray, or other types of contaminants. Used chiefly by law enforcement agencies, military forces, firefighters, and hazardous material workers.

ZEPPELIN, FERDINAND VON. 1838-1917. German aeronautic pioneer, German military officer, Union military observer (American Civil War), and inventor: The noted airship designer is credited with inventing the world's first successful, gas-filled rigid airship (1900). Known after its creator as a "Zeppelin," its strong, stable, metal framework made it ideal for military, commercial, and civilian transportation. His *Graf Zeppelin* became famous for being the first airship to circumnavigate the globe (1929). His company, Luftschiffbau Zeppelin (founded 1908), built the ill-fated hydrogen-filled *Hindenburg* airship, which crashed and burned in Lakehurst, New Jersey, May 6, 1937. Modern airships based on Zeppelin's designs still inhabit the world's airspace, one of the more familiar being the "Goodyear Blimp."

Ferdinand von Zeppelin.

ZERNIKE, FRITS. 1888-1966. Dutch scientist, physicist, chemist, mathematician, educator, and inventor: Discovered phase-contrast microscopy (1932), which led to his invention of the phase-contrast microscope (1938), a high resolution microscope used to observe and study living cells without the need of dyes or fixation. Nobel Prize winner.

ZHONGMING, TANG. 1897-1980. Chinese engineer, businessman, entrepreneur, and inventor: Invented an automobile equipped with a charcoal-powered internal combustion engine (1931). Zhongming's unique car was a commercial success in China for the following 20 years, but was eventually replaced by less expensive gasoline-powered vehicles.

ZHOU, JIAN. 1957-1999. Chinese scientist, immunologist, virologist, and researcher. Co-developed the HPV vaccine (1991), or human papillomavirus vaccine, a non-infectious vaccine for cervical cancer.

ZHUKOVSKY, NIKOLAI Y. 1847-1921. Russian scientist, physicist, aerospace innovator, mechanical engineer, and inventor: Developed numerous early theories involving aerodynamics, along with a number of innovations, his most famous being the eponymous "Joukowsky Airfoil" (1910)—an idealized hypothetical model used to study and understand complicated air flow principles. Zhukovsky is known as the "father of Russian aviation."

ZIEGLER, KARL. 1898-1973. German scientist, chemist, and inventor: Co-invented the Ziegler-Natta Catalyst (1955), a substance that permits alpha-olefins, such as ethylene and propylene, to be polymerized. The resulting product, high-density polyethylene (HDPE), has numerous applications in a wide variety of fields. These include: plumbing pipes, boxes, boat parts, food products, cutting boards, toys, gasoline tanks, plastic bottles and containers, motor oil, film wrapping, outdoor furniture, crates, detergents, marine buoys, housewares, chemical storage tanks, cosmetics, drainage systems, and shoes. Nobel Prize winner.

ZIEHL, FRANZ. 1857-1926. German bacteriologist, biologist, and inventor: Co-developed (with Friedrich Neelsen) the Ziehl-Neelsen Stain (1882), an acid-fast staining technique used to detect and identify acid-fast bacilli. The ZN stain is useful primarily for diagnosing tuberculosis and leprosy.

ZUSE, KONRAD. 1910-1995. German computer scientist, civil engineer, businessman, entrepreneur, author, and inventor: Invented the first programmable computer in the world (1941). Known as the "Z3," Zuse's computer revolutionized electronic technology and laid the foundation for the coming computer age.

ZVYOZDOCHKIN, VÄSILII. 1876-1956. Russian woodturner, doll maker, artist, and wood carver: Credited with co-inventing (with Sergei V. Malyutin) the Matryoshka Doll. More popularly known as Babushka ("Grandmother") Dolls or Russian nesting dolls, this set of hollowed out, painted wooden figures, which gradually decrease in size, "nest" inside one another, with the last or innermost doll (usually depicted as a baby), being the smallest.

ZWORYKIN, VLADIMIR. 1888-1982. Russian scientist, physicist, engineer, and inventor: Invented 1) the iconoscope (1923), the world's first electric TV camera tube, 2) the kinescope (1929), a cathode-ray based picture tube, and 3) an all-electric TV system (1932). Zworykin's innovations paved the way for a host of new technologies in the fields of electron microscopes, infrared-sensitive instruments, broadcasting, and digital media.

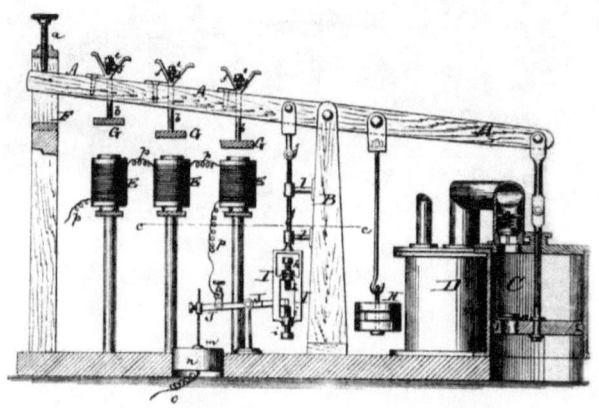

Electro Magnetic Engine, 1872

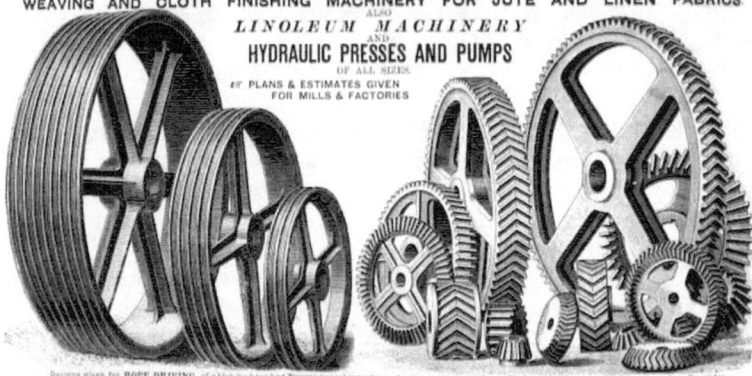

Ads from *Engineering Magazine*, Great Britain, 1888.

NOTES

ALL FOOTNOTES, ENDNOTES, & NOTES IN GENERAL ARE MINE, UNLESS OTHERWISE INDICATED. L.S.
(ALL BIBLE CITATIONS ARE FROM THE KJV.)

1. Karl G. Wilcox, "Job, His Daughters and His Wife," *Journal for the Study of the Old Testament*, Vol. 42, Issue 3, April 10, 2018, pp. 303-305.
2. For more on the American Civil War, or what I refer to as the War for the Constitution, see my many books on the topic, especially: *Everything You Were Taught About the Civil War is Wrong, Ask a Southerner!*
3. I own and play each of the three guitars listed in this entry.
4. For more on these topics, see my books: *Confederacy 101: Amazing Facts You Never Knew About America's Oldest Political Tradition*; *The Articles of Confederation Explained: A Clause-by-Clause Study of America's First Constitution*; *The Constitution of the Confederate States of America Explained: A Clause-by-Clause Study of the South's Magna Carta*; *America's Three Constitutions: Complete Texts of the Articles of Confederation, U.S. Constitution, and C.S. Constitution*; *All We Ask is to Be Let Alone: The Southern Secession Fact Book*; *Abraham Lincoln Was a Liberal, Jefferson Davis Was a Conservative: The Missing Key to Understanding the American Civil War*; *I, Confederate: Why Dixie Seceded and Fought in the Words of Southern Soldiers*.
5. According to ancient tradition, radiation can be health-giving in small doses. See our Sea Raven Press book, *The Hormesis Effect: The Miraculous Healing Power of Radioactive Stones*, by Jane G. Goldberg, PhD, and Jay Gutierrez, edited by Lochlainn Seabrook.
6. For more information on this topic see our Sea Raven Press book, *The Hormesis Effect: The Miraculous Healing Power of Radioactive Stones*, by Jane G. Goldberg, PhD, and Jay Gutierrez, edited by Lochlainn Seabrook.
7. I have spent many wonderful hours photographing and hiking in and around Hayden Valley.
8. For more information on these two battles, see my books: *Encyclopedia of the Battle of Franklin: A Comprehensive Guide to the Conflict that Changed the Civil War*; *The Battle of Franklin: Recollections of Confederate and Union Soldiers*; *The Battle of Spring Hill: Recollections of Confederate and Union Soldiers*; *The McGavocks of Carnton Plantation: A Southern History*.
9. In 1984, as a touring singer-songwriter-musician and recording artist, I was one of the first to purchase a Kurzweil K250 synthesizer.
10. For more on General Lee see my books, *The Old Rebel: Robert E. Lee as He was Seen by His Contemporaries*, and *The Quotable Robert E Lee: Selections From the Writings and Speeches of the South's Most Beloved Civil War General*.
11. For more on ancient theology, female spirituality, and goddess-worship see my academic books, *Britannia Rules: Goddess-Worship in Ancient Anglo-Celtic Society*; *The Book of Kelle: An Introduction to Goddess-Worship and the Great Celtic Mother-Goddess Kelle*; *The Goddess Dictionary of Words and Phrases: Introducing a New Core Vocabulary for the Women's Spirituality Movement*.
12. My books, including this one, have all been submitted to the Library of Congress.
13. I owned and performed on one of the earliest digital sampling synthesizers.
14. For more on the authentic history of America's two-party system and the progressive presidency of big government Left-winger Lincoln, see my book, *Abraham Lincoln Was a Liberal, Jefferson Davis Was a Conservative: The Missing Key to Understanding the American Civil War*.
15. For more on the true history of Abraham Lincoln see my books, *Abraham Lincoln: The Southern View*, *The Unquotable Abraham Lincoln*, *The Great Impersonator*, *Lincolnology: The Real Abraham Lincoln Revealed in His Own Words*, *The Unholy Crusade: Lincoln's Legacy of Destruction in the American South*, and *Twelve Years in Hell: Victorian Southerners Expose the Myth of Reconstruction, 1865-1877*.
16. These are some of the same reasons the American South (Conservative) seceded and took up arms against the Union (Liberal) in the mid 19th Century. For more on this topic see my books, in particular: *All We Ask is to Be Let Alone: The Southern Secession Fact Book*; *I, Confederate: Why Dixie Seceded and Fought in the Words of Southern Soldiers*; and *Lincoln's War: The Real Cause, the Real Winner, the Real Loser*.
17. "Electricity Explained: How Electricity is Generated," U.S. Energy Information Administration (EIA), October 31, 2023.
18. Carol Hodanbosi, "Pascal's Principle and Hydraulics," NASA, circa August 1996.

19. I have written over 50 evidence-based books on what has been misrepresented by the uninformed as the "Civil War," but which I more accurately term the "War for the Constitution." In particular see my books: *Abraham Lincoln Was a Liberal, Jefferson Davis Was a Conservative: The Missing Key to Understanding the American Civil War*; *Lincoln's War: The Real Cause, the Real Winner, the Real Loser*; and *Everything You Were Taught About the Civil War is Wrong, Ask a Southerner*.

20. For more on this topic see my books, *Everything You Were Taught About the Civil War is Wrong, Ask a Southerner!*; *Lincoln's War: The Real Cause, the Real Winner, and the Real Loser*; *Abraham Lincoln Was a Liberal, Jefferson Davis Was a Conservative: The Missing Key to Understanding the American Civil War*; and *Everything You Were Taught About American Slavery is Wrong, Ask a Southerner!*

21. Debate continues between mainstream science and alternative health science as to the benefits and detriments of consuming artificial vitamin C.

22. As a performing and recording artist throughout the mid to late 20th Century, I am quite familiar with Rickenbacker guitars, and have myself owned and played a number of different models.

23. For more on this topic see my books: *UFOs and Aliens: The Complete Guidebook*; *Mysterious Invaders: 12 Famous 20th-Century Scientists Confront the UFO Phenomenon*; and *The Martian Anomalies: A Photographic Search for Intelligent Life on Mars*.

24. For more on this topic see my books, *Jesus and The Law of Attraction: The Bible-Based Guide to Creating Health, Wealth, and Happiness Following Christ's Simple Formula*, and *Vitamin D: The Miracle Treatment for Nearly Every Disease and Health Issue*.

25. I was not able to find any definitive confirmation for or against this view.

26. Although Washington is regarded as "America's first president," this is not historically true. For more information see my book, *Confederacy 101: Amazing Facts You Never Knew About America's Oldest Political Tradition*.

BIBLIOGRAPHY

And Suggested Reading

"That as we enjoy great advantages from the inventions of others, we should be glad of an opportunity to serve others by any invention of ours; and this we should do freely and generously." — Benjamin Franklin

Alderman, Edwin Anderson, and Joel Chandler Harris (eds.). *Library of Southern Literature: Biographical Dictionary of Authors*. Atlanta, GA: The Martin and Hoyt Co., 1907.
Allen, James T. *Automobile Patents From 1789 to July 1, 1899*. Washington, D.C.: H. B. Russell and Co., 1900.
Allen, William. *An American Biographical and Historical Dictionary*. Boston, MA: William Hyde and Co., 1832.
Aminoff, Michael J., François Boller, and Dick F. Swaab (eds.). *Handbook of Clinical Neurology — Vol. 175: Sex Differences in Neurology and Psychiatry*. Amsterdam, Netherlands: Elsevier, 2020.
A New Biographical Dictionary, of 3000 Contemporary Public Characters, British and Foreign, of All Ranks and Professions. London, UK: George B. Whittaker, 1825.
Appleyard, Rollo. *Pioneers of Electrical Communication*. Freeport, NY: Books for Libraries Press, 1930.
Archer, Mark. *William Hedley: The Inventor of Railway Locomotion on the Present Principle*. Newcastle-Upon-Tyne, UK: self-published, 1882.
Bachman, Frank P. *Great Inventors and Their Inventions*. New York: American Book Company, 1918.
Ball, Robert S. *Great Astronomers*. London, UK: Isbister and Co., 1895.
Bally, Edward. *Industry and Manufactures in the United States: Look Out for Yourselves!* Boston, MA: Beacon Press, 1878.
Barlow, Charles, and Philip Le Capelain (eds.). *Patent Journal, and Inventors' Magazine*. Vol. 1, May 30, 1846, through November 21, 1846. London, UK: self-published, 1840.
Baron-Cohen, Simon. *The Essential Difference: Male and Female Brains and the Truth about Autism*. New York: Perseus Books Group, 2003.
Beale, Thomas William. *The Oriental Biographical Dictionary*. Calcutta, India: self-published, 1881.
Bellchambers, Edmund. *A General Biographical Dictionary: Containing Lives of the Most Eminent Persons of All Ages and Nations*. 4 vols. London, UK: Allan Bell and Co., 1835.
Beltzhoover, George M. *James Rumsey, the Inventor of the Steamboat*. Charleston, WV: West Virginia Historical and Antiquarian Society, 1900.
Bidwell, Benson. *Benson Bidwell, Inventor of the Trolley Car, Electric Fan and Cold Motor*. Chicago, IL: The Henneberry Press, 1907.
Blake, John L. *A Biographical Dictionary: Comprising a Summary Account of the Lives of the Most Distinguished Persons of All Ages, Nations, and Professions*. Philadelphia, PA: H. Cowperthwait, 1859.
Brewster, David. *The Life of Sir Isaac Newton*. London, UK: John Murray, 1831.
Brown, John Howard (ed.). *Lamb's Biographical Dictionary of the United States*. Boston, MA: Federal Book Company of Boston, 1903
Browne, Walter, and Frederick A. Austin (eds.). *Who's who on the Stage: The Dramatic*

Reference Book and Biographical Dictionary of the Theatre. New York: self-published, 1906.

Bryan, Michael. *Dictionary of Painters and Engravers: Biographical and Critical*. London, UK: George Bell and Sons, 1889.

Buhle, Mari Jo. *Women and American Socialism, 1870-1920*. Chicago, IL: University of Illinois Press, 1981.

Burke, Bernard. *A Genealogical and Heraldic History of the Peerage and Baronetage*. London, UK: Harrison and Sons, 1885.

Burt, Horace. *William Austin Burt, Inventor of the Typewriter, First Constructed in Any Country*. Chicago, IL: R.R. Donnelley and Sons Company, 1920.

Carnegie, Andrew. *James Watt*. New York: Doubleday, Page and Co., 1905.

Casson, Herbert N. *The History of the Telephone*. Chicago, IL: A. C. McClurg, 1910.

Chalmers, Alexander. *The General Biographical Dictionary: Containing an Historical and Critical Account of the Lives and Writings of the Most Eminent Persons in Every Nation; Particularly the British and the Irish, From the Earliest Accounts to the Present Time*. London, UK: self-published, 1812.

Chalmers, Patrick. *James Chalmers, the Inventor of the "Adhesive Stamp," Not Sir Rowland Hill*. London, UK: Effingham Wilson, 1884.

Chambers, Robert. *Lives of Illustrious and Distinguished Scotsman, Forming a Complete Scottish Biographical Dictionary*. 4 vols. Glasgow, Scotland: Blackie and Son, 1834.

Cleghorn, Thomas. *The Hydro-aeronaut, Or, Navigator's Life-buoy*. London, UK: J. M. Richardson, 1810.

Cochrane, Robert. *Heroes of Invention and Discovery: Lives of Eminent Inventors and Pioneers*. Edinburgh, Scotland: W. P. Nimmo, Hay, and Mitchell, 1897.

Cook, Frederick Francis. *Bygone Days in Chicago: Recollections of the "Garden City" of the Sixties*. Chicago, IL: A. C. McClurg and Co., 1910.

Cooper, Thompson. *A Biographical Dictionary: Containing Concise Notices of Eminent Persons of All Ages and Countries*. London, UK: George Bell and Sons, 1890.

Cooper, William Ricketts. *An Archaic Dictionary: Biographical, Historical, and Mythological; From the Egyptian, Assyrian, and Etruscan Monuments and Papyri*. London, UK: Samuel Bagster and Sons, 1876.

Daniel, Francis Colombine. *Memoir of Sir F. C. Daniel, Knt., M.D., Inventor of the Life Preserver, Used in Cases of Shipwreck, Bathing Etc*. London, UK: self-published, 1821.

Davies, George, Edmund Hunt, and Bristow Hunt. *Handbook for Inventors and Patentees: Containing Every Information as to the Protection of Inventions, the Registration of Designs, and the Obtainment of British, Colonial, and Foreign Patents*. London, UK: British Government, Office of Patents, 1868.

Deitch, JoAnne Weisman (ed.). *A Nation of Inventors: Researching American History*. Carlisle, MA: Discovery Enterprises, Ltd., 2001.

Derby, George. *The National Cyclopaedia of American Biography: Being an Analytical Study of American History and Biography*. New York: James T. White and Co., 1906.

Dickson, William Kennedy Laurie, and Antonia Dickson. *The Life and Inventions of Thomas Alva Edison*. London, UK: Chatto and Windus, 1894.

Dircks, Henry. *Inventors and Inventions*. London, UK: E. and F. N. Spon, 1867.

Doubleday, Russell. *Stories of Inventors: The Adventures of Inventors and Engineers—True Incidents and Personal Experiences*. New York: Doubleday Page and Company, 1904.

Evans, Joseph. *Biographical Dictionary of Ministers and Preachers of the Welsh Calvinistic Methodist Body*. Caernarfon, Wales: self-published, 1907.

Fairbairn, Patrick. *The Imperial Bible-Dictionary: Historical, Biographical, Geographical, and Doctrinal*. London, UK: Blackie and Son, 1866.

Faunce, Cy Q. *The Airliner and Its Inventor, Alfred W. Lawson, With a Summary of the*

Entire Aeronautical Movement. Columbus, OH: Rockcastel Publishing, 1921.
Forrer, Leonard. *Biographical Dictionary of Medallists*. London, UK: A. H. Baldwin and Sons, Ltd. 1904.
Foucaud, Édourd. *Les Artisans Illustres*. Paris, France: Bethune et Plon, 1841.
Foss, Edward. *Biographia Juridica. A Biographical Dictionary of the Judges of England, From the Conquest to the Present Time, 1066-1870*. London, UK: John Murray, 1870.
French, Gilbert J. *The Life and Times of Samuel Crompton: Inventor of the Spinning Machine Called the Mule*. London, UK: Simpkin, Marshall and Company, 1859.
Garbit, Frederick J. *The Phonograph and Its Inventor, Thomas Alvah Edison: Being a Description of the Invention and a Memoir of Its Inventor*. Boston, MA: Gunn, Bliss and Company, 1878.
Geffroy, Gustave. *Claude Monet: Sa vie, son Œuvre*. Paris, France: G. Cres and Co., 1924.
Gilbert, Frank. *Jethro Wood, Inventor of the Modern Plow: A Brief Account of His Life, Services, and Trials*. Chicago, IL: Rhodes and McClure, 1882.
Giles, Herbert A. *A Chinese Biographical Dictionary*. London, UK: Bernard Quaritch, 1898.
Glaisher, James, Camille Flammarion, W. De Fonvielle, and Gaston Tissandier. *Travels in the Air*. London, UK: Richard Bentley and Son, 1871.
Godwin, Parke. *The Cyclopædia of Biography: A Record of the Lives of Eminent Persons*. New York: G. P. Putnam's Sons, 1878.
Gordon, Timothy J. *The Case for Patriarchy*. Manchester, NH: Crisis Publications, 2021.
Gould, John. *Biographical Dictionary of Eminent Artists: Comprising Painters, Sculptors, Engravers, and Architects*. 2 vols. London, UK: Effingham Wilson, 1835.
Green, B. M. (ed.). *Who's Who and Why: 1919-1920*. Vancouver, CAN: International Press Limited, 1912.
——. (ed.). *Who's Who in Canada: An Illustrated Biographical Record of Men and Women of the Time: 1925-1926*. Vancouver, CAN: International Press Limited, 1925.
Gress, Carrie. *The End of Woman: How Smashing the Patriarchy Has Destroyed Us*. Washington, D.C.: Regnery, 2024.
Gruber, Helmut, and Pamela Graves (eds). *Women and Socialism, Socialism and Women: Europe Between the Two World Wars*. Oxford, UK: Berghahn Books, 1998.
Hale, Edward E. *Stories of Invention Told by Inventors and Their Friends*. Boston, MA: Roberts Brothers, 1889.
Hall, Julius. *Practical Suggestions to All Inventors and Patentees and Purchasers of Patents*. London, UK: self-published, 1872.
Hardie, James. *The New Universal Biographical Dictionary, and American Remembrancer of Departed Merit*. New York: Johnson and Stryker, 1801.
Hart, Charles. *Agriculture v. The Cotton Trade*. London, UK: self-published, 1852.
Hays, Will H. *Motion Pictures: An Outline of the History and Achievements of the Screen From its Earliest Beginnings to the Present Day*. Garden City, NY: Doubleday, Doran and Co., 1929.
Herbermann, Charles G. (ed.). *The Catholic Encyclopedia*. New York: The Encyclopedia Press, 1914.
Hessels, Jan Hendrik. *Gutenberg: Was He the Inventor of Printing?: An Historical Investigation*. London, UK: Bernard Quaritch, 1882.
Hole, Charles. *A Brief Biographical Dictionary*. New York: Hurd and Houghton, 1866.
Holweck, Frederick George. *A Biographical Dictionary of the Saints, With a General Introduction on Hagiology*. St. Louis, MO: B. Herder Book Co., 1924.
Hopkins, George M. *Inventor's Manual: How to Work a Patent to Make it Pay*. New York: Norman W. Henley Publishing, 1924.
Howson, Charles, and Henry Howson. *Patents: A Manual Relating to United States Patents for Inventors and Patentees*. Philadelphia, PA: self-published, 1904.

Hubert, Jr., Philip G. *Inventors.* New York: Charles Scribner's Sons, 1893.
Iles, George. *Leading American Inventors.* New York: Henry Holt and Company, 1912.
Ingram, William H. (ed.). *Who's Who in Paris Anglo-American Colony: A Biographical Dictionary.* Paris, France: The American Register, 1905.
Ishikawa, Yasujiro. *The Who's Who in Japan.* Tokyo, Japan: The Who's Who in Japan Publishing Dept., 1916.
Jewitt, Llewellynn. *The Wedgwoods: Being a Life of Josiah Wedgwood.* London, UK: Virtue Brothers and Co., 1865.
Johnson, Clifton. *The Rise of an American Inventor: Hudson Maxim's Life Story.* Garden City, NY: Doubleday, Page and Company, 1907.
Johnson, John Henry. *Hints to Inventors Desirous of Obtaining Letters Patent for Their Inventions.* London, UK: British Government, Office for Patents and the Registry of Designs, 1870.
Johnson, Rossiter (ed.). *The Twentieth Century Biographical Dictionary of Notable Americans.* Boston, MA: The Biographical Society, 1904.
Jones, Francis Arthur. *Thomas Alva Edison: Sixty Years of an Inventor's Life.* New York: Thomas Y. Crowell and Co., 1907.
Jones, Stephen (ed.). *A New Biographical Dictionary; Or, Pocket Compendium: Containing a Brief Account of the Lives and Writings of the Most Eminent Persons in Every Age and Nation.* London, UK: G. G. Robinson, 1794.
Jordan, John C. *An Historical Citizen: Career of Phineas Davis, the Noted Inventor.* York, PA: Historical Society of York County, 1904.
Kellogg, John Henry. *Plain Facts for Young and Old.* Burlington, IA: Segner and Condit, 1881.
Kennedy, John. *The History of Steam Navigation.* Liverpool, UK: Charles Birchall, Ltd., 1903.
Konvalinka, Joseph G. *Memoirs of an Inventor and Scientist: Sketches of Original Ingenious Ideas.* New York: self-published, 1897.
Kress, Wilhelm. *Aviatek: Wie der Vogel fliegt und wie der Mensch Fliegen Wird.* Wien, Austria: Spielhagen and Schurich, 1905.
Landman, Isaac (ed.). *The Universal Jewish Encyclopedia: An Authoritative and Popular Presentation of Jews and Judaism Since the Earliest Times.* New York: The Universal Jewish Encyclopedia, 1943.
Leonard, John William. (ed.). *Who's Who in Pennsylvania: A Biographical Dictionary of Contemporaries.* New York: L. R. Hamersly and Co., 1908.
———. *Who's Who in Finance, Banking and Insurance: A Biographical Dictionary of Contemporary Bankers, Capitalists and Others Engaged in Financial Activities in the United States and Canada.* New York: Joseph and Sefton, 1911.
———. *Woman's Who's Who of America: 1914-1915.* New York: The American Commonwealth Co., 1914.
Lethbridge, Roper. *The Golden Book of India: A Genealogical and Biographical Dictionary.* London, UK: Macmillan and Company, 1893.
Lidstone, Thomas. *Some Account of the Residence of the Inventor of the Steam-engine.* London, UK: Longmans, Green, Reader and Dyer, 1869.
Linton, William C. *The Inventor's Adviser, and Manufacturer's Handbook to Patents, Trade-marks, Designs, Copyrights, Prints and Labels.* Montreal, CAN: self-published, 1921.
Lynch, Joseph H. *Tips to Inventors.* New York: Hamilton Publishing and Reporting Company, 1911.
Macfarlane, Robert. *History of Propellers and Steam Navigation: With Biographical Sketches of the Early Inventors.* New York: George P. Putnam, 1851.
Macleod, Henry Dunning. *A Dictionary of Political Economy: Biographical, Bibliographical, Historical, and Practical.* London, UK: Longman, Green, Longman, Roberts, and Green, 1863.

Marquis, Albert Nelson. *The Book of Chicagoans: A Biographical Dictionary of Leading Living Men of the City of Chicago.* Chicago, IL: A. N. Marquis and Co., 1911.

———. (ed.) *Who's Who in New England: A Biographical Dictionary of Leading Living Men and Women.* Chicago, IL: A. N. Marquis and Co., 1916 ed.

Martin, Gloria. *Socialist Feminism: The First Decade, 1966-76.* Seattle, WA: Freedom Socialist Publications, 1978.

Martin, Thomas Commerford. *The Inventions, Researches and Writings of Nikola Tesla: With Special Reference to His Work in Polyphase Currents and High Potential Lighting.* New York: The Electrical Engineer, 1894.

Mather, Frank Lincoln (ed.). *Who's Who of the Colored Race: A General Biographical Dictionary of Men and Women of African Descent.* Chicago, IL: self-published, 1915.

Maunder, Samuel. *The Biographical Treasury: A Dictionary of Universal Biography.* London, UK: Longman, Brown, Green, Longmans, and Roberts, 1856.

McCabe, Joseph (ed.). *A Biographical Dictionary of Modern Rationalists.* London, UK: Watts and Co., 1920.

McClure, James Baird (ed.). *Edison and His Inventions: Including the Many Incidents, Anecdotes, and Interesting Particulars Connected With the Life of the Great Inventor.* Chicago, IL: Rhodes and McClure, 1879.

McLean, John. *James Evans, Inventor of the Syllabic System of the Cree Language.* Toronto, CAN: William Briggs, 1890.

Metallurgical and Chemical Engineering. Vol. 10, January-December 1912. New York: Electrochemical Publishing Company, 1912.

Miguelez, Ernest, and Carsten Fink. *Measuring the International Mobility of Inventors: A New Database.* Geneva, Switzerland: World Intellectual Property Organization, 2013.

Mish, Frederick (ed.). *Webster's Ninth New Collegiate Dictionary.* Springfield, MA: 1984.

Moore, James. *Memoirs of James Moore, Inventor of the New Life Steam Vessel.* London, UK: T. Goode, 1839.

Mottlelay, Paul Fleury. *Bibliographical History of Electricity and Magnetism: Chronologically Arranged.* London, UK: Charles Griffin and Co., 1922.

Mowry, William A., and Arthur May Mowry. *American Inventions and Their Inventors.* New York: Silver, Burdett and Company, 1900.

Muir, John. *The Story of My Boyhood and Youth.* Boston, MA: Houghton Mifflin Co., 1913.

Murray-Aynsley, Harriet. *Symbolism of the East and West.* London, UK: George Redway, 1900.

Myers, Gustavus. *History of the Great American Fortunes.* New York: Modern Library, 1907.

Neill, Jo C., and Jayashri Kulkarni (eds.). *Biological Basis of Sex Differences in Psychopharmacology.* Heidelberg, Germany: Springer-Verlag, 2011.

Neilson, William (ed.). *Webster's Biographical Dictionary: A Dictionary of Names of Noteworthy Persons With Pronunciations and Concise Biographies* (1st edition). Springfield, MA: G. and C. Merriam Company, 1943.

Nolan, Melanie (ed.). *Australian Dictionary of Biography.* Acton, Australia: The Australian National University Press, 2021.

Oberlin, J. P. *Pitfalls or How an Invention May Be Lost.* A paper read December 3, 1914, before the Examining Corp of the U.S. Patent Office. Washington, D.C.: U.S. Government Printing Office, 1914.

O'Byrne, William R. *A Naval Biographical Dictionary: Comprising the Life and Services of Every Living Officer in Her Majesty's Navy.* London, UK: John Murray, 1849.

O'Donoghue, David J. *The Poets of Ireland: A Biographical Dictionary with Bibliographical Particulars.* London, UK: self-published, 1892.

Ottley, Henry. *A Biographical and Critical Dictionary of Recent and Living Painters and*

Engravers. London, UK: Chatto and Windus, 1875.

Page, George. *George Page, Machinist, Inventor, and Manufacturer, West Baltimore*. Baltimore, MD: self-published, 1850.

Palmer, Daniel David. *The Science of Chiropractic: Its Principles and Adjustments*. Davenport, IA: The Palmer School of Chiropractic, 1906.

Palmer, Samuel. *A General History of Printing*. London, UK: A. Bettesworth, 1733.

Papini, Mauricio R. *Comparative Psychology: Evolution and Development of Brain and Behavior*. 2001. New York: Routledge, 3rd ed., 2021.

Patrick, David, and Francis Hindes Groome (eds.). *Chamber's Biographical Dictionary: The Great of All Times and Nations*. Philadelphia, PA: J. B. Lippincott Co., 1900.

Peirce, Bradford Kinney. *Trials of an Inventor: Life and Discoveries of Charles Goodyear*. New York: Carlton and Porter, 1866.

Perky, Henry D. *Wisdom Vs. Foolishness*. Worcester, MA: Perky Publishing Co., 1902.

Perry, Frances M. *Four American Inventors: Robert Fulton, Eli Whitney, Samuel F. B. Morse, Thomas A. Edison*. New York: American Book Company, 1901.

Plummer, John. *The Story of a Blind Inventor; Being Some Account of the Life and Labours of Dr. James Gale*. London, UK: William Tweedie, 1868.

Polovtsov, Aleksandr Aleksandrovich (ed.). *Russian Biographical Dictionary*. Moscow, Russia: Imperatorskago Russkago Istoricheskago Obshchestva, 1914.

Powel, Jr., Harford. *Walter Camp: The Father of American Football—An Authorized Biography*. Boston, MA: Little, Brown, and Co., 1926.

Pupin, Michael. *From Immigrant to Inventor*. New York: Charles Scribner's Sons, 1926.

Ramsaye, Terry. *A Million and One Nights: A History of the Motion Picture*. New York: Simon and Schuster, 1926.

Rankine, J., and W. H. Rankine. *Biography of William Symington, Civil Engineer: Inventor of Steam Locomotion*. Falkirk, Scotland: A. Johnston, 1862.

Reed, Thomas Allen. *A Biography of Isaac Pitman (Inventor of Phonography)*. London, UK: Griffith, Farran, Okeden and Welsh, 1890.

Remy, Alfred. *Baker's Biographical Dictionary of Musicians*. New York: G. Schirmer, 1919.

Richardson, Alexander. *The Evolution of the Parsons Steam Engine*. London, UK: Offices of Engineering, 1911.

Rigaud, Stephen Peter. *Biographical Account of John Hadley, the Inventor of the Quadrant*. London, UK: Fisher, Son, and Co., 1835.

Roberts, T. R. *Eminent Welshmen: A Short Biographical Dictionary of Welshmen*. Cardiff, Wales: Cardiff and Merthyr Tydfil, 1908.

Robinson, Henry. *Inventors and Inventions*. New York: self-published, 1911.

Robinson, Paschal. *The Writings of Saint Francis of Assisi*. Philadelphia, PA: The Dolphin Press, 1905.

Rogers, Thomas J. (ed.). *A New American Biographical Dictionary*. Easton, PA: self-published, 1824.

Rolt-Wheeler, Francis. *The Boy With the U.S. Inventors*. Boston, MA: Lothrop, Lee and Shepard Company, 1920.

Root, Henry. *Henry Root, Surveyor, Engineer and Inventor: Personal History and Reminiscences*. San Francisco, CA: self-published, 1921.

Rorschach, Hermann. *Psychodiagnostik: Methodik und Ergebnisse eines Warhrnehmungsdiagnostischen Experiments*. Bern, Switzerland: Hans Huber, 1937.

Rose, Hugh James. *New General Biographical Dictionary*. 12 vols. London, UK: B. Fellowes, 1848.

Russell, Bertrand. *The Natural Superiority of Women*. New York: Collier Books, 1974.

Ryan, Richard. *Biographia Hibernica: A Biographical Dictionary of the Worthies of Ireland*. London, UK: self-published, 1819.

Sainsbury, John S. *A Dictionary of Musicians: From the Earliest Ages to the Present Time*. London, UK: Sainsbury and Company, 1824.

Sanderson, John. *Biography of the Signers to the Declaration of Independence.* Philadelphia, PA: R. W. Pomeroy, 1823.
Savic, Ivanka (ed.). *Sex Differences in the Human Brain, Their Underpinnings and Implications.* Amsterdam, Netherlands: Elsevier, 2010.
Seabrook, Lochlainn. *Aphrodite's Trade: The Hidden History of Prostitution Unveiled.* 1994. Franklin, TN: Sea Raven Press, 2011 ed.
——. *The Goddess Dictionary of Words and Phrases: Introducing a New Core Vocabulary for the Women's Spirituality Movement.* 1997. Franklin, TN: Sea Raven Press, 2010 ed.
——. *Britannia Rules: Goddess-Worship in Ancient Anglo-Celtic Society - An Academic Look at the United Kingdom's Matricentric Spiritual Past.* 1999. Franklin, TN: Sea Raven Press, 2010 ed.
——. *The Book of Kelle: An Introduction to Goddess-Worship and the Great Celtic Mother-Goddess Kelle, Original Blessed Lady of Ireland.* 1999. Franklin, TN: Sea Raven Press, 2010 ed.
——. *Carnton Plantation Ghost Stories: True Tales of the Unexplained from Tennessee's Most Haunted Civil War House!* 2005. Franklin, TN, 2016 ed.
——. *Nathan Bedford Forrest: Southern Hero, American Patriot.* 2007. Franklin, TN, 2010 ed.
——. *Abraham Lincoln: The Southern View.* 2007. Franklin, TN: Sea Raven Press, 2013 ed.
——. *The McGavocks of Carnton Plantation: A Southern History - Celebrating One of Dixie's Most Noble Confederate Families and Their Tennessee Home.* 2008. Franklin, TN, 2011 ed.
——. *Christmas Before Christianity: How the Birthday of the "Sun" Became the Birthday of the "Son."* Franklin, TN: Sea Raven Press, 2010.
——. *A Rebel Born: A Defense of Nathan Bedford Forrest.* 2010. Franklin, TN: Sea Raven Press, 2011 ed.
——. *Everything You Were Taught About the Civil War is Wrong, Ask a Southerner!* 2010. Franklin, TN: Sea Raven Press, 2024 ed.
——. *The Quotable Jefferson Davis: Selections From the Writings and Speeches of the Confederacy's First President.* Franklin, TN: Sea Raven Press, 2011.
——. *The Quotable Robert E. Lee: Selections From the Writings and Speeches of the South's Most Beloved Civil War General.* Franklin, TN: Sea Raven Press, 2011 Sesquicentennial Civil War Edition.
——. *Lincolnology: The Real Abraham Lincoln Revealed In His Own Words.* Franklin, TN: Sea Raven Press, 2011.
——. *The Unquotable Abraham Lincoln: The President's Quotes They Don't Want You To Know!* Franklin, TN: Sea Raven Press, 2011.
——. *Honest Jeff and Dishonest Abe: A Southern Children's Guide to the Civil War.* Franklin, TN: Sea Raven Press, 2012.
——. *Encyclopedia of the Battle of Franklin - A Comprehensive Guide to the Conflict that Changed the Civil War.* Franklin, TN: Sea Raven Press, 2012.
——. *The Quotable Nathan Bedford Forrest: Selections From the Writings and Speeches of the Confederacy's Most Brilliant Cavalryman.* Spring Hill, TN: Sea Raven Press, 2012.
——. *Forrest! 99 Reasons to Love Nathan Bedford Forrest.* Spring Hill, TN: Sea Raven Press, 2012.
——. *Give 'Em Hell Boys! The Complete Military Correspondence of Nathan Bedford Forrest.* Spring Hill, TN: Sea Raven Press, 2012.
——. *The Constitution of the Confederate States of America Explained: A Clause-by-Clause Study of the South's Magna Carta.* Spring Hill, TN: Sea Raven Press, 2012 Sesquicentennial Civil War Edition.
——. *The Great Impersonator: 99 Reasons to Dislike Abraham Lincoln.* Spring Hill, TN: Sea Raven Press, 2012.
——. *The Old Rebel: Robert E. Lee As He Was Seen By His Contemporaries.* Spring Hill, TN:

Sea Raven Press, 2012 Sesquicentennial Civil War Edition.

——. *The Quotable Stonewall Jackson: Selections From the Writings and Speeches of the South's Most Famous General*. Spring Hill, TN: Sea Raven Press, 2012 Sesquicentennial Civil War Edition.

——. *Saddle, Sword, and Gun: A Biography of Nathan Bedford Forrest for Teens*. Spring Hill, TN: Sea Raven Press, 2013.

——. *Jesus and the Law of Attraction: The Bible-Based Guide to Creating Perfect Health, Wealth, and Happiness Following Christ's Simple Formula*. Franklin, TN: Sea Raven Press, 2013.

——. *The Bible and the Law of Attraction: 99 Teachings of Jesus, the Apostles, and the Prophets*. Franklin, TN: Sea Raven Press, 2013.

——. *The Alexander H. Stephens Reader: Excerpts From the Works of a Confederate Founding Father*. Spring Hill, TN: Sea Raven Press, 2013.

——. *The Quotable Alexander H. Stephens: Selections From the Writings and Speeches of the Confederacy's First Vice President*. Spring Hill, TN: Sea Raven Press, 2013 Sesquicentennial Civil War Edition.

——. *Christ Is All and In All: Rediscovering Your Divine Nature and the Kingdom Within*. Franklin, TN: Sea Raven Press, 2014.

——. *Jesus and the Gospel of Q: Christ's Pre-Christian Teachings as Recorded in the New Testament*. Franklin, TN: Sea Raven Press, 2014.

——. *Give This Book to a Yankee! A Southern Guide to the Civil War for Northerners*. Spring Hill, TN: Sea Raven Press, 2014.

——. *The Articles of Confederation Explained: A Clause-by-Clause Study of America's First Constitution*. Spring Hill, TN: Sea Raven Press, 2014.

——. *Confederate Blood and Treasure: An Interview With Lochlainn Seabrook*. Spring Hill, TN: Sea Raven Press, 2015.

——. *Nathan Bedford Forrest and the Battle of Fort Pillow: Yankee Myth, Confederate Fact*. Spring Hill, TN: Sea Raven Press, 2015.

——. *Everything You Were Taught About American Slavery War is Wrong, Ask a Southerner!* Spring Hill, TN: Sea Raven Press, 2015.

——. *Confederacy 101: Amazing Facts You Never Knew About America's Oldest Political Tradition*. Spring Hill, TN: Sea Raven Press, 2015.

——. *The Great Yankee Coverup: What the North Doesn't Want You to Know About Lincoln's War!* Spring Hill, TN: Sea Raven Press, 2015.

——. *Slavery 101: Amazing Facts You Never Knew About America's "Peculiar Institution."* Spring Hill, TN: Sea Raven Press, 2015.

——. *Confederate Flag Facts: What Every American Should Know About Dixie's Southern Cross*. Spring Hill, TN: Sea Raven Press, 2016.

——. *Nathan Bedford Forrest and the Ku Klux Klan: Yankee Myth, Confederate Fact*. Spring Hill, TN: Sea Raven Press, 2016.

——. *Seabrook's Bible Dictionary of Traditional and Mystical Christian Doctrines*. Spring Hill, TN: Sea Raven Press, 2016.

——. *Everything You Were Taught About African-Americans and the Civil War is Wrong, Ask a Southerner!* Spring Hill, TN: Sea Raven Press, 2016.

——. *Nathan Bedford Forrest and African-Americans: Yankee Myth, Confederate Fact*. Spring Hill, TN: Sea Raven Press, 2016.

——. *Women in Gray: A Tribute to the Ladies Who Supported the Southern Confederacy*. Spring Hill, TN: Sea Raven Press, 2016.

——. *Lincoln's War: The Real Cause, the Real Winner, the Real Loser*. Spring Hill, TN: Sea Raven Press, 2016.

——. *The Unholy Crusade: Lincoln's Legacy of Destruction in the American South*. Spring Hill, TN: Sea Raven Press, 2017.

——. *Abraham Lincoln Was a Liberal, Jefferson Davis Was a Conservative: The Missing Key to Understanding the American Civil War*. Spring Hill, TN: Sea Raven Press, 2017.

——. *All We Ask is to be Let Alone: The Southern Secession Fact Book*. Spring Hill, TN: Sea Raven Press, 2017.
——. *The Ultimate Civil War Quiz Book: How Much Do You Really Know About America's Most Misunderstood Conflict?* Spring Hill, TN: Sea Raven Press, 2017.
——. *Rise Up and Call Them Blessed: Victorian Tributes to the Confederate Soldier, 1861-1901*. Spring Hill, TN: Sea Raven Press, 2017.
——. *Victorian Confederate Poetry: The Southern Cause in Verse, 1861-1901*. Spring Hill, TN: Sea Raven Press, 2018.
——. *Confederate Monuments: Why Every American Should Honor Confederate Soldiers and Their Memorials*. Spring Hill, TN: Sea Raven Press, 2018.
——. *The God of War: Nathan Bedford Forrest as He Was Seen by His Contemporaries*. Spring Hill, TN: Sea Raven Press, 2018.
——. *The Battle of Spring Hill: Recollections of Confederate and Union Soldiers*. Spring Hill, TN: Sea Raven Press, 2018.
——. *I Rode With Forrest! Confederate Soldiers Who Served With the World's Greatest Cavalry Leader*. Spring Hill, TN: Sea Raven Press, 2018.
——. *The Battle of Nashville: Recollections of Confederate and Union Soldiers*. Spring Hill, TN: Sea Raven Press, 2018.
——. *The Battle of Franklin: Recollections of Confederate and Union Soldiers*. Spring Hill, TN: Sea Raven Press, 2018.
——. *A Rebel Born: The Screenplay* (for the film). Written 2011. Franklin, TN: Sea Raven Press, 2020.
——. (ed.) *A Short History of the Confederate States of America* (Jefferson Davis, Belford Company, NY, 1890). A Sea Raven Press Reprint. Spring Hill, TN: Sea Raven Press, 2020.
——. (ed.) *Prison Life of Jefferson Davis: Embracing Details and Incidents in his Captivity, With Conversations on Topics of Public Interest* (John J. Craven, Sampson, Low, Son, and Marston, London, UK, 1866). A Sea Raven Press Reprint. Spring Hill, TN: Sea Raven Press, 2020.
——. *What the Confederate Flag Means to Me: Americans Speak Out in Defense of Southern Honor, Heritage, and History*. Spring Hill, TN: Sea Raven Press, 2021.
——. *Heroes of the Southern Confederacy: The Illustrated Book of Confederate Officials, Soldiers, and Civilians*. Spring Hill, TN: Sea Raven Press, 2021.
——. *Support Your Local Confederate: Wit and Humor in the Southern Confederacy*. Spring Hill, TN: Sea Raven Press, 2021.
——. *America's Three Constitutions: Complete Texts of the Articles of Confederation, Constitution of the United States of America, and Constitution of the Confederate States of America*. Spring Hill, TN: Sea Raven Press, 2021.
——. *Vintage Southern Cookbook: 2,000 Delicious Dishes From Dixie*. Spring Hill, TN: Sea Raven Press, 2021.
——. *The Bittersweet Bond: Race Relations in the Old South as Described by White and Black Southerners*. Spring Hill, TN: Sea Raven Press, 2022.
——. (ed.) *The Rise and Fall of the Confederate Government* (Jefferson Davis, D. Appleton, New York, 1881). 2 vols. A Sea Raven Press Facsimile Reprint. Spring Hill, TN: Sea Raven Press, 2022.
——. *Secrets of Celebrity Surnames: An Onomastic Dictionary of Famous People*. Cody, WY: Sea Raven Press, 2023.
——. *I, Confederate: Why Dixie Seceded and Fought in the Words of Southern Soldiers*. Spring Hill, TN: Sea Raven Press, 2023.
——. *Twelve Years in Hell: Victorian Southerners Expose the Myth of Reconstruction, 1865-1877*. Cody, WY: Sea Raven Press, 2023.
——. *Seabrook's Complete Battle Book: The War Between the States, 1861-1865*. Cody, WY: Sea Raven Press, 2023.
——. *The Hampton Roads Conference: The Southern View*. Cody, WY: Sea Raven Press,

2024.
———. *Rocky Mountain Equines: A Photographic Collection of Horses, Donkeys, and Mules of the American West.* Cody, WY: Sea Raven Press, 2024.
———. *Rocky Mountain Bison: A Photographic Collection of Bison of the American West.* Cody, WY: Sea Raven Press, 2024.
———. *Mysterious Invaders: Twelve Famous 20th-Century Scientists Confront the UFO Phenomenon.* Cody, WY: Sea Raven Press, 2024.
———. *We Called Him Jeb: James Ewell Brown Stuart as He Was Seen by His Contemporaries.* Cody, WY: Sea Raven Press, 2024.
———. *Your Soul Lives Forever: Documented Victorian Case Studies Proving Consciousness Survives Death.* Cody, WY: Sea Raven Press, 2024.
———. *Authentic Victorian Ghost Stories: Genuine Early Reports of Apparitions, Wraiths, Poltergeists, and Haunted Houses.* Cody, WY: Sea Raven Press, 2024.
———. *The Greatest Jesus Mystery of All Time: Where Was Christ Between the Ages of 12 and 30?* Cody, WY: Sea Raven Press, 2024.
———. *Vitamin D: The Miracle Treatment for Nearly Every Disease and Health Issue.* Cody, WY: Sea Raven Press, 2024.
———. *Jesus and the Gospel of Thomas: A Christian Mystic's View of Christianity's Most Important Ancient Text.* Cody, WY: Sea Raven Press, 2025.
———. *The Way of Holiness: The Story of Religion and Mythology, from the Cave Bear Cult to Christianity—A Study of the Origins, Development, Functions, Symbols, and Themes of Spiritual Thought.* Unpublished manuscript.
———. *Mothers and Bachelors: Ending the Battle of the Sexes—A New Approach to Marriage and the Family Based on the Sciences of Anthropology, Primatology, and Sociobiology.* Unpublished manuscript.
———. *Seabrook's Encyclopedia of Religion and Myth: A Comparative Guide to the Major Beliefs, Deities, People, and Legends of the World's Religions.* Unpublished manuscript.
———. *Families Around the World: A Children's Guidebook to the Marriages and Families of Different Cultures.* Unpublished manuscript.
———. *The True Legend of King Arthur: The Magical Story of Britain's Most Famous Ruler.* Unpublished manuscript.
———. *Glimpses of Heaven: A Guidebook to the Near-Death Experience.* Unpublished manuscript.
———. *Rowena: The Cat Who Wanted to be a Person.* Unpublished manuscript.
———. *Blackie: The Crow Who Was Afraid of Heights.* Unpublished manuscript.
Seitz, Frederick. *The Cosmic Inventor: Reginald Aubrey Fessenden (1866-1932).* Philadelphia, PA: American Philosophical Society, 1999.
Shore, Thomas T. (ed.). *Cassell's Biographical Dictionary.* London, UK: Caseel, Petter, and Galpin, 1867.
Simpson, Walter Grindlay. *The Art of Golf.* Edinburgh, Scotland: David Douglas, 1892.
Sinclair, Angus. *Development of the Locomotive Engine.* New York: Angus Sinclair Publishing Co., 1907.
Smith, Goodwin B. *How to Succeed as an Inventor: Showing the Wonderful Possibilities in the Field of Invention.* Philadelphia, PA: Inventors and Investors Corporation, 1909.
Smith, Sigurd. *Søren Hjorth: Inventor of the Dynamo-electric Principle.* Copenhagen, Denmark: J. Jørgensen and Company, 1912.
Smith, William, and Henry Wace (eds.). *A Dictionary of Christian Biography, Literature, Sects and Doctrines.* Boston, MA: Little, Brown, and Company, 1877.
Stark, John. *Biographia Scotica: Or Scottish Biographical Dictionary; Containing a Short Account of the Lives and Writings of the Most Eminent Persons and Remarkable Characters, Natives of Scotland, From the Earliest Ages to the Present Time.* Edinburgh,

Scotland: 1805.
Stephen, Leslie. *Dictionary of National Biography.* New York: Macmillan and Co., 1886.
Stuart, Robert. *Historical and Descriptive Anecdotes of Steam-engines: And of Their Inventors and Improvers.* London, UK: Wightman and Cramp, 1829.
Sturgis, Russell. *A Dictionary of Architecture and Building: Biographical, Historical, and Descriptive.* New York: Macmillan and Co., 1905.
Tarbell, Ida Minerva. *The History of the Standard Oil Company.* New York: S. S. McClure Co., 1902.
Tesla, Nikola. *Experiments With Alternate Currents of High Potential and High Frequency.* New York: W. J. Johnston Co., 1892.
The American Inventor: An Illustrated Monthly Journal Devoted to Inventions, Mechanics and Industrial Progress. Vol. 16, January-December, 1907. New York: Gough and Gough, 1907.
The Biographical Dictionary of the Society for the Diffusion of Useful Knowledge. London, UK: Longman, Brown, Green, and Longmans, 1843.
The Electrical Engineer: An Illustrated Record and Review of Electrical Progress. Vol. 44, July 2, 1909, through December 31, 1909. London, UK: The Electrical Engineer, Ltd., 1909.
The Encyclopedia Britannica: A Dictionary of Arts, Sciences, Literature and General Information. 11th ed. New York: The Encyclopedia Britannica Company, 1911.
The Implement Age: A Journal Published in the Interest of the Farm Implement and Wagon Trade. Vol. 33, No. 1, issue. Philadelphia, PA: July 9 1908.
The Inventors' Advocate, and Journal of Industry; A British and Foreign Miscellany of Inventions, Manufactures, Trade, Science, and the Arts. Vol. 2, Jan. 4 through June 30, 1840. London, UK: William Kidd, 1840.
The Leisure Hour: A Family Journal of Instruction and Recreation. October 1, 1868, issue. London, UK: Religious Tract Society, 1868.
The Story of the Sewing Machine. Boston, MA: Singer Manufacturing Company, 1851.
The United States Biographical Dictionary and Portrait Gallery of Eminent and Self-Made Men. Chicago, IL: American Biographical Publishing Co., 1877.
Thompson, Edward P. *How to Make Inventions: Or, Inventing as a Science and an Art.* New York: D. Van Nostrand Company, 1891.
Thompson, Silvanus P. *Philipp Reis: Inventor of the Telephone: A Biographical Sketch, with Documentary Testimony.* London, UK: E. and F. N. Spon, 1883.
Timbs, John. *Stories of Inventors and Discoverers in Science and the Useful Arts. A Book for Old and Young.* London, UK: Kent and Company, 1860.
Toner, Joseph M. *George Washington as an Inventor and Promoter of the Useful Arts.* An Address Delivered at Mount Vernon, April 10, 1891. Washington, D.C.: U.S. Government Printing Office, 1891.
Tower, Oswald (ed.). *Spalding's Official Basketball Guide Containing the Official Rules: 1914-1915.* New York: American Sports Publishing Co., 1914.
Traub, Hamilton P. *The American Literary Yearbook: A Biographical and Bibliographical Dictionary of Living North American Authors.* Henning, MN: self-published, 1919.
Turner, Thomas. *Counsel to Inventors of Improvements in the Useful Arts.* London, UK: F. Elsworth, 1850.
United States Government. *Small Inventors Program.* Congressional Hearing Before the Subcommittee on Energy Development and Applications of the Committee on Science and Technology, U.S. House of Representatives, 96th Congress, 1st Session, November 13, 1979, Vol. 14. Washington, D.C.: U.S. Government Printing Office, 1980.
———. *Patents and Inventions: An Information Aid for Inventors.* Washington, D.C.: U.S. Dept. of Commerce, Patent and Trademark Office, 1977.
———. *American Inventors Protection Act of 1999.* U.S. House of Representatives, 106th

Congress, 1ˢᵗ Session, August 3, 1999. Washington, D.C.: Committee on the Judiciary, 1999.
——. United States Patent and Trademark Office Website: www.uspto.gov/patents.
Verwey, E. J. (ed.). *New Dictionary of South African Biography*. Pretoria, South Africa: HSRC Publishers, 1995.
Walker, William. *The Inventor's Guide Book; or, Plain Directions for Obtaining Letters Patent*. London, UK: George Philip and Son, 1860.
Watkins, John. *Universal Biographical Dictionary*. London, UK: Longman, Rees, Orme, Brown, and Green, 1830.
Webb, Gerald B. *René Théophile Hyacinthe Laennec: A Memoir*. New York: Paul B. Hoeber, 1928.
Wedderburn, John. *How to Get a Patent: A Complete Compendium of Useful Information for Inventors*. Washington, D.C.: John Wedderburn and Co., 1896.
Weigand, Kate. *Red Feminism: American Communism and the Making of Women's Liberation*. Baltimore, MD: Johns Hopkins University Press, 2001.
Weller, Charles Edward. *The Early History of the Typewriter*. La Porte, IN: self-published, 1918.
Westscott, Thompson. *Life of John Fitch: The Inventor of the Steam-boat*. Philadelphia, PA: J, B. Lippincott and Company, 1857.
Wheeler, Joseph M. *A Biographical Dictionary of Freethinkers of All Ages and Nations*. London, UK: Progressive Publishing Co., 1889.
Wile, Frederic William. *A Century of Industrial Progress*. Garden City, NY: Doubleday, Doran and Co., 1928.
Wilkes, John. *A Christian Biographical Dictionary*. London, UK: Longman, Hurst, Rees, Orme, and Brown, 1821.
Wilson, E. O. *Sociobiology: The New Synthesis*. Cambridge, MA: Harvard University Press, 1975.
Wilson, Rachel. *Occult Feminism: The Secret History of Women's Liberation*. N.p. Independently published, 2021.
Woodcroft, Bennet. *Brief Biographies of Inventors of Machines for the Manufacture of Textile Fabrics*. London, UK: Longman, Green, Longman, Roberts, and Green, 1863.
——. *Alphabetical Index of Patentees and Applicants for Patents of Invention, for the Year 1870*. London, UK: Office of the Commissioners of Patents for Inventions, 1871.
Woodward, Horace L. *Patents: A Talk to the Inventor*. Washington, D.C.: George E. Howard Press, 1910.
Wylson, James. *The Mechanical Inventor's Guide; Comprising Familiar Treatises on the Laws of Motion*. London, UK: Simpkin, Marshall, and Co., 1859.
Yale, Rodney Horace. *Yale Genealogy and the History of Wales*. Beatrice, NE: self-published, 1908.

INDEX

INCLUDES TOPICS, PEOPLE, KEYWORDS, KEY PHRASES, & SPELLING VARIATIONS

(THE MALE INVENTORS ALPHABETICALLY CHRONICLED IN THIS BOOK ARE NOT LISTED IN THE INDEX)

"Upon every invention of value, we erect a statue to the inventor, and give him a liberal and honourable reward." — Lord Bacon

1960s, 20, 45, 46, 55, 57, 63, 68, 80, 128, 132, 172, 185, 208, 220, 221, 232, 235, 246, 249, 272
1970s, 46, 63, 161, 176, 194, 201, 208, 220, 221, 235, 242
3Com Corporation, 174
3D model, 69
3D movie glasses, 23, 152
3D optical imager, 176
3D printer, 15, 125
3D printing, 69, 144, 186
3D Systems, 125
3D technology, 125
3D volume holography, 74
405-line television waveform, 50
A Radiant Sea of Energy (Moray), 179
A Song of Ice and Fire (Martin), 169
A-22 (guitar), 46
AB Separator, 154
Abaca, 144
Abalakov thread, 39
Abbe condenser, 39
ABC, 42
abdomen, 82, 116
absolute zero temperature, 248
absorption refrigeration, 182, 209
abstract algebra, 42
AC electrical system, 247, 248
AC radio signals, 90
academic institutions, 114, 169
academician, 151
Academy Award, 231
acceleration, 77
accordion, 16, 57, 74, 264
account, 281-284, 286, 290
acetate disks, 50
acetone, 75
acetylcholine, 203
acetylcholinesterase inhibitor, 203
acetylene, 49, 71
acid catalyst, 75

acid-fast bacilli, 277
acoustic suspension loudspeaker, 256
acoustical tile, 49
acoustician, 163
acoustics, 62, 117, 118, 142, 163
acrylic paint, 28, 51, 104
acrylic polymer emulsion, 51, 104
actinometer, 118
action, 12, 52, 73, 78, 98, 219, 227, 266, 267
activist, 103, 104, 162, 225, 236
actor, 40, 45, 47, 50, 63, 65, 73, 75, 76, 102, 123, 125, 127, 134, 139, 144, 146, 162, 184, 200, 201, 230, 235, 236, 241, 265, 267
actors, 229
ACV, 66
Ad.ly, 217
ADD, 33, 203
addiction recovery organization, 236, 265
adding, 15, 66, 73, 113, 168, 179, 203, 240
adding machine, 15, 203
ADHD, 185
adhesive, 27, 67, 95, 144
adhesive substance, 144
adhesives, 155
adjustable pipe wrench, 30
adjustable spanner, 30, 132
ADS, 51, 278
Advanced Research Projects Agency Network, 176
advertising, 97, 204
advertising and marketing, 97
AED, 201
aeolipile, 118
aerial firefighting, 205
aerial propulsion, 53
aerial reconnaissance, 172
aerial screw, 73
aerial steam carriage, 241

aerodynamic advancements, 264
aerodynamic car, 20, 206
aerodynamic innovations, 206
aerodynamicist, 133
aerodynamics, 66, 138, 276
aeronaut, 92, 227, 282
aeronautical engineer, 91, 133, 155, 176, 264
aeronautical pioneer, 231
aeronautical principle, 264
aeronautical principles, 205
aeronautical researcher, 178
aeronautical scientist, 207
aeronautical site, 45
aeronautics, 62, 92, 118, 145, 177, 178, 210, 211, 241
aeronautics engineer, 62, 210
aeronautics pioneer, 211
Aerosmith, 161, 162
aerosol spray paint, 28, 231
aerospace, 55, 63-65, 72, 86, 92, 103, 125, 130, 132, 138, 147, 154, 161, 166, 174, 176, 184, 194, 210, 211, 234, 242, 244, 251, 259, 264, 276
aerospace designer, 63
aerospace engineer, 55, 64, 65, 72, 92, 103, 138, 161, 166, 176, 194, 210, 211, 242, 244, 251, 259, 264
aerospace engineering, 86
aerospace industry, 154
aerospace innovator, 276
aerospace science, 147
aetherphone, 248
AFib, 121
AFM, 100, 215
Africa, 292
African-Americans, 288
AGA cooker, 71
agamassan, 71
agar, 207
age, 45, 54, 109, 147, 156, 171, 189, 222, 229, 233, 253, 277, 284, 291
agricultural advancement, 261
agricultural calendar, 232
agricultural economics, 217
agricultural fertilizers, 111
agricultural implement, 226
agricultural research, 61
agricultural science center, 154
agricultural scientist, 61
agriculture, 48, 79, 101, 157, 171, 202, 283
agronomist, 154, 157, 210

AHW, 153
AI, 124, 167, 176, 212
AI pioneer, 176
aide-de-camp, 237
AIDS, 76, 94, 144, 187, 226, 229, 231, 249
aileron, 53
ailments, 109, 145, 151, 174, 203, 243
air capacity, 59
air compressor, 147
air conditioner, 15, 86, 153
air conditioning system, 205
air consumption, 59
air flight, 157, 235
air flow, 65, 242, 276
air flow amplification technology, 65
air flow principles, 276
air gliders, 241
air pressure, 115, 147, 177
air pump, 26, 107, 133
air traffic control computers, 133
air-conditioning, 175
air-cooled machinery and vehicle engines, 141
air-cushion vehicle, 66
air-filled boxing gloves, 147
air-filled tire, 248
air-pumping machine, 116
air-sensitive chemical research, 229
airbag, 134
airborne surveys, 45
airborne toxins, 276
aircraft, 18, 20, 25, 29, 45, 64, 67, 77, 92, 94, 95, 102, 108, 113, 133, 134, 146, 147, 153, 157, 160, 163, 165, 168, 176, 191, 196, 197, 205, 207, 210, 211, 227, 229, 231, 234, 238, 241, 242, 248, 249, 251, 252, 256, 259, 260, 264, 268
aircraft accidents, 260
aircraft builder, 242
aircraft control stick, 146
aircraft design and construction, 191
aircraft designer, 94, 108, 134, 153, 176, 207, 210, 211, 231, 234, 242, 251, 256, 268
aircraft designs, 94, 251
aircraft engineer, 242, 249, 256
aircraft industry, 241
aircraft parts, 196
aircraft pilot, 64, 67, 92, 134, 197, 210, 211, 234, 268
aircraft technology, 231
aircraft with boat hull, 205

aircrafts, 40
airflow, 108
airplane designer, 130
airplane technology, 227
airplane wing, 53
airplanes, 78, 82, 113
airports, 89
airship, 102, 109, 227, 229, 276
airship designer, 227, 276
airsickness bag, 22, 228
airspeed, 260
AIRTRAC, 45
aiuton, 65
AIV fodder, 19, 256
AIV method, 256
AK-47, 137, 138
alarm, 22, 24, 72, 169, 183, 253
alarm clock, 169
Alba, 181
Albany Plan, 93
Alberti cipher, 40
Alcantara, 194
alchemist, 142, 229
alchemy, 235
alcohol, 52, 114, 134, 206
alcohol meter, 52, 114, 134
alcoholics, 19, 135, 236, 265
Alcoholics Anonymous, 19, 135, 236, 265
alcoholism, 236, 265
aldehydes, 83
Aldrin, Edwin, Jr., 115
alexandrite, 256
Alfa Laval, 154
algebraic theory, 42
algebraist, 42
algorithmic computations, 172
alizarin, 206
alkali groups, 57
alkaline battery, 25, 253
alkaline potassium carbonates, 124
alkaline solution, 49
All in the Family (TV series), 28, 154
all-electric TV system, 277
all-terrain land train, 156
all-terrain vehicles, 52
allergies, 228
allergy, 228
Allied Powers, 91
alloy scientist, 177
alloys, 105, 263
AllVoices, 217
Almaz, 63
alpha-olefins, 187, 276
alternating-current, 247
alternating-current dynamos, 247

alternative health, 219, 385
alternative health and fitness, 385
alternative medicine, 220
altitude, 111, 207, 211, 249, 260
aluminum, 9, 29, 105, 112, 118, 178, 187, 196
aluminum chloride, 178
aluminum foil, 29, 187
Alzheimer's disease, 49, 203
AM radio, 26, 88
AM-38F engine, 176
AMA, 132
Amaryllidaceae plant family, 203
amateur singers, 128, 187
amateurs, 79
ambassador, 93, 238
AMBS-1 two-stroke engine, 176
America, 6, 21, 67, 81, 93, 131, 170, 222, 284, 287, 289, 382
American Academy of Pediatrics, 228
American Civil War, 3, 83, 93, 174, 205, 213, 288, 385
American colonies, 93, 115
American confederacy, 93
American Flyer Trains, 27, 102
American history, 3, 282
American Jewish Joint Distribution Committee, 101
American patriots, 205
American politics, 3
American press, 117
American Revolutionary War, 223
American slavery, 213, 288
American Society for Artificial Internal Organs, 145
American South, 3, 174, 184, 288
American Southerners, 213
American Telegraphone Company, 211
American time scheduling, 153
Americans, 284, 288, 289
America's nuclear program, 225
Ames Test, 40
Amici roof prism, 40
amino acid, 80
amino acids, 206
ammeters, 263
ammo, 227, 233
ammonia, 49, 108, 111, 212
ammonia engine, 108
ammunition, 56, 75, 92, 142, 227
ammunition designer, 227
AMP MP3 Playback Engine, 254
amphibious aircraft, 20, 146, 231
amphibious plane, 205

amplification, 65, 87, 137, 145, 244
amplification systems designer, 87
amplifiers, 235, 246, 249
Amplitron, 56
amplitude modulation, 88
amputation techniques, 208
anaesthetic, 109, 208, 230
anaesthetist, 54
analog recording, 77, 207
analog storage, 137, 244
analysis, 46, 98, 100, 146, 154, 215, 228, 266, 267
analytical chemistry, 49
anatomical diseases, 246
anatomical features, 208
anatomist, 73, 98, 112, 115, 117, 151, 208, 230
anchor escapement, 122
ancient Egypt, 88, 212, 220, 272
ancient history, 3
ancient peoples, 109
André, Jean C., 125
Andreas Stihl AG and Company, 241
Androgynous Peripheral Attach System, 244
anemia, 43
anemometer, 30, 39, 73, 122
anesthesia, 54
anesthesiologist, 230
angel, 125
Angel Light, 125
angels, 24, 236
anger, 41
Anglomania, 17, 131
angular velocity, 51
aniline purple, 206
animal, 29, 68, 74, 83, 98, 104, 115, 122, 129, 168, 184, 232, 236
animal behavior, 236
animal magnetism, 173
animal models, 83, 236
animal powered mills, 68
animal rights activist, 104
animal tissue culture, 115
animal tissues, 98
animals, 19, 29, 53, 98, 104, 143, 155, 158, 183, 226, 230, 256
animation, 40, 50, 76, 123, 128, 239, 240
animation films, 40
animation technology, 123, 239
animator, 40, 50, 102, 107, 128, 184, 201, 240, 241
ANS synthesizer, 183
answering machine, 24, 115, 211

Antarctica, 385
antenna, 28, 55, 82, 130, 175, 248, 253, 260, 271
antenna designs, 82
antennas, 203
anthrax, 30, 145
Anthropocene Epoch, 262
anthropologist, 148, 257
anthropology, 24, 131, 156, 182, 290
anthropomorphic device, 40
anti-disease vaccines, 120
antibacterial compounds, 228
antibacterial drugs, 228
antiballistic missiles, 63
antibiotic, 17, 99
antibiotics, 17, 81, 90
antibodies, 67, 245
antifreeze, 257
antilock brakes, 20
antilock braking systems, 121
antimatter, 265
antimicrobial, 260
antioxidant protection, 219
antipersonnel projectile, 233
antiperspirant, 178, 183
antiperspirant deodorant, 178
antiseptics, 22, 219
anxiety, 30, 204
API, 152
apiarist, 200, 213
apiculturist, 213
aplanatic optical system, 167
apochromatic lens system, 39
Apollo 11, 115
apothecary, 143
Appalachia, 385
Appalachian heritage, 385
apparel, 100, 191
apparitions, 290
Apple Computer company, 262, 267
Apple I computer, 132
Apple II computer, 132
Apple Inc., 132
appliance industry, 194, 238
appliances, 79, 85, 101, 120, 121, 196, 202, 218
application programming interface, 152
applications, 15, 51, 55, 61, 66, 69, 74, 87-89, 94, 97, 98, 100, 101, 105, 109, 111, 112, 119-122, 124, 128, 130-133, 135, 137, 142, 144, 153, 160, 166-169, 172, 174, 182, 185-187, 202-204, 206, 209,

212, 215, 218, 220, 225, 226,
  229, 231, 232, 235, 242-244,
  247, 250, 267, 272, 277, 291
APS, 235
APS Underwater Assault Rifle, 235
aqualung, 68
aquarist, 385
Aquinas, Thomas, 57
Arabs, 143
arborite, 146
arc converter, 211
arc furnace, 118
arc lamp, 21, 238, 271
arc light regulator, 110
arc lights, 234, 248
arc welding, 23, 47, 153, 236
archaeological record, 167
archaeology, 24, 156, 212
archetypal origins, 167
archetypes, 31, 135
Archimedes' screw, 41
architect, 39, 55, 58, 60, 61, 73, 78,
  82, 95, 102, 105, 113, 122,
  129, 145, 155, 166, 167, 183,
  223, 226, 233, 243, 246, 267,
  273
architects, 205, 283
architectural engineer, 238
architectural industry, 138
architecture, 188, 265, 291
archival storage, 207
area rule, 264
Argand burner, 41
Argand lamp, 41
Argentina, 50
argon ion laser, 48
Arithmometer, 66
Arkin, Adam P., 144
Arlandes, Marquis d', 223
armaments manufacturer, 190
armies, 219
armillary sphere, 25, 117, 148, 272
armor-piercing shells, 166
armored car, 73
armored gun transport, 154, 176
armored turret warship, 83
arms, 46, 56, 63, 67, 116, 137, 166,
  184, 219, 227, 266
Armstrong, Neil, 115
army, 15, 17, 35, 82, 95, 109, 116,
  133, 155, 174, 195, 205, 242
army colonel, 155
army officer, 133
Arnott stove, 42
Arnott ventilator, 42
aroma, 113

aromatic compounds, 49
aromatic scents, 143
ARPANET, 26, 144, 176
Arrested Development (TV series),
  28, 125
arrhythmias, 121
art, 6, 21, 24, 27, 41, 44, 50, 51,
  66, 78, 79, 82, 104, 134, 148,
  167, 168, 177, 196, 200, 217,
  231, 239, 265, 290, 291
art critic, 44
art form, 50, 196, 239
art historian, 41
art industry, 231
Art Technology Group, 134
art world, 51, 104
arterial blood pressure, 145
arteries, 67, 145
Articles of Confederation, 93, 288,
  289
artificial cell, 63
artificial cement, 42
artificial eyebrows, 18, 187
artificial eyelashes, 18, 187
artificial gum massager, 34
artificial heart, 31, 74, 130, 145, 265
artificial intelligence, 177
Artificial Intelligence Project, 177
artificial kidney, 31, 145
artificial language, 275
artificial limb, 18, 241
artificial pearl, 228
artificial refrigeration, 70
artificial rivers, 176
artificial satellite, 103, 141, 145, 249
artificial silk, 190
artificial sweetener, 81
artificial wearable kidney, 145
artificially generated precipitation,
  228
artillery, 67, 70, 97, 145, 167, 233
artillery officer, 67, 145, 233
artillery weapon, 167
artilleryman, 103
Artin rings, 42
artisan, 385, 388
artist, 5, 28, 39, 40, 44, 45, 50, 58,
  73, 76, 88, 95, 117, 122, 123,
  125, 130, 141, 144, 147, 149,
  167, 175, 177, 179-181, 183,
  184, 186, 190, 210, 213, 217,
  220, 232, 237, 265, 277, 380,
  385, 388
artistic innovator, 177
artistic possibilities, 261
artists, 58, 161, 162, 205, 217, 283,

385
arts and crafts, 128
ascorbic acid, 219
ash, 77
Asia, 243
asphalt, 19, 155, 171, 236
assault rifle, 24, 96, 137, 235
assault vehicle, 95
assembly line, 22, 91, 194, 198
astrobotanist, 249
astrocolorimetrist, 249
astrolabe, 120, 142
astrologer, 60, 89, 117, 140, 169, 173, 184, 186, 218, 273
astrological sun signs, 259
astrology, 143, 235
astronautic engineer, 138
astronomer, 40, 41, 46, 51, 62, 64, 73, 92, 97, 103, 107, 111, 117, 118, 120, 122, 124, 125, 130, 140, 142, 143, 148, 153, 159, 166, 169, 173, 184, 186, 188, 195, 197, 199, 215, 217, 228, 229, 232, 238, 239, 248, 249, 272
astronomical clock, 169, 238
astronomical clock tower, 238
astronomical measurements, 120
astronomical observatory, 169
astronomical sextant, 142
astronomical tables, 142
astronomical views, 259
astronomy, 25, 85, 92, 94, 119, 130, 148, 158, 186, 206, 218, 235, 385
astrophysicist, 116, 153, 239, 249
astrophysics, 64
at sign (@), 250
Atanasoff-Berry Computer, 42
Atari Hotz Box, 123
athlete, 19, 52, 59, 68, 102
Atkins, Chet, 107, 385
Atlanta, GA, 205
ATM, 19, 133, 235
ATM machines, 133
atmospheric contamination, 153
atmospheric data, 57, 177
atmospheric electricity, 159
atmospheric pressure, 250
atomic bomb, 62, 141, 148, 154, 195, 244
atomic clocks, 81
atomic energy, 156
atomic force microscope, 31, 50, 100, 215
atomic hydrogen welding, 153

atomic level, 100, 215
atomic number, 173
atomic physics, 51
atomic research, 255
atomic weight, 173
atomizer, 157
atoms, 40, 49, 147, 156, 222
atrial fibrillation, 121
attention-deficit hyperactivity disorder, 185
ATV, 20, 211
audio, 51, 54, 76, 85, 125, 128, 155, 183, 197, 207, 230, 231, 234, 249, 255, 256
Audio Distribution System, 51
audio engineer, 183, 230, 256
audio equipment, 231
audio industry, 207, 256
audio products, 249
audio recording, 155
audio storage, 207
audio technology, 128
audiologist, 123
audiometer, 47
audiophile, 48
auditoriums, 49
Auerlicht, 263
aura, 143, 219
auscultatory device, 151
auscultatory method, 31, 145
Australia, 180, 285
Australian military, 250
Austria, 284
author, 3, 5, 9-12, 33, 39-41, 49, 52, 54, 55, 57, 59, 66, 67, 72, 80, 82, 86, 89, 90, 93-95, 101-106, 108, 109, 111, 114-120, 122, 123, 125, 128-135, 137-140, 142-149, 151-154, 156-161, 163, 166, 167, 169-171, 173, 174, 178-182, 184-187, 189, 190, 195-197, 202, 204, 206, 210-213, 215, 218, 219, 221-223, 225, 227-232, 235, 236, 238-241, 243, 244, 246, 248, 250, 255-257, 259-261, 266, 267, 273, 275, 277, 380-382, 385
autism, 281
auto industry, 194
auto world, 231
auto-gyroscope stabilization, 113
auto-intoxication, 129
autochenile, 139
autogiro, 64

autogyro, 64
automated door system, 176
automated external defibrillator, 201
automated industrial robotics, 155
automated tabulating machine, 121
automated telegraph, 156
automated teller machine, 235
automatic (cow) milking machine, 86
automatic bread turner, 67
automatic cooking device, 128
automatic lubricators, 171
automatic message accounting, 132
automatic sun valve, 71
automatic tapping machines, 124
automatic telegraph, 80
automatic telephone billing equipment, 132
automatic telephone exchange, 242
automatic weapon, 96
automobile, 20, 48, 51, 60, 64, 76, 78, 86, 91, 119, 139, 141, 148, 155, 163, 194, 196, 204-206, 217, 232, 276, 281
automobile air conditioning, 205
automobile bumper, 163
automobile designer, 119, 206
automobile differential, 204
automobile engineer, 60
automobile engines, 86
automobile manufacturer, 194
automobile parts, 196, 205
automobile sled, 139
automobiles, 79, 94, 101, 120, 141, 202, 205, 238, 251
automotive, 48, 100, 132, 139, 147, 148, 155, 174, 175, 177, 179, 180, 184, 191, 204
automotive engineer, 48, 155, 204
automotive history, 148
automotive industry, 147, 175
automotive parts, 180
autonomous robots, 172
autopilot, 113
avalanche photodiodes, 189
avant-garde artist, 44
avant-garde musicians, 183
Avelo Dive System, 59
average man, 215
aviation, 44, 64, 79, 89, 101, 108, 146, 157, 175, 191, 202, 205, 219, 223, 241, 264, 268, 276
aviation accident, 223
aviation engineer, 64
aviation fuel, 219
aviation history, 264
aviation pioneer, 223, 264, 268

aviation researcher, 108
aviation science, 157
aviation technology, 175, 205
aviator, 117, 123, 130, 134, 157, 170, 205, 212, 223, 227, 268
aviator, first, 227
Avicenna, 235
Avogadro's number, 81
Avtomat Kalashnikova model 1947, 137
axis, 92
Ayurveda, 243
Azerbaijan, 190
Aztecs, 139
B programming language, 26, 221, 248
B. F. Goodrich, 134
BAAS, 55
Babesia, 43
Babushka Dolls, 167, 277
baby, 101, 167, 202, 231, 277
baby monitors, 231
baby rattles, 202
bacilli, 172, 277
Bacillus lactis aërogenes, 56
back, 6, 28, 33, 53, 57, 76, 77, 88, 95, 116, 138, 148, 163, 167, 173, 178, 180, 181, 209, 212, 220, 223, 226, 237, 240
background noise, 77
backup storage, 207
bacon, 293
bacteria, 60, 90, 106, 144, 145, 155, 172, 207
bacteria cultures, 144, 145
bacteria staining methods, 144
bacteria-based infections, 90
bacterial filter, 62
bacterial infections, 81
bacteriological work, 144
bacteriologist, 56, 65, 67, 90, 101, 106, 111, 144, 207, 226, 266, 267, 277
bacteriology, 101
bacteriophage therapy, 81
bacterium, 56, 144, 213
badminton, 22, 212, 237
Badminton (British estate), 237
Badminton House, 237
Baer, Theodor, 211
bagless vacuum cleaner, 17, 78
Bahco, 132
bake oven, 196
Bakelite, 43
baking paper, 88, 212
balance spring regulation, 250

balanced ternary numeral system, 93
balances, 221
ball bearing, 160
ballistic missile, 103, 145, 166, 185
ballistic missiles, 63, 249
ballistics, 62, 118, 125
balloon, 20, 31, 57, 91, 109, 145, 178, 200, 207, 223, 227
balloon design, 207
balloon flight, 178, 207, 223
balloonist, 207, 223, 227
balloonists, 223
balloons, 146, 177, 178, 241
ballpoint pen, 25, 50, 160
banana grower, 227
band leader, 385
bandleader, 230
bands, 46, 97, 161, 162, 220
banker, 173, 180, 273
banking, 79, 85, 101, 152, 202, 235, 284
banking industry, 235
banks cards, 180
bar code readers, 53
bar-code scanners, 130
barber, 44, 176, 187, 221
barbituric acid, 44
barcode, 23, 92, 114, 153, 185, 234, 267
barcode readers, 92
barcode technology, 234, 267
bareback rigging, 45
barge, 148
Barnes and Mortlake Regatta, 180
Barnes Football Club, 180
barograph, 30, 256
barometer, 30, 122, 250, 256
barometric pressure, 256
Barrow, Joseph, 184
Barsanti-Matteucci engine, 170
baseball, 22, 61
BASIC, 26, 66, 99, 116, 138, 140, 148, 162, 188, 211, 267
bass player, 385
bass viol, 25, 40
batch distiller, 66
bathyscaphe, 25, 207
batteries, 9, 16, 85, 201, 243
battery, 25, 46, 57, 72, 73, 82, 98, 107, 140, 157, 164, 170, 201, 209, 211, 251, 253, 257, 275
battery operated radios, 46
battery technologies, 275
battery technology, 72
battery-powered, 82, 140, 157, 170
battle, 83, 134, 139, 287-290

Battle of Dunbar, 187
Battle of Edgehill, 187
Battle of Franklin, 134, 287, 289
Battle of Marston Moor, 187
Battle of Spring Hill, 134, 289
Baudot code, 45
Bayerische Flugzeugwerke, 197
Bayerische Motoren Werke, 197
BB gun, 24, 168
Beagle (British naval vessel), 72, 114
beam antenna, 130
beam communication systems, 168
beam ray, 220
beam stations, 168
beams, 56, 66, 94, 135, 149, 177, 189, 244
beard, 56
Beatles, the, 46, 107, 161, 162, 220
Beaufort cipher, 46
Beaufort scale, 46
bedding, 194, 243
bedroom, 34
bee brood disease, 213
beef tallow, 172
beehive frame, 24, 213
beekeeping, 213
beeping tone, 204
beers, 113
beeswax, 24
Beginner's All-purpose Symbolic Instruction Code, 140, 148
behavior, 236, 286
behavioral psychological treatment, 204
behavioral psychologist, 30
behaviorist, 235
Bekhterev's Band, 47
Bekhterev's Disease, 47
Bekhterev's Mixture, 47
Bekhterev's Nucleus, 47
Bekhterev's Reflexes, 47
Bekhterev's Sign, 47
Bekhterev's Symptom, 47
Bekhterev's Test, 47
Bekhterev's Tract, 47
bell, 8, 23, 47, 73, 86, 153, 174, 182, 204, 212, 249, 273, 281, 282
Bell 47 (helicopter), 273
Bell Company, 47
Bell Helicopter, 23, 273
Bell Helicopter series, 273
Bell, Alexander G., 86, 153, 174, 249
Bell, Bill, 212
bellmaker, 182

bellows, 74, 109
belts, 155, 180, 210, 238, 248
benevolent patriarchy, 32
benzene, 49, 86
Bergius Process, 48
Berthelot's reagent, 49
beverage cans, 196
beverages, 166
Bible, 260, 279, 282, 288, 381, 385
Bible authority, 385
Bible studies, 385
Bic, 50
Bic Cristal pen, 50
bichloride of mercury, 149
bicycle, 18, 21, 60, 62, 77, 82, 105, 148, 159, 170, 175, 232, 240, 248, 250, 271
bicycle derailleur, 60
bicycle designer, 82
bicycle locks, 271
bicycle maker, 159
bicycle parts manufacturer, 60
bicycle performance, 60
bicycle racer, 82
bicycle wheel, 62
bicycles, 9, 17, 125, 147, 166, 229, 235, 238, 251
bicycling, 78, 240
Biefeld–Brown Effect, 55
Bielschowsky Silver Stain Technique, 49
bifocal glasses, 93
big government, 174
Bill of Rights, 17, 166
billiard balls, 202
binary data, 74
binoculars, 26, 158, 236
binomial nomenclature, 19, 158
biochemist, 40, 65, 80, 157, 175, 182, 226, 236, 238, 251, 256, 266
biodegradable, 106, 195
bioengineered product, 53
biography, 3, 15, 282, 283, 285-288, 290-292, 385
Biologic Living, 139
biological functions, 63
biological hierarchical groupings, 158
biological male, 32
biological men, 33
biologist, 50, 60, 63, 98, 99, 104, 115, 122, 129, 142, 144, 172, 175, 182, 188, 190, 201, 208, 226, 228, 229, 236, 238, 244, 277
biologists, 158, 255

biology, 24, 63, 64, 79, 80, 98, 100, 101, 131, 144, 155-157, 166, 182, 202, 215, 226, 275
biology researcher, 80
biomass, 89, 250
biomedical engineer, 229
biomedical engineering, 229
biomedical scientist, 130, 225
biomedical surgeon, 272
biomedicine, 67, 189
biomimicry, 174
biophysical method, 251
biophysicist, 41, 71, 88, 121, 229, 251
BIOS system, 223
biotechnological wonder, 91
biotechnologist, 53, 219
biotechnology, 53, 63, 100, 215
biotechnology industries, 53
biplane, 210
bird, 52, 109, 188, 209
bird's nest, 188
birds, 389
Biro, 50
Bíró Pens of Argentina, 50
Birome, 50
birth, 9, 13, 28, 29, 34, 48, 63, 173, 188, 208, 217, 221, 257
birth control, 9, 28, 29, 63, 173, 208, 221, 257
birth control industry, 63, 208, 221
birth control sponge, 29, 257
bison, 290
bisphenol A, 75
Black and Decker Workmate, 119
black box, 20, 260
black hole mechanics, 116
black holes, 116
black peppercorn plant, 196
black powder explosive, 230
black reaction, 104
Black Sabbath, 161, 162
blacksmith, 73, 109, 170, 224
bladeless fan, 78
Blenkinsop rack railway system, 50
blimp, 20, 102, 276
Blissymbolics, 50
Blissymbols, 16, 50
blocked airway, 54
blood, 17, 43, 68, 91, 133, 145, 155, 158, 267, 288
blood analysis, 267
blood clots, 91
blood pressure cuff, 145
blotting paper, 162
blowers, 66

blowpipe, 108
blowtorch, 23, 191
Blu-ray, 128
blue denim, 18
blue jeans, 9, 18, 241
blue laser, 186
blue light, 251
blue semiconductor LED, 186
blues, 58
Bluetooth, 113
BMI, 31, 189, 215, 216
BMW, 22, 197
boar, 170
Boar War, 170
board game, 58, 72, 119, 173, 202
board game designer, 72, 173
board game developer, 119
board game pieces, 202
boat, 20, 21, 29, 30, 60, 86, 98, 115, 129, 147, 180, 187, 191, 196, 205, 238, 244, 251, 259, 277, 292
boat engines, 86
boat hulls, 196
boat owners, 115
boat parts, 187, 277
boat propellers, 191
boat-builder, 98
boating, 83
boats, 113, 122, 166, 168, 220, 232, 242
bodies, 29, 104, 112, 202, 218
bodily disequilibrium, 152
body, 15, 16, 29, 31, 53, 70, 72, 83, 106, 124, 129, 138, 140, 151, 158, 203, 206, 215, 219, 226, 230, 243, 260, 282, 388
Body Mass Index, 31, 215
body temperature, 53
body weight, 138, 215
Boem System, 51
Bohr Model, 24, 51
Bohr, Niels, 249
boilers, 53, 113
boiling point, 62, 86
boiling point of water, 86
Bolling, Edith, 385
bolometer, 153
bolt-action rifle, 219
bolt-action rifle of the century, 266
bomber geodetic designs, 259
bombs, 233
bombsight, 231
bone, 72, 127, 156
bone lengthening medical procedure, 127

bones, 127, 208
Bono, 107
book binding, 144
book burning, 218
book cover designer, 385
book designer, 385
book formatter, 385
book stand, 131
books, 3, 6, 9, 12, 26, 83, 91, 93, 109, 114, 134, 143, 155, 167, 169, 174, 184, 205, 213, 218, 236, 247, 260, 281, 283, 286, 380-382, 385, 388, 389
Boone, Pat, 385
boring machine, 106
boring machines, 56
Bose-Einstein condensate, 81
Boston Conservatory of Elocution, Oratory, and Dramatic Art, 82
Boston Technology Corporation, 134
botanist, 19, 40, 61, 72, 73, 118, 148, 158, 169, 183, 211, 215, 218, 219, 238, 251, 266
botany, 235
bottle cap, 22, 183, 199
bottle neck, 199
bottle opener, 22, 199
bottles, 41, 75, 95, 187, 277
Bottom Line, 132
Boulton and Watt, 262
Bouly, Léon G., 162
bouncing bomb, 34, 259
bow and arrow, 33
bowel, 129
Bowes, Walter, 208
boxes, 187, 220, 242, 277
boy, 286
Boyle, Charles, 105
boys, 122, 287
BPA, 75
Bragg gratings, 237
braid theory, 42
Braille writing machine, 112
Braille writing system, 44
brain, 31, 48, 54, 76, 85, 104, 189, 272, 286, 287
brain tissue neurons, 104
brains, 281
Brandeis University, 23, 104, 225
Brandeis, Louis D., 104, 225
Branobel, 190
brass cylinder, 211
Braun Tube, 55
Brazil, 227
bread, 25, 67, 188, 204
bread buckle, 204

bread clip, 204
bread manufacturers, 204
bread tab, 204
bread tag, 25, 204
bread tie, 204
breakfast, 19, 88, 139, 206
Breaking Bad (TV series), 28, 102
breaking glass, 237
breathable, 16, 105, 194
Breathalyzer, 21, 52, 114, 134
breech, 92
Brice, Fanny, 194
brick manufacturer, 227
bridge, 25, 64, 72, 73, 113, 142, 148, 152, 230, 241
bridge construction, 148
bridge designer, 241
bridge maintenance and repair, 142
bridge-builder, 113
bridges, 9, 25, 66, 177
Britain, 50, 91, 100, 108, 259, 261, 278
Britain's war efforts, 223
British Association for the Advancement of Science, 55
British cinema industry, 255
British Interplanetary Society, 161
British peer, 237
British Royal Navy, 114
broadband networks, 237
broadcast media, 28, 229
broadcast signals, 82
broadcast television, 60
broadcasting, 74, 131, 189, 277
broadcasting technology, 189
broiling food, 30
broken bones, 127, 208
bromide paper, 243
bronc saddle, 45
bronze statue, 73
brother, 64, 126, 159, 161, 162, 173, 175, 178, 189, 190, 230, 257, 268
brothers, 130, 157, 178, 227, 283, 284
Brown University, 23, 55
Brownian movement, 81
Brownie camera, 79
Browning Arms Company, 56
browsing the Internet, 152
Broxodent, 34
brush strokes, 177
bubble chamber, 103
bubonic plague, 111
Buchanan Windmill, 33
Buchanan, Patrick J., 385

Buddhist compassion-goddess, 177, 273
Buddhist monk, 269
Bude-light, 108
Buick, 78
building block toy, 64
building material, 152
buildings, 77, 89, 124, 172, 247, 271
Bull Engine, 57
bulldozer, 18, 156
bulletins, 61
bullets, 125, 233
bumper designs, 163
bumper sticker, 29, 102
Bunsen battery, 57
Bunsen Burner, 29, 57
buoyancy, 59
Buran, 161
burns, 57
buses, 113, 229, 231, 246
business efficiency, 233
business magnate, 91
businesses, 12, 98, 170, 174, 190, 207, 235, 273
businessman, 40-42, 45, 46, 50, 52, 54-58, 60, 61, 64-66, 68, 69, 71, 74, 76, 78-80, 82, 83, 85-89, 91, 92, 94, 95, 98, 101-107, 109, 112, 113, 115, 119, 120, 122-125, 127, 128, 130-134, 138-141, 149, 152-163, 165, 167, 168, 170-183, 187, 188, 190, 191, 193-197, 199, 200, 202, 204-206, 208, 210-213, 217-219, 221, 223, 224, 226-229, 231-241, 243, 245, 246, 248-251, 254, 255, 257, 259-264, 266-268, 271-273, 275-277, 385
butter alternative, 172
buttons, 202
C programming language, 26, 221
C.S. commander, 155
C++ programming language, 26, 242
cabinets, 89, 271, 272
cable car, 22, 113
cable cars, 113
cable TV, 9, 27, 28, 124, 253, 259, 260, 271
cable TV technology, 124
cables, 73, 119, 159, 234
cacao, 124
Cadillac, 78

cadmium cell, 263
cakes, 19, 190
calcium, 128
calcium chlorate, 128
calculations, 186, 193, 228
calculator, 15, 66, 143, 203, 228, 272
calculators, 87, 120, 121, 235
calculus, 28, 188, 203, 250
calendars, 142, 148
calibrating, 132
California, 113, 239, 241
caller-ID system, 115
calligraphy, 237
Callisto, 97
camera, 15, 24, 28, 40, 41, 50, 53, 68, 75, 76, 79, 94, 97, 115, 128, 152, 173, 189, 211, 218, 229, 243, 249, 262, 275, 277
camera bodies, 218
camera miniaturization technology, 275
camera obscura, 40
camera pioneer, 275
camera technology, 115
cameras, 15, 41, 53, 80, 92, 169, 236, 267, 272
camming devices, 39
camoufleur, 265
Campbell, Joseph, 385
campers, 67, 98, 263
camphor, 201
camping, 16, 82, 128, 218, 263
camping equipment, 16, 218
camping expeditions, 82
camping lanterns, 263
campstool, 131
Canada, 61, 180, 283, 284
canal builder, 220
canal lock engineer, 210
canal navigation, 220
cancer, 30, 40, 94, 99, 109, 120, 129, 152, 201, 220, 276
cancer cells, 220
cancer patients, 99
cancer-causing chemicals, 40
candle, 17, 271
candles, 22, 174, 219
candy maker, 52
canning technology, 169
cannon, 103, 195
cannon balls, 195
cannonball, 233
Canon, 27, 73, 177, 273
Canon, Inc., 273
canvas, 147, 149

canvas catamaran, 147
capacitance, 86, 189
capacitor, 50, 184
capers, 195
capital, 108
capitalists, 284
capron, 144
Caprotti valve gear, 60
Captain Kangaroo (TV series), 28, 139
car, 16, 20, 22, 26, 29, 44, 48, 64, 73, 92, 113, 147, 148, 153, 159, 193, 194, 198, 201, 205, 206, 208, 213, 227, 231-233, 248, 250, 251, 276, 281
car industry, 147, 233
car wheels, 159
carbon, 29, 31, 47, 48, 80, 83, 89, 91, 100, 111, 125, 143, 153, 156, 159, 191, 194, 203, 210, 212, 250, 263
carbon arc welding, 47
carbon atom, 83
carbon buildup, 194
carbon dioxide emissions, 111
carbon dioxide laser, 31, 203
carbon emissions, 48, 91, 194
carbon filament, 153, 263
carbon filament light system, 263
carbon filaments, 153
carbon microphone, 80, 125
carbon monoxide, 31, 89, 203, 212, 250
carbon monoxide laser, 31, 203
carbon paper, 29, 91
carbon pigment printing, 210
carbon transmitter, 80, 125
carbon wire, 159
carbon-14 atoms, 156
carbonated water, 19, 212
carbonyl chloride, 75
carburetor, 193, 194
Cardan Grille, 60
cardboard disk, 209, 239
cardiac defibrillator, 31, 146
cardiac monitoring, 121
cardiologist, 46, 201
cards, 9, 23, 26, 121, 124, 128, 146, 167, 180, 202, 212, 218
cargo, 91, 121
cargo ships, 121
caricaturist, 76
Carlyle, Thomas, 15
Carnegie Corporation of New York, 24, 61
Carnegie Endowment for Peace, 24,

61
Carnegie Hall, 24, 61
Carnegie Institution for Science, 24, 60
Carnegie Mellon University, 23, 60, 173
Carnegie Technical Schools, 60
Carnton Plantation, 134, 287
carpenter, 50, 64, 106, 114, 115, 140, 159, 226, 261
carpenters, 161, 162
Carpenters, the, 161, 162
Carrel-Lindbergh Perfusion Pump, 158
carrots, 195
Carryall Scrapers, 156
cars, 20, 78, 112, 113, 119, 122, 125, 198, 209, 221, 229, 231, 235
Carson, Martha, 385
cartographer, 39, 73, 95, 117, 135, 148, 196, 228, 230, 238
cartons, 187
cartoonist, 76, 107, 128, 217, 385
cascading style sheets, 157
Cash, Johnny, 385
cassette technology, 207
cast iron, 53
cast-iron heating radiator, 226
castle, 113, 237
cat, 29, 161, 290
cat litter, 29, 161
catadioptric telescope, 24, 229
catalytic converter, 20, 139, 179
catapult, 69, 161, 162
cataract surgery, 31, 243
caterpillar tractor, 18, 62
catheter, 31, 32, 91, 93
cathode ray tube, 69, 249
cathode ray tubes, 60, 81
cathode-ray based picture tube, 277
cathode-ray tube, 55
cathodes, 69
Catholic cleric, 59
cattlemen, 256
cavalry, 289
cavity structural effect, 106
CCD, 53
CD, 15, 57, 76, 112, 118, 121, 147, 197
CD players, 147
CD readers, 112
CD technology, 118
CD-ROM, 76
CED, 68
ceiling fan, 29, 76

ceilings, 44
celestial navigation, 115
cell, 122
cell biology, 98
cell phone, 15, 24, 26, 86, 168, 233
cell phones, 9, 15, 81, 87, 119, 120, 133, 143, 169, 196, 245
cello, 25, 40
cellophane, 9, 19, 54, 78
cellophane tape, 19, 78
cells, 25, 32, 73, 74, 88, 104, 115, 122, 129, 137, 152, 155, 164, 172, 201, 220, 244, 245, 249, 251, 276
cellular abnormalities, 88
cellular ceramic substrate, 44
cellular distribution, 98
cellular phone, 67
cellular processes, 251
cellular regenerator, 152
celluloid, 75, 126, 201, 202
cellulose, 54, 162, 266
Celsius scale, 86
Celsius temperature scale, 62
cement, 19, 42, 152, 171, 243
cement reinforcement, 243
cementless concrete, 77
Cenocell, 77
censorship, 218
centigrade scale, 62
central heating, 22, 226
central nervous system, 47, 49, 185
central nervous system stimulant, 185
central processing unit, 85
centrifugal cream separator, 154
centrifugal fan, 30, 225
ceramic engineer, 44
ceremonial bodyguard, 266
cerium, 262, 263
cervical cancer, 94, 201, 276
cervical cap, 173
cervix, 201
cesium, 57
cesspool, 114
ChaCha, 134
chaffcutter, 19, 226
chain drive, 18, 238, 240
chainsaws, 74
chair, 16, 131, 238
Chalfie, Martin, 251
Chamberland filter, 62
Chapman Stick, 63
character, 17, 76, 92, 104, 128, 149, 245, 265
character actor, 265

character designer, 128
charcoal, 75, 276
charge carriers, 118
charge-coupled device, 53
charged particles, 64, 103, 265
Charlotte Dundas (steamboat), 244
Cheers (TV series), 28, 63
cheese, 9, 22, 50
cheese slicer, 22, 50
chef, 41, 195
chemical, 19, 41, 46, 50, 73, 80, 87, 89, 93, 98, 141, 144, 146, 147, 154, 158, 159, 179, 187, 196, 203, 206, 210, 212, 228, 229, 245, 253, 257, 259, 277, 285
chemical action, 73, 98
chemical analysis, 46, 228
chemical and water use, 41
chemical engineer, 41, 50, 93, 146, 179, 210, 245, 253
chemical fertilizer, 19, 154
chemical manufacturing, 257
chemical manure, 157
chemical reaction, 87, 259
chemical storage tanks, 187, 277
chemical warfare, 159
chemical weapons program, 144
chemicals, 31, 40, 76, 87, 219, 276
chemist, 39, 41, 43, 44, 46, 48, 49, 54, 55, 57, 60-62, 65, 67, 69-73, 75-77, 80, 83, 85-90, 92-94, 99, 101, 104, 106-109, 111, 112, 114, 118-122, 124, 125, 128, 134, 135, 139, 143, 144, 147, 153, 154, 156, 157, 159, 160, 165, 172-174, 178, 185, 187, 188, 190, 194-196, 201, 203-206, 210, 212, 213, 218, 219, 221, 223, 228, 229, 232, 236, 243, 245, 248-250, 256, 257, 259-263, 265, 275, 276
chemistry, 49, 94, 143, 144, 157, 165, 190, 210, 212, 221, 275
chemists, 255, 259
chemotherapeutic drugs, 144
Cherenkov effect, 63, 64
chess, 119
Chevrolet, 22, 64, 78
Chevrolet Motor Car Company, 64
Chevy, 64
chewing gum, 144
chi, 218
chicken ramen, 41
chickenpox, 30, 120
chicks, 188

child, 22, 24, 183, 385
child prodigy, 385
child-proof bottle cap, 183
children, 35, 168
children's books, 385
children's construction toy set, 199, 267
Children's Flour, 188
children's nutrition, 228
children's toy, 102, 112
China, 162, 178, 181, 230, 276
Chinese, 63, 117, 135, 148, 152, 157, 162, 238, 244, 269, 276, 283
Chinese typing, 157
chips, 19, 147, 190
chiropractic, 31, 200, 286
chiropractor, 200
Chlamydia, 101
chloric ether, 109
chloroform, 27, 109
chocolate, 9, 19, 88, 124, 188, 206, 207
chocolate egg, 88
chocolate factory, 124
chocolate maker, 124
chocolate milk, 19, 124, 206, 207
chocolate products, 124, 188, 207
chocolates, 207
chocolatier, 52, 88, 124, 188, 206, 207
choking blockages, 116
cholera, 30, 65, 111, 145
Christ, 288, 290, 389
Christian, 57, 62, 106, 156, 181, 195, 288, 290, 292, 385
Christian history, 385
Christianity, 287
Christmas, 9, 23, 66, 121, 287
Christmas card, 23, 66
Christmas lights, 121
chromatic French horn, 65
chromatic trumpet, 65
chromatography, 31, 251
chromolithographer, 107
chromosomal analysis, 98
chronic, multi-site, multi-electrode recordings, 189
church, 120
church clocks, 120
cigar, 76
cigarette, 18, 76, 144, 157
cigarette paper, 144
cigarettes, 157, 180
cinema, 162
cinematic art form, 50

Cinématographe, 162
cinematographer, 50, 75, 88, 94, 184, 239
cinematography, 94, 131, 168
cipher wheel, 131
cipher-breaking methods, 143
circles, 255
circuit, 17, 64, 81, 118, 120, 139, 143, 193
cistern, 114
citrus-scented, 86
city life, 138
city planner, 78
civil engineer, 56, 64, 67, 90, 93, 106, 141, 163, 171, 210, 230, 233, 240, 248, 277, 286
Civil War, 3, 83, 93, 134, 155, 174, 205, 213, 287-289, 385, 389
civilian transportation, 276
clairvoyant, 129
Clapton, Eric, 107
Clark electrode, 65
Clark Equipment Company, 65
Clark, Barney, 130
Clarke, Arthur C., 13
Clasp Locker, 134
class, 99, 158
classical composer, 385
classical conditioning, 30, 204
classical music, 9, 16, 17, 228
classification, 68, 114, 158, 169, 249
clavichord, 16, 238
clay, 161
clean energy, 90, 92
cleaning, 22, 27, 101, 103, 109, 132, 194, 195, 219, 231, 232, 249
cleaning products, 22, 194, 219
cleaning surfaces, 249
cleps, 211
clergyman, 41, 45, 57, 75, 82, 92, 115, 131, 183, 185, 190, 197, 208, 212, 223, 228, 241, 255, 264, 266, 275
cleric, 59, 62, 63, 155
Clermont (steamboat), 95
climatology, 24, 156
clinical innovations, 151
clinical psychologist, 204
clinical research, 246
clock, 69, 73, 89, 91, 93, 115, 125, 131, 148, 169, 173, 204, 210, 217, 227, 238, 246, 269, 272
clock design, 246
clock maker, 173, 210
clock pulse, 91

clock tower, 238
clockmaking, 246, 250
clocks, 81, 91, 97, 120, 122, 176
closed iron core transformer, 75
cloth, 149, 162, 170
cloth printing, 170
clothes, 30, 46, 131
clothes rack, 131
clothing, 18, 22, 27, 30, 34, 105, 132, 174, 194, 232, 241, 243, 265
clothing designer, 132
clothing industry, 194, 232, 241
clotted lymph, 115
cloud chamber, 18, 265
cloud computing, 26, 81
cloud computing technology, 81
cloud seeding, 19, 228, 257
cloud storage, 233
Cloudbuster, 218
Club des Sous l'Eau, 212
CMOS sensor, 92
CO_2, 89, 203
coach, 59, 185, 196, 230, 239
coal, 49, 77, 183, 250, 276, 385
coal gas mask, 276
coal-burning process, 77
Coalite, 201
Coanda Effect, 94
Coandă effect, 65
coat, 275
coated guitar-string technology, 105
coaxial rotor, 160
Coca-Cola, 205
Cochran, Eddie, 107
cocklebur plant, 174
cockpit, 20, 231, 260
cockpit voice recorder, 20, 260
cocoa butter, 19, 124
cocoa powder, 19, 124
cocoa press, 124
code book, 63
code-breaking computer, 91
coding systems, 128
Coelostat, 158
coffee, 3, 19, 227
coffee table books, 3
Coffey still, 66
cog rail, 169
cog railway, 168
cog wheel, 168, 169
cog wheel train technology, 169
cog wheels, 221, 242
cognitive functions, 203
cognitive scientist, 176
cogway, 159, 169

cogway railroad, 159
coherent optics, 74
coherer, 52, 54
coiled springs, 251
cold fusion, 90, 272
cold fusion systems, 272
Cold War, 226
collapsible metal paint tube, 217
collectible items, 122
collective unconscious, 31, 135
College of New Jersey, 57, 75, 266
Collegiate School, 41, 208, 223
colliding-beam storage ring, 195
collier, 286
Collins, Phil, 107
colloidal solutions, 195
collotype, 210
cologne, 86, 87
color cinematography, 94
color film scanning, 240
color film slides, 213
color films, 213
color image development process, 158
color management printing, 240
color motion picture projection, 213
color negative photographic paper, 262
color photograms, 262
color photograph, 152, 171, 243
color photography, 171, 213
color photography pioneer, 213
color prints, 107
color scanner, 15, 117
color television, 39, 44, 104
colored inks, 261
colored race, 285
colorings, 172
Colossus, 73, 91
Colt 1911, 67
Colt Manufacturing Company, 67
Colt Patent Fire-Arms Manufacturing Company, 67
Colt revolver, 66
Columbia University, 23, 100
columns, 66, 177
combat, 67, 103, 113, 154, 176, 196, 207, 239
combat engineer, 103
combat medics, 67
Combat Sambo, 196, 239
combs, 180, 202, 385
Combs, Bertram T., 385
combustion, 44, 45, 48, 49, 67, 74, 76, 122, 123, 138, 139, 155, 160, 170, 179, 180, 189, 200, 231, 240, 241, 276
combustion chamber, 138
combustion engine, 44, 45, 48, 67, 74, 76, 123, 155, 170, 180, 189, 200, 240, 276
comedian, 40, 63, 65, 73, 102, 127, 134, 200, 230, 265
comets, 122
commander-in-chief, 261
commerce, 152, 234, 267, 291
commerce industry, 234, 267
commercial beekeeper, 213
commercial demolition, 55
commercial fisherman, 98
commercial honey production, 213
commercial printing industry, 235
commercial productivity, 233
commercial products, 79, 101, 202
commercial transportation, 276
commercialization of beekeeping, 213
communication, 50, 54, 55, 60, 74, 152, 168, 170, 174, 181, 194, 196, 211, 281
communication devices, 196
communication issues, 50
communication technology, 211
communications, 22, 30, 52, 63, 65, 81, 85, 101, 112, 115, 119-121, 130, 145, 166, 168, 177, 183, 186, 189, 196, 203, 207, 218, 228, 233, 267
communications industry, 207
communications satellite, 22, 65, 228
communications systems, 267
communications technique, 63
communications technology, 65, 218
communism, 292
commutators, 131
commuting vehicle, 73
compact disk, 48, 76, 197
compact disk player, 48
compacted substructure, 171
comparative anatomist, 98
comparative embryologist, 172
comparative mythology, 385
comparative pathologist, 172
comparative religion, 385
compass, 26, 68, 97, 142, 200
competition, 144, 196, 239
competitive exclusion principle, 99
competitors, 268
complex compounds, 251
composer, 41, 46, 51, 65, 73, 118, 128, 139, 144, 146, 157, 183,

230, 385
composers, 157, 183
composite, 152
composition, 47, 69, 238
compositions, 47
compound, 67, 94, 122, 130, 158, 196, 210
compound microscope, 122, 130, 158
compound microscope and illumination system, 122
compounds, 31, 49, 135, 189, 206, 219, 228, 245, 251, 266
compressed air, 66, 69, 93, 147, 158, 183, 212, 264
compressed air applications, 66
compressed air steam guns, 183
compressed air tank, 212
compressed air-powered catapult, 69
computational possibilities, 251
computer, 15, 26-28, 40, 42, 43, 45, 49, 51, 54-56, 59, 62-64, 74, 76, 79-82, 85, 87, 91, 93, 105, 112, 113, 116, 119, 120, 123, 124, 132, 133, 137-140, 144, 146-149, 152, 156, 157, 163, 170-174, 176, 177, 179, 188, 189, 194, 196, 199, 200, 217, 221, 222, 230, 232, 233, 235, 241, 245, 246, 248, 250, 251, 255, 262, 266, 267, 277
computer age, 277
computer bulletin board service, 15, 64
computer chips, 147
computer clocks, 176
computer data, 76, 245
computer design model, 188
computer files, 82
computer hardware, 223
computer hobbyist, 64
computer industry, 85
computer markup language, 157
computer memory, 217, 246
computer mouse, 15, 82, 144, 163
computer ports, 119
computer programmer, 43, 49, 59, 80, 123, 139, 156, 171, 250
computer programming language, 140, 148
computer programming languages, 139
computer science, 51, 93, 146, 177, 189, 255
computer scientist, 40, 43, 49, 51, 54-56, 59, 62, 63, 74, 79, 80,
85, 91, 105, 113, 116, 120, 132, 133, 137, 138, 140, 144, 148, 149, 152, 156, 157, 171-174, 176, 179, 188, 194, 199, 200, 221, 230, 232, 235, 241, 245, 248, 250, 251, 266, 267, 277
computer screen, 26, 144, 163
computer software, 15, 123, 149
computer tablet, 139
computer technologist, 123
computer technology, 74, 81, 132, 157, 170
computer tomography, 123
computer-assisted tomography, 87
computer-information age, 54, 233
computer-like data analysis, 146
computer-to-computer email, 250
computers, 9, 15, 17, 75, 80, 81, 87, 116, 120, 124, 133, 139, 143, 147, 155, 167, 169, 174, 185, 188, 193, 196, 212, 231, 235, 245, 246, 267
computing process, 81
Computing-Tabulating-Recording Company, 121
concentrated gas, 71
concept car, 147
conceptual artist, 144
conceptualist, 132
concert halls, 163
concerto, 17, 47
concrete, 16, 66, 77, 85, 152, 156, 158, 177
concrete construction industry, 66, 177
concrete floors, 85
condensed milk, 188
condensed soup, 19, 77
condensing chambers, 222
condensing process, 77
conditioned reflex, 29, 204
conditions, 12, 69, 138, 151, 172, 203, 222, 272
condom, 29, 94, 119
condoms, 94
conducted energy device, 68
conductor, 100, 114, 191, 230
conductors, 77, 92
confectionary, 52, 207
confectionary candy, 52
confectionary scientist, 52
confectioner, 41, 146
Confederacy, 93, 261, 288, 289, 385
Confederate, 83, 93, 134, 155, 180, 184, 205, 213, 287-289, 385

Confederate flag, 288, 289
Confederate general, 180, 205
Confederate government, 289
Confederate military, 213
Confederate soldiers, 289, 385
Confederate States, 93, 155, 205, 287, 289
Confederate States army veteran, 205
Confederate States of America, 93, 287, 289
Confederate veterans, 385
confederation, 93, 288, 289
confocal scanning microscopy, 176
confusion, 153
Congress, 6, 29, 39, 114, 166, 169, 291, 292
Congreve rocket, 67
conical V beam TV antenna, 82
connective tissue, 102
consciousness, 290
conservation of energy, 107, 117
conservationist, 125, 182, 385
conservationists, 100
Conservatives, 385
conspiracy, 220
conspiracy theories, 92, 148
conspiracy theorists, 90, 95, 247
constant current electricity, 98
constantan, 263
constitution, 17, 39, 83, 93, 166, 205, 287-289, 385
Constitution of the Confederate States, 93, 287, 289
constitutional principles, 93
construction, 42, 60, 66, 69, 76, 78, 79, 89, 91, 100-103, 107, 112, 113, 118, 122, 130, 135, 142, 144, 148, 152, 154, 156, 166, 172, 174, 177, 183, 190, 191, 196, 199, 202, 218, 226, 229, 231, 234, 239, 242, 247, 267
construction builder, 103, 226
construction engineer, 113
construction industry, 66, 89, 172, 177, 231
construction sites, 242
consultant, 123
consumable electrode, 236
consumer goods, 112, 196
consumer industry, 187
consumers, 67, 68, 103
contact information, 114, 185
contact lens, 88, 99, 265
contact lens technology, 99
contact lenses, 16, 88, 99, 265

containers, 75, 187, 277
containment barrier, 204
contaminants, 276
content addressed memory, 56
content creator, 385
Continental Army, 195
continuous still, 66
continuously variable transmission, 77
continuum mechanics, 67
contraception, 273
contraceptive device, 220, 257
contraceptive pill, 63, 208, 221
contradictions, 12
Control Pad, 27, 272
controlled flight, 268
controversies, 247
conventional aircraft, 95
conventional engine, 175
conventional health organizations, 221
conventions, 6
conveyor belts, 155, 210, 238
cookies, 19, 190, 207
cooking, 3, 119, 128, 238, 257
cooking industry, 238
cookless breakfast food, 206
cooks, 238
cookstove, 242
cookware, 18, 19, 147, 210
cooling, 83, 113, 133, 182, 209, 225, 234
cooling process, 182, 209
cooling system, 133
Copeman Electric Stove Company, 67
Copernican science, 97
copper, 73, 117, 186, 242
copper production, 186
copper rivets, 73
copper wire, 242
cordite, 75
cordless portable riveter, 251
cords, 214
cork, 122, 199, 260
corn, 232
corn grinder, 232
cornea, 88, 99
corneal contacts, 88
Cornell University, 23, 68, 264
Cornell, Ezra, 264
corona discharge, 143
corporate communications, 52
corrective lens, 89
corrosion, 105, 135, 210, 264
cosmetic chemist, 178

cosmetics, 22, 187, 219, 277
cosmic energy, 179, 218
cosmodrome, 45
cosmologist, 116, 140, 143
cosmology, 81, 116
cosmos, 179
Cosray Receiver, 179
COTRAN, 45
cotton, 19, 101, 114, 155, 181, 243, 264, 283
cotton candy machine, 181
cotton gin, 264
cotton production, 264
cotton replacement, 243
cotton seeds, 264
cotton trade, 283
cotton wadding, 101
Coulter Principle, 68
counterfeit money detecting machine, 199
country music, 58
courmarin, 206
courtier, 114
cowboy, 45
cowcatcher, 43
cows, 256
CPU, 85, 171, 188
CPUs, 124, 167, 212
crackers, 19, 88, 190
craftsman, 109, 182
crane, 18, 156
cranes, 238, 248
crankless engine, 20, 175
crankshaft-powered engines, 175
crash simulations, 40
crash test dummy, 40
crates, 187, 277
Crawford, Cindy, 385
Cream, 18, 19, 154, 161, 162, 180
cream separator, 19, 154
creationism, 72
creative, 11, 34, 67, 128, 208, 385
credit card, 26, 141, 152
credit cards, 9, 26, 202, 218
Creedence Clearwater Revival, 46, 220
creeds, 385
Creek, 139
creosote, 22, 149, 219
crescent wrench, 132
crescograph, 52
crewed spacecraft, 145
Crick, Francis, 175
crime, 24, 183, 236
criminal investigations, 131
criminals, 257
criminologist, 215, 257
Crisco Vegetable Shortening, 190
critic, 44, 94
crochet, 174
Crookes dark space, 69
Crookes radiometer, 69
Crookes tube, 69
crop duster, 210
crop nutrients, 256
crop rotation, 62
crop yields, 217
crops, 61, 100
cross field amplifier, 56
cross-sectional images, 124
crossword puzzle, 27, 268
Crown Cork and Seal Company of Baltimore, 199
crowns, 67
CRT, 55, 81, 117, 222
CRT television, 222
CRT type-setter, 117
cruise missiles, 63
Cruise, Tom, 385
crushed stone, 171
cryogenics, 31, 195
Cryotron, 56
cryptanalysis, 143
cryptocurrency, 63
cryptographer, 39, 63, 211, 251
cryptographic code, 46
cryptography, 40, 60, 143
crystal, 15, 55, 87, 116, 152, 179, 201, 272
crystal rectifier, 55
crystallization rate of metals, 70
crystals, 70, 219
CSA, 93
CSS, 26, 80, 157
CT, 31, 87, 123, 124
CT technology, 87, 124
Cube, 22, 27, 67, 132, 223
cultivation, 61, 93
cultivator, 19
cultivators, 226
cultured pearl industry, 228
cupboards, 89
Curb Your Enthusiasm (TV series), 73
cure development, 145
cure for cancer, 260
current, 33, 45, 46, 77, 85, 86, 92, 98, 106, 111, 114, 118, 121, 125, 131, 148, 169, 187, 196, 199, 208, 226, 247, 260, 263
cursor, 82, 144, 163
curtain rods, 89

curtains, 16
curve, 65
cut-off valve, 53
cutler, 82
cutting boards, 187, 277
CVR, 260
CVT, 77
cyanoacrylate adhesive, 67
Cyberdyne (2004), 226
cyborg robots, 226
cycling shorts, 233
cyclist, 60
cyclone separator, 78
cyclonic action, 78
cyclostyle duplicating process, 101
cyclotron-produced radionuclides, 246
cylinder lock, 26, 272
Cyrus, Miley, 385
cytology, 101
cytopathology, 104, 201
Czochralski Process, 70
da Vinci, Leonardo, 86, 89
dactyloscopy, 257
Daedaleum, 123
daffodils, 203
Daguerre, Louis, 71, 189
daguerreotype camera, 53
daguerreotype photography, 71, 189
daguerreotypist, 210
dairy, 19, 78, 159
dairy industry, 78, 159
Dalén light, 71
dam, 25, 142, 152, 190, 210, 259
dam construction, 152
dam engineer, 210
dam maintenance and repair, 142
Dandy Horse, 77
dangerous surgeries, 130
Daniell cell, 72
dark matter, 103
dark space, 69
darkroom, 262
DARPA, 81, 116
Dartmouth College, 23, 264
Darwin, Charles, 114, 249, 262
dasymeter, 107
data, 11, 12, 15, 16, 20, 57, 74, 76, 79, 98, 101, 113, 114, 121, 124, 128, 130, 133, 146, 155, 169, 176, 177, 185, 202, 203, 207, 214, 218, 222, 233, 245, 255, 260
data analysis, 146
data archiving, 222
data compression, 222

data management, 222
data recording, 128, 155
data recording technology, 128
data structure, 176
data tables, 146
data transmission, 113, 130
data/voice combination recording device, 260
dating application, 172, 217, 260
Davy miners lamp, 73
Daylight Savings Time, 29, 93
Dayton Engineering Laboratories Company, 141
Dayton-Wright Airplane Company, 141
dazzle camo, 265
dazzle camouflage, 27, 265
DC, 90, 106, 118, 131, 197, 199, 247
DC dynamo, 199
DC electrical system, 247
DC motors, 131
De Laval Nozzle, 154
Dead Sea Scrolls, 156
Dearborn, MI, 147
death, 3, 13, 112, 186, 190, 223, 231, 290
death ray, 247
debit cards, 180, 202
decarbonization process, 89
Declaration of Independence, 131, 287
deep sea exploration, 207
defense, 68, 196, 239, 287, 289
defibrillator, 31, 46, 146, 201
defibrillators, 45, 212
deformed limb bones, 127
degreasing, 153
Delco, 141
Dell, 27, 74
Dell Computer Corporation, 74
Delta faucets, 141
delta wing concept, 133
delta-flying hang glider, 146
dementia, 203
demolition, 22, 55
demonstration, 211, 251
denim clothing, 241
Denisyuk hologram, 74
Denmark, 290
dental equipment, 186
dental implants, 147
dental motor, 76
dentist, 95, 181, 238
dentists, 34, 67
dentures, 67

deodorant, 178, 183
deodorant industry, 183
Department of Government
    Efficiency, 183
depression, 30, 174, 204
design engineer, 64
designer, 40, 46, 48, 52, 56, 57, 63,
    64, 66, 67, 71, 72, 78, 82, 87,
    88, 94, 95, 101-103, 105, 108,
    119, 122-124, 128, 130, 132,
    134, 137, 138, 141, 143, 145,
    146, 153, 154, 159, 167, 171,
    173, 176, 177, 179, 181, 185,
    186, 199, 200, 206, 207,
    210-213, 217, 220, 223, 227,
    229, 231, 234, 237, 240-242,
    251, 254, 256, 268, 272, 273,
    276, 385
desk, 131, 381
desktop computers, 87
dessert spread, 88
detachable pneumatic tire, 175
detecting disease, 49, 72
detergents, 187, 277
deuterium oxide, 99
Deutsch, Herb, 179
development, 28, 39, 45, 48-50, 52,
    53, 55, 57, 60, 63, 75, 78, 90,
    92, 94, 105, 113, 114, 116,
    117, 124, 125, 128, 129, 132,
    133, 141-146, 148, 149, 151,
    152, 154-159, 165, 168, 169,
    172, 180, 183, 184, 189, 194,
    202, 205, 207-209, 212, 213,
    221, 223, 225, 227, 228, 230,
    232, 234, 237, 239-241, 243,
    246, 249, 250, 253, 257, 266,
    286, 290, 291
Devil, 268
Dewar flask, 75
diabetes, 44, 49, 243
diabetes mellitus, 44, 49
diabetic retinopathy, 48
diagnosis, 101, 124, 151, 222
diagnostic cameras, 41
diagnostic equipment, 231
diagnostic tools, 81
diamagnetism, 86
diamond scalpel, 88
diaphragm, 29, 122, 173, 249
dichromate relief, 210
Diddley, Bo, 107
Dideoxy Technique, 226
Diehl incandescent lamp, 76
Diehl Manufacturing Company, 76
dielectric constants, 195

diesel engine, 21, 76, 108, 141, 149
diesel fuel, 21, 49, 99
diesel internal combustion
    locomotive engine, 160
diesel-powered subs, 166
diet, 3, 34, 119, 243, 385
diet and nutrition, 3, 243, 385
diet and sex pills, 119
dietary supplement, 219
differential gear train, 20, 205
diffraction, 122
digestion, 205
digestive aid, 206
digestive tract, 129
digital audio, 54
digital cameras, 53
digital cash, 63
digital computer, 80, 170
digital data, 207
digital data storage, 207
digital delay, 17, 203
digital effects, 240
digital film editing, 231
digital imaging, 53
digital information, 76
digital media, 277
digital meteorological instruments,
    256
digital photography, 134
digital pioneer, 128
digital potentiometers, 137, 244
digital switching systems, 214
digital technology services, 134
digital telephone answering machine,
    24, 115
digital television industry, 113
digital thermometer, 63, 169, 182
digital thermometers, 86
digital voltmeter, 139
digital watches, 87
digitized research tool, 57
diglyceride, 49
dill, 195
dimension, 147
dimensions, 132, 147
Dinky line, 122
diode rectifier, 112
dioptra, 142
diorama, 71
diorama theater, 71
dip pens, 261
diphtheria, 228
diphtheria test, 228
diplomat, 39, 93, 228, 232, 264
dipole TV antenna, 175
direct current (DC) motor, 118

direct currents, 90
Direct Sequence Spread Spectrum, 107
direct-current, 106, 131
director, 50, 55, 63, 65, 75, 76, 94, 102, 103, 116, 123, 128, 134, 148, 162, 195, 201, 232, 239, 240, 262
dirigible, 20, 102
dirigibles, 92, 146
dirty dishes, 123
disaster, 91, 168
disaster relief, 91
disclaimer, 9, 11, 12
disease, 34, 43, 47, 49, 72, 98, 104, 114, 120, 172, 189, 203, 207, 213, 219, 222, 249, 260, 267, 290, 389
disease control, 114
disease prevention, 219
disease-carrying protozoans, 43
diseases, 83, 129, 139, 145, 151, 152, 179, 209, 225, 236, 246
dishwasher, 25, 123
disk, 9, 15, 17, 18, 48, 49, 51, 76, 94, 104, 133, 181, 186, 189, 197, 209, 217, 233, 239, 242, 245, 246, 255
disk data storage medium, 76
disk record gramophone, 49
Disney company, 76, 128, 141
Disney theme park attractions, 128
Disneyland Theme Park, 76
disorder, 30, 185, 204
dispersive optical prism, 40
displacement lubricator, 171
displacement pump, 41
display devices, 272
disposable bottle cap, 22, 199
disposable hypodermic syringe, 31, 182
disposable safety razor, 18
dissection, 243
distilled coal, 183
distilling science, 66
distraction osteogenesis, 31, 127
dital harp, 157
dive bomber aircraft, 207
dive helmet, 234
dive mask, 22, 89, 92, 212
dive suit, 123
divergent infinite series, 107
dividing, 66, 215
diving apparatuses, 141
diving bell, 73
diving gear, 59, 234

division, 48, 197, 210
division of mass labor, 210
Dixie, 93, 184, 289
DNA, 29, 53, 60, 88, 129, 131, 144, 175, 176, 182, 226, 236, 238, 245
DNA fingerprinting, 29, 131
DNA molecules, 60, 226
DNA sequences, 129
DNA typing, 131
Döbereiner's lamp, 76
docent, 172, 385
dock maintenance and repair, 142
docking mechanism, 244
docks, 271
doctor, 28, 29, 56, 65, 139, 142, 146, 151, 188, 201, 243
Doctor Who (TV series), 28, 188
doctors, 143
dog, 74, 204
DOGE, 184
Dolby noise-reduction system, 77
doll maker, 277
dolls, 167, 277
dominoes, 119
door, 16, 25, 48, 51, 81, 86, 123, 138, 143, 157, 166, 176, 194, 200, 204, 218, 222, 228, 240, 249, 253, 264
door frames, 16, 218
doors, 22, 131, 138, 161, 162, 271
Doors, the, 161, 162
dot matrix display, 246
dot-matrix impact teleprinter, 117
double clad fiber laser, 237
double stranded plasmid DNA, 236
double-armed weapon, 187
doughnuts, 19, 190
Dow Jones Industrial Average, 77
dowsing, 219
Drachenflieger (sea plane), 146
draftsman, 153, 213, 224, 231, 264
drag chute, 146
drag coefficient, 206
Dragon-flier, 146
drainage, 21, 171, 187, 267, 277
drainage systems, 187, 277
Draisienne, 77
DRAM, 74
drawing paper, 162
dream, 121
dream machine, 148
drill machine, 128
Drinker Respirator, 31, 78, 232
drinking, 18, 237
drinkware, 236

driver's licences, 180
driving, 70, 186
drogue parachute, 146
drone, 20, 161, 229, 247, 249
drone military aircraft, 249
drone package delivery technology, 229
drones, 9, 19, 122, 168, 232, 242
drop hammer, 186
drug testing and development, 172, 249
drugs, 135, 144, 228
drum machine, 16, 248
drum manufacturer, 161
drum memory, 246
drummer, 385
drums, 9, 16, 161, 162, 272
Drunkometer, 114
dry breakfast cereals, 139
dry cleaning, 132
dry cleaning process, 132
dry dock engineer, 210
dry ice, 228
dry scouring, 132
drying food, 30
DSSC, 106, 195
DSSS, 107
dual-clutch transmission, 20, 139
Ducie, Bill, 134
Duco paint, 21, 141
Duke of Beaufort (9th), 237
dumb waiters, 131
durability, 105, 122, 138, 162, 168, 171, 206, 232, 242
durable, 66, 152-154, 177, 194, 196, 201, 232, 236, 241, 260
Dutch, 68, 77, 81, 86, 124, 125, 127, 130, 143, 145, 155, 184, 195, 196, 222, 231, 246, 266, 276
Dutch processing, 124
Duvall, Robert, 385
DVD, 15, 76, 121, 128
DVD players, 121
DVDs, 75
DVM, 139
dwarf planet, 250
dye, 44, 101, 106, 195, 206
dye industry, 44
dye solution, 101
dye-sensitized solar cell, 106, 195
dyes, 67, 206, 276
Dymaxion House, 95
Dymaxion Vehicles, 95
Dynabook, 139
dynamic random-access memory, 74

dynamic range, 69
dynamics engineer, 206
dynamite, 190
dynamo, 86, 106, 131, 142, 199, 201, 238, 263, 273, 290
dynamos, 197, 199, 247, 248
Dyson Air Multiplier, 78
e-cigarette, 157
e-commerce technology, 152
E-phase, 45
eagle scout, 385
ear mufflers, 107
ear protectors, 107
Earhart, Amelia, 140
Earl of Orrery, 105
Earl of Oxford, 385
early keyboard instrument, 52
early man, 34
earmuffs, 107
ears, 81
earth, 24, 30, 33, 92, 115, 140, 141, 156, 161, 169, 178, 179, 189, 229, 248
earth days, 140
earth globe, 169
earth orbit, 30, 141
earth-digging equipment, 156
earth-free manned flight, 178
earth-free unmanned flight, 178
earthmovers, 156
earthquake activity, 98
earthquake early warning systems, 98
earthquake maker, 247
earthquakes, 109, 117, 122, 200, 220
earth's crust, 98
earth's magnetic fields, 117
earth's orbit, 140
earth's rotational speed, 65
Eastman 910, 67
Eau de Cologne, 87
ebook, 27, 57
ecash, 63
ECG, 31, 81
Echo I, 228
ecological habitat, 99
ecologist, 95, 99
ecology, 68, 99
economic sciences, 190
economics, 188, 217
economist, 43, 142, 148, 187, 211, 273
Ecsaine, 194
ECT, 49, 62
edaphology, 61
eddy currents, 92

Edison, Thomas A., 27, 86, 88, 103, 115, 140, 153, 243, 247-249, 253
editor, 28, 50, 76, 93, 139, 157, 180, 204, 213, 219, 231, 233, 239, 243, 268, 385
editorial designer, 385
Edman degradation, 80
EDS, 281, 283, 285, 286, 290
education, 3, 11, 62, 69, 97, 125, 168, 247
educational device, 223
educational programs, 61
educators, 11, 265, 385
EEG, 31, 48
eels, 155
EFB, 213
effervescent drinks, 174
efficiency, 42, 112, 118, 119, 122, 124, 138, 168, 170, 175, 180, 186, 206, 217, 232, 233, 242, 245, 247, 256, 261, 264, 267
egg, 80, 88, 148
egg-shaped clock, 148
eggs, 195
Egypt, 88, 212, 220, 272
Egypt, ancient, 272
eidouranion, 259
Eiffel Tower, 102
Eiffel, Gustave, 102
Einstein, Albert, 165, 171, 209, 211, 249
ejection seat, 20, 77
Elastane, 16, 232
elasticity, 105, 122, 206
electric arc, 48, 118, 236
electric arc furnace, 118
electric arc welding, 236
electric automobiles, 238
electric battery, 209, 257
electric bicycles, 235
electric boat, 129
electric car, 44, 201, 251
electric cars, 235
electric cash register, 21, 141
electric chain saw, 241
electric chair, 238
electric charge, 143
electric conduction, 108
electric copy machine, 80
electric current, 86, 125, 196, 199
electric current induction, 86
electric displays, 185
electric dynamo, 86
electric elevator, 22, 234
electric elevators, 239

electric extension cord, 16, 42
electric fan, 242, 281
electric fence, 17, 98
electric field, 60
electric fire alarm, 253
electric garage door opener, 25, 86
electric generator, 21, 26, 107, 138, 142
electric generators, 121
electric guitar, 16, 25, 86, 203, 220
electric guitars, 46, 87
electric light, 26, 65, 103, 263
electric light bulb, 26
electric lighting, 104, 141, 201
electric motor, 50, 73, 86, 129, 131, 199, 239, 251
electric motor technologies, 199
electric motors, 121
electric pH meter, 46
electric plants, 201
electric potential, 227
electric power stations, 148
electric printing press, 73
electric probes, 68
electric railway, 73, 234, 239
electric railway locomotive, 73
electric self-starter automobile ignition, 141
electric shock, 201
electric speedometer, 47
electric stove, 15, 67
electric street car, 208
electric street railway, 239
electric switch, 142
electric telegraph, 129, 181
electric telegraph technology, 129
electric thermostat, 67
electric toothbrush, 25, 34, 86, 267
electric tramways, 201
electric transformer, 86
electric TV camera tube, 277
electric vehicle, 44, 79, 138, 245, 247, 250
electric wall outlet, 124
electric wall plug, 124
electric water pump, 25, 86
electric waves, 86
electric wheels, 156
electric wire, 238
electric-guided rocket, 161
electric-powered gyroscope, 161
electrical battery cell, 98
electrical cables, 234
electrical charge, 74
electrical charges, 190
electrical circuit, 64

electrical currents, 17, 98
electrical discharges, 69
electrical energy, 86, 197
electrical engineer, 41, 42, 45-47, 49, 51, 54-56, 60, 65, 68, 74, 76, 77, 79, 81, 85, 87-89, 97, 106, 116, 120, 121, 123-125, 131, 132, 137, 142-147, 151, 154, 163, 167, 172, 174, 186, 193, 196, 199, 201, 203, 210, 212, 213, 215, 218, 220, 223, 228, 231, 232, 234, 237-239, 242-244, 246, 248-250, 261, 263, 267, 271, 285, 291
electrical engineering, 48
electrical equipment, 124, 218
electrical experimenter, 131
electrical generator, 74, 131
electrical generators, 208
electrical instrument, 124
electrical light systems, 239
electrical property analysis, 100, 215
electrical science, 142
electrical scientist, 59, 119, 124
electrical sewing machine motor, 76
electrical shock, 204
electrical signals, 16, 45, 54, 233
electrical switches, 250
electrical telegraphy, 72
electrical units, 60
electrical wiring, 196
electrically charged signals, 53
electrically-powered door opener, 176
electrician, 106, 114, 125, 133, 246, 263
electricity, 17, 19, 21, 25, 40, 62, 70, 72, 77, 86, 98, 100, 102, 106, 107, 131, 159, 179, 182, 184, 189, 191, 195, 196, 202, 208, 209, 211, 212, 239, 241, 242, 247, 248, 250, 257, 263, 285
electricity distribution technology, 189
electricity producing industry, 209
electricity production, 19, 159, 257
electricity revolution, 98
electrified overhead cables, 73, 159, 234
electrified railway system, 208
electro-chemical action, 73
electro-optic effects, 116
electro-phonetic receiver, 124
electrocardiograph, 81
electrocardiographical instrument, 158
electrocardiography, 81
electrochemist, 111, 142
electrochemistry, 86
electroconvulsive therapy, 49, 62
electrodes, 69, 246
electrodynamics, 45, 117
electroencephalogram, 48
electroepitaxy, 189
electrogravitics, 55
electrohydrodynamics, 55
electroluminescence, 108, 160, 223
electroluminescence technology, 160
electrolysis, 86, 112
electrolytic method, 118
electromagnet, 18, 81, 211, 242
electromagnet technology, 81
electromagnetic actuators, 121
electromagnetic frequencies, 220
electromagnetic induction, 247
electromagnetic radiation, 168, 171, 222, 261
electromagnetic technology, 45
electromagnetic theory, 196
electromagnetic transmitter-receivers, 152
electromagnetic waves, 125, 238
electromagnetics, 131
electromagnetism, 86, 117, 118, 196
electromagnets, 242, 246
electromechanical Chinese language typewriter, 157
electromechanical engineer, 142
electromechanical relay, 118
electrometer, 17, 102, 158, 227
Electromote, 234
electromotive force, 118, 257
electron, 31, 56, 57, 62, 88, 223, 275, 277
electron beam, 56, 223
electron beam lithography, 56
electron microscope, 31, 88, 223
electron microscopes, 277
electron paramagnetic resonance, 31, 275
electronic balances, 221
electronic book, 57
electronic camera tube, 50
electronic cigarette, 18, 157
electronic communications industry, 196
electronic computer, 42, 91, 120
electronic device, 88, 120, 137, 244
electronic devices, 174, 272
electronic displays, 121
electronic equipment pioneer, 234

electronic games, 120
electronic industry, 194
electronic instrument, 248
electronic links, 187
electronic measuring devices, 263
electronic motion pictures, 44
electronic music, 123, 179, 230
electronic music genre, 179
electronic music visionary, 123
Electronic Numerical Integrator and Computer, 80, 170
electronic products, 81
electronic recipe cards, 128
electronic scales, 221
electronic sensors, 196
electronic signals, 189
electronic surveillance technology, 265
electronic technology, 171, 193, 277
electronic telephone switching systems, 132
electronic television, 87, 105
electronic ternary computer, 56
electronic textual work, 57
electronic TV system, 249
electronics, 45, 50, 51, 79, 85, 90, 100, 101, 112, 119, 120, 125, 147, 151, 166, 168, 177, 191, 195, 196, 202, 210, 213, 223, 249, 253, 262, 266
electronics engineer, 50, 223, 266
electronics industry, 85, 210, 253, 262
electronics manufacturing, 195
electrons, 51, 53, 57, 64
electrophotographist, 143
electrophotography, 61, 143
electrophysiological method, 187, 226
electrophysiology, 98
electroscope, 102, 190
electrospray ionization, 87
electrostatic battery, 275
electrostatic precipitator, 19, 231
electrotyping, 129
elemento-organic chemistry, 144
elements, 14, 35, 99, 157, 173, 196, 212, 239
elevator, 19, 22, 148, 176, 196, 224, 234
elevator technology, 176
elevators, 9, 22, 176, 239
Eleven Fifty Academy, 134
Elixir guitar strings, 105
Elizabeth I, Queen, 114
Elwell, Paul B., 201

Elwell-Parker car, 201
email, 15, 51, 250
embolectomy catheter, 91
emeralds, 256
emergency contraception, 273
emergency device, 77
emergency response, 99
emergency services, 79, 101, 202
emergency situations, 128
Emerson College, 23, 82
Emerson respirator, 82
Emerson, Lake and Palmer, 161, 162
Emitron, 50
empire, 42
emulsifiers, 172
Enceladus, 118
encryption code, 152
Encyclopedia Britannica, 291
encyclopedist, 385
endoscopes, 122
endoscopy, 53
Endy, Andrew D., 144
energy, 40-42, 48, 67, 68, 81, 85, 86, 89, 90, 92, 100, 103, 106, 107, 112, 117, 121, 130, 131, 147, 156, 172, 175, 179, 180, 191, 195, 197, 200-202, 209, 211, 218, 226, 247, 249, 265, 291
energy industry, 89
energy management, 41
energy producing technologies, 90
energy production, 201
energy technologies, 112, 147
Enfield rifles, 213
engine, 17, 20, 21, 26, 44, 45, 48, 55, 57, 58, 60, 62, 65, 67, 70, 74, 76, 93, 95, 102, 103, 108, 112, 113, 118, 122, 123, 138, 141, 149, 155, 156, 159, 160, 163, 168-170, 175, 176, 180, 188, 189, 194, 199, 200, 210, 220, 222, 228, 232, 240, 241, 244, 249, 251, 254, 261, 263-265, 268, 276, 277, 284, 286, 290
engine design, 48, 60, 95, 159, 241
engine designer, 48, 138, 220
engine efficiency, 261
engine rocket, 249
engine technologies, 102
engine technology, 156, 163, 170
engineer, 39-51, 54-57, 59-79, 81, 82, 85, 87-98, 100-108, 111-125, 127-135, 137-149, 151-163, 166-176, 178-191,

193-197, 199-215, 217-235, 237-246, 248-251, 253, 255, 256, 259-267, 271-273, 276, 277, 285, 286, 291, 385
engineering, 33, 48, 53, 86, 100, 120, 140, 141, 147, 161, 191, 229, 232, 236, 256, 261, 278, 285, 286
engineering firms, 256
engineers, 102, 141, 205
Engineers Club of Dayton, 141
engines, 18, 48, 49, 53, 70, 86, 103, 113, 139, 141, 150, 157, 163, 171, 175, 176, 179, 183, 184, 186, 191, 211, 230, 231, 238, 241, 242, 263, 291
England, 60, 61, 67, 82, 101, 106, 114, 173, 188, 283, 285
English, 42-44, 46, 49, 50, 53, 54, 56, 57, 60, 62-69, 72, 73, 78, 80, 82, 83, 86, 87, 90, 91, 93, 94, 98, 100, 102-107, 111-116, 118-120, 122, 123, 125, 127, 131, 134, 154, 155, 157, 158, 161, 167, 170, 171, 180, 184, 187, 188, 197, 199-202, 206, 212, 213, 217, 218, 220, 221, 223, 226, 228, 233, 235, 236, 238, 240-244, 246, 248, 250, 251, 255, 259, 260, 262-265, 268
English Channel, 223
English Civil Wars, 187
English postal system, 43
engraver, 53, 228
ENIAC, 170
enjoyableness, 261
Enköpings Mekaniska Verkstad, 132
entertainer, 40, 76, 82
entertainment, 11, 27, 28, 97, 130, 207, 229, 272, 385
entertainment industry, 207
entertainment media, 229
entomologist, 106
entomology, 158
entrepreneur, 40, 44, 47, 49, 50, 52, 53, 55-58, 60, 64, 66-68, 71, 73, 74, 76-83, 87-89, 91, 94, 98, 102, 103, 105, 107, 109, 113, 119, 120, 122, 124, 125, 127, 128, 130, 132-135, 138, 139, 141, 144, 149, 152-155, 157-163, 165, 167, 168, 170, 172-174, 176, 177, 179-183, 187, 190, 191, 193-197, 199, 200, 202,
204-206, 208, 210, 213, 217, 219, 221, 223, 226-241, 245, 246, 248-250, 254, 255, 257, 259-264, 266-268, 271-273, 275-277, 385
envelopes, 144
environment, 87, 100, 119, 123, 262
environmental conservation, 154
environmental impact reduction, 41
environmental practices, 62
environmental protection, 79, 101, 142, 202
environmental research, 99, 166
environmental science, 24, 156, 186
environmentalist, 61, 182
environmentally friendly, 92, 242, 250, 260
environmentally friendly car, 92
environmentally friendly hemp car, 92
enzyme, 219
Epcot Center, Walt Disney World, 95
ephedrine, 185
epidemiologist, 111
epigenetic reprogramming, 129
epigeneticist, 129
epilepsy, 47, 189
epistemology, 117, 143
epoxy, 88
EPR, 275
equinoxes, 120
era of silent movies, 249
Erector Set, 27, 102
Erlenmeyer flask, 83
Erlenmeyer rule, 83
Ermak, 166
Eroica (Beethoven), 47
error-free TV signals, 113
errors, 12, 113
escape artist, 123
eschatology, 143
Escient Technology, 134
escort fighter aircraft, 207
Esperanto, 28, 275
essayist, 94, 135, 147
essential nutrient, 219
essential oils, 86
establishment conspiracy, 220
etcher, 265
ethanol, 91
Ethernet, 15, 51, 163, 174
ethnic studies, 3
Ethyl gasoline, 21, 175
ethylene, 187, 276

etymologist, 385
etymology, 3, 167, 385
eudiometer, 133
Europa, 97, 119
Europe, 55, 102, 188, 227, 269, 283
European foulbrood, 213
European history, 3
European linguistics, 275
European mainland, 106
European pottery, 262
European royalty, 385
EV, 44, 201, 250, 251
evacuation, 91
evolution, 24, 51, 66, 72, 80, 122, 143, 154, 249, 286
evolutionary embryologist, 172
evolutionist, 99
EVs, 44
excavation, 56
excavators, 229
exciseman, 66
execution, 108, 161, 238
execution devices, 108, 161
exercise, 33, 243
exhaust gases, 138
exoskeleton, 31, 226
expansion, 62
expansion-air engine, 62
experience, 174, 290
experiences, 282
experimental cropland, 154
experimental physicist, 190
experimental physics, 133
experimental psychology, 49
experimental scientist, 122
experimentalist, 92, 98, 99, 103
experimenter, 125, 131, 202
exploding device, 233
exploration tower, 101
explorer, 57, 65, 68, 72, 116, 157, 158, 166, 182, 207, 227, 250
explosions, 138
explosive, 22, 27, 55, 75, 103, 128, 135, 138, 175, 190, 229, 230, 259
explosive fuels, 229
explosive shells, 103
explosives, 67, 109, 111, 141, 228, 248
explosives researcher, 109
expressways, 119
extension arm, 93
extension cord, 16, 42
external cardiac defibrillator, 146
external heart defibrillator, 46
extraction, 147, 225

extraction techniques, 147
extraterrestrials, 3, 247
extreme weather conditions, 138
extruded polystyrene foam, 182
extrusion-based 3D printing, 69
eye, 76, 88, 96, 117, 140, 382
eyebrows, 18, 187
eyeglass frames, 202
eyeglass lenses, 75
eyeglass maker, 158
eyeglasses, 9, 16, 34, 72, 76
eyelashes, 18, 187
eyes, 81, 103
eyewear, 72, 88
F-number system, 85
FA, 180
fabric, 105, 209, 262
fabrics, 155, 292
Fabry-Pérot Interferometer, 85, 206
face, 85, 212, 276
Face floor profile numbering system, 85
faceted drinking glass, 237
faceted drinkware, 236
facial recognition, 15, 155
facsimile device, 62
factories, 114, 119, 183, 210, 234, 271
factory, 27, 124, 171, 224
Fahlberg, Constantin, 219
Fahrenheit degrees, 86
Fahrenheit scale, 86
Fairlight Instruments, 235
family, 3, 28, 55, 76, 124, 154, 158, 203, 290, 291, 385, 388
family entertainer, 76
family histories, 3
fan-maker, 91
fans, 9, 29, 66, 86, 122, 168, 232, 242, 385
fantasy toy figures, 46
FAR file manager, 26, 222
farad, 86
Faraday Cage, 86
Faraday Effect, 86
Faraday, Michael, 208, 249
farm elevator, 19, 224
farm machinery, 224
farmer, 19, 39, 68, 101, 108, 140, 154, 261, 266
farmers, 61, 98, 106, 228, 256
farming, 61, 62, 76, 86, 111, 140, 217, 226, 238, 264
farming equipment, 226, 238
farming industry, 111
farming methods, 217

farms, 89, 100
fashion, 61, 241, 253, 265, 271
fashion industry, 241
fastest ship, 202
fat, 19, 190, 215
father, 39, 75, 93, 94, 119, 131, 166, 266, 271, 286, 288, 385
father of all manure spreaders, 140
father of American movie animation, 50
father of American technology, 265
father of artificial organs, 145
father of aviation, 89
father of basketball, 186
father of Bengali science fiction, 53
father of cardiopulmonary resuscitation, 146
father of chiropractic, 200
father of cinema, 209, 239
father of cryptography, 143
father of cytopathology, 201
father of digital word processing, 117
father of electric traction, 239
father of electricity, 247
father of electrophysiology, 98
father of emergency medicine, 201
father of English clockmaking, 250
father of food science, 41
father of home video games, 43
father of hydrostatics, 41
father of ice hockey, 69
father of inherent immunity, 172
father of Japanese microelectronics, 189
father of Java, 105
father of lawn tennis, 266
father of mathematics, 41
father of medical auscultation, 151
father of microbiology, 145, 155
father of microneurosurgery, 272
father of microwave power transmission, 56
father of modern cell biology, 98
father of modern day agriculture, 171
father of modern electrocardiography, 81
father of modern football, 180
father of modern golf, 181
father of modern hypnotism, 54
father of modern music, 203
father of modern optics, 40, 141
father of modern rocket propulsion, 103
father of modern rodeo, 45
father of modern Russian art, 167

father of modern soccer, 180
father of modern stencil duplicating, 101
father of modern vaccines, 120
father of objective psychology, 47
father of orgone therapy, 218
father of our national park system, 182
father of Peeps, 52
father of perfumery, 143
father of PHP, 156
father of playmobil, 46
father of Poland's oil industry, 162
father of positron emission tomography, 246
father of probiotics, 56
father of PVC, 218
father of radar, 119
father of radio control, 114
father of radio guidance systems, 161
father of refrigerated transportation, 133
father of relativity, 81
father of Russian aviation, 276
father of Russian battlefield surgery, 208
father of Russian shipbuilding, 147
father of serotherapy, 43
father of stereophony, 51
father of stop-motion animation, 240
father of supersonic flight, 138
father of surgery, 243
father of taxonomy, 158
father of television, 55
father of the American clock industry, 246
father of the atomic bomb, 195
father of the chainsaw, 241
father of the computer, 43
father of the Constitution, 166
father of the electroencephalogram, 48
father of the first successful artificial heart, 130
father of the fluorescent lamp, 100
father of the frozen foods industry, 50
father of the glass laser, 237
father of the ground effect vehicle, 40
father of the H-Bomb, 246
father of the handheld cellular phone, 67
father of the hard disk drive, 133
father of the information age, 109
father of the interstate highway

system, 81
father of the manmade diamond, 112
father of the modern battery, 253
father of the modern bicycle, 240
father of the modern liquid crystal industry, 87
father of the modern railway, 240
father of the plastics industry, 43
father of the postage meter, 208
father of the radio, 74, 168
father of the runway, 130
father of the Soviet atomic bomb, 148
father of the Soviet hydrogen bomb, 226
father of the telephone, 47
father of the theory of sonics, 67
father of the typewriter, 233
father of the wireless remote control, 210
father of thermosonic bonding, 68
father of voice radio, 88
father of water parks, 176
father of Yellowstone National Park, 116
father of Yosemite National Park, 182
Father Time of the Internet, 176
fathers of photography, 71
fathers of Soviet particle physics, 40
fathers of the Internet, 62, 137
fatigue, 261
fats, 190
fax machine, 15, 62, 117
fax machines, 155
FDA, 67, 218
FDR, 260
FDR/CVR, 260
feathering spectrograph, 249
federal government, 184
federal spending, 184
Federalist Papers, 166
Fedorov Avtomat, 96
female, 29, 32, 35, 119, 167, 225, 281
female brains, 281
female condom, 29, 119
Fender Manufacturing, 87
Fender Musical Instruments Corporation, 87
Fender Precision Bass, 87
Fender Stratocaster, 87
Fender Telecaster, 87
Fenton's reagent, 87
fermentation, 256
fermium, 88

Fernez-Le Prieur Diving Apparatus, 212
Ferrari, Gaudenzio, 40
ferro rod, 263
ferrocement, 152
ferrocerium, 16, 263
ferrocerium rod, 16, 263
ferroelectric memory, 56
Ferrone, Steve, 107
ferry, 241
ferrying, 66
fertility, 167, 256
fertility symbol, 167
fertilizer, 9, 19, 111, 140, 154, 157
fertilizer pioneer, 154
fertilizers, 111, 257
Feulgen reaction, 31, 88
Feulgen stain, 88
fever, 104
Fewa, 49
FFF, 69
FGMOS, 137, 244
FGT, 137, 244
fiber amplifier, 237
fiber optic cables, 119
fiber optical laser, 237
fiber optics, 122, 237
fiber-optic communications, 112
fiberglass, 81
Fidelio (Beethoven), 47
field, 33, 34, 42, 44, 49, 53, 56, 60, 62, 63, 66-68, 79, 81, 98, 103, 105, 106, 114, 118, 124, 125, 130, 132, 137, 144, 146, 147, 151, 152, 155, 158, 161, 163, 167, 196, 200, 201, 204, 208, 212-214, 218, 221, 223, 228, 229, 238, 241, 242, 244, 246, 267, 290
field surgeon, 44
film, 3, 23, 28, 47, 50, 54, 75, 76, 79, 80, 94, 106, 122, 123, 128, 132, 133, 137, 143, 162, 166, 184, 187, 195, 201, 207, 213, 218, 230, 231, 236, 239, 240, 244, 249, 255, 259, 265, 267, 277, 289, 385
film actor, 47, 76, 265
film art director, 240
film coating, 236
film composer, 385
film director, 50, 75, 76, 94, 123, 128, 239
film editing, 231
film editing machine, 231
film editing technology, 231

film editor, 28, 50, 231, 239
film industry, 80, 128, 184, 207, 231, 240, 255
film producer, 50, 75, 80, 123, 128, 132, 162, 239, 267
film projector, 162
film score composer, 230
film scores, 230
film sheets, 143
film wrapping, 187, 277
filmmaker, 40, 57, 68, 102, 162, 168, 184, 187, 240, 385
films, 29, 40, 74, 168, 213, 243, 248
filter, 28, 44, 50, 57, 62, 126, 139, 173, 179
filter pump, 57
filtering coal gas mask, 276
filters, 44, 85, 134, 152, 194, 243
finance, 77, 180, 207, 284
financial services, 273
financial world, 180
financier, 132, 180
finger-driven touchscreen interface, 132
fingerprinting, 29, 131, 257
fire, 16, 17, 22, 29, 66, 67, 72, 77, 93, 99, 128, 148, 160, 167, 169, 223, 230, 253, 260
fire alarm, 22, 72, 253
fire extinguisher, 17, 99, 160, 167
fire fighters, 99
fire prevention, 128
firearms, 34, 56, 67, 103, 142, 200, 219, 227, 266
firearms development, 142
firearms manufacturer, 200, 219
firebombs, 73
firefighters, 276
firefighting, 20, 30, 91, 119, 160, 205
firefighting foam, 160
firelock, 142
fireman, 171
Firepaste, 125
fireplaces, 128
first law of thermodynamics, 180
first photos of earth from the moon, 24, 115
first successful flight, 153
first-aid technique, 116
Fischer assay, 89
Fischer wall plug, 89
Fischer, Franz J. E., 250
Fischer-Tropsch Process, 21, 250
Fischer-Tropsch synthesis, 89

fish, 382
fishbone antenna, 253, 271
fishbone TV antenna, 260
Fisher space pen, 89
fishing, 83, 243
fishing nets, 243
fitness innovator, 139
five-note air-horns, 244
fixation, 276
fixator, 127
fixed-wing aircraft, 45
flaked cereal industry, 139
flame, 73, 159, 191, 260, 263
flammable, 99, 109, 202, 203
flash boiler, 231
flash freezing, 50
flash memory, 137, 169, 244, 245
flash steam boiler, 231
flash stick, 245
Flash-Matic, 210
flasks, 229
flat homeoscope, 146
flat screen displays, 116
flat-panel displays, 245
flat-screen TVs, 87, 89
flatbed scanner, 15, 149
flautist, 51
flavor, 113
flavorings, 172, 206
Fleetwood Mac, 161, 162
Fleming valve, 90
flexible belts, 248
flexible endoscopes, 122
Flexor-Line clothesline, 67
flight data recorder, 20, 260
flight engineer, 133, 210
flight industry, 138
flight parameters, 260
flight science, 133
flight scientists, 109
flint mechanism, 76
flintlock, 24, 53, 92, 109, 142
flintlock firearm mechanism, 92
flintlock gun mechanism, 53
flip-disk display, 17, 246
flip-dot display, 246
floating, 137, 244
floating gate transistor, 137, 244
flooding issues, 267
floppy disk, 15, 186, 233, 245
floppy disks, 233
Florida, 57, 95
flu, 30, 120
fluctuation-dissipation theorem, 81
fluid, 68, 107-109, 117, 122, 138, 168, 195, 202, 232

fluid dynamics, 117, 122, 168, 232
fluid flow theory, 138
fluorescein, 44
fluorescent dyes, 67
fluorescent light, 26, 100, 119
fluorescent X-rays, 119
fluorocarbon chemistry school, 144
FLUOROSCAN, 45
flush toilet, 30, 114
flute, 25, 51
flute fingering technique, 51
fly base, 73
Flyin-Saucer, 181
flying machine, 73, 89, 108, 130, 247
flying machines, 205
FM radio, 26, 42
foam fire extinguisher, 17, 99, 160
focal-plane shutter SLR camera, 115
Fogarty balloon catheter, 91
foil electret microphone, 17, 231
fold-up bed, 183
folding campstool, 131
folding footstool, 131
Foley, Charles, 217
foliage string trimmer, 44
folk music, 58
folk wrestling, 196, 239
Foo Fighters, the, 161, 162
food, 16, 17, 19, 26, 29, 30, 41, 50, 52, 75, 78, 79, 100, 101, 111, 119, 133, 139, 140, 144, 154, 166, 169, 172, 187, 190, 191, 194, 202, 204, 206, 238, 261, 272, 277
food colorings, 172
food containers, 75
food delivery, 79, 101, 202
food industry, 190, 194
food items, 272
food packaging, 144
food preservation, 169, 187
food preservation industry, 169
food producers, 204
food production, 119, 140, 154
food products, 187, 277
food-drying technology, 78
foods, 34, 50, 119, 128, 133, 152, 190, 219
foot pedals, 82, 250
football, 22, 59, 60, 82, 180, 185
Football Association, 180
Foote, Shelby, 385
footstool, 131
footwear, 16, 27, 155, 210, 218
force, 31, 49, 68, 69, 79, 95, 100, 101, 118, 138, 202, 215, 218, 219, 257
Ford Gyron, 232
Ford II, Henry, 148
Ford Motor Company, 91, 147, 216
Ford Rotunda, 147
Ford Thunderbird, 141
Ford, Henry, 140
Ford-Welton, Hannah, 107
forehead strip thermometer, 201
forensic authorities, 131
forensic science, 24, 156
forensics, 166, 182
forest, 74, 77, 100, 102
forest belt, 100
forester, 100, 220
forging metal, 186
forklift, 18, 65
forklift truck, 65
forklifts, 127, 229, 238
Forlanini airships, 92
formic acid, 49
fossil fuel industry, 201
fossil remains, 209
fossils, 122, 148, 156
Foucault pendulum, 92
foundational components, 62
foundations, 66, 167, 177
Founders, 13, 14, 55, 236, 265
founders of modern holography, 74
Founding Father, 39, 75, 93, 131, 166, 266, 288
Founding Fathers, 68, 93
fountain pen, 25, 29, 34, 73, 91, 160, 205, 210, 261
four-wheel steering, 201
fourth state of matter, 69
Foveon, 173
fragile nervous system structures, 272
fragrances, 206
frame beehive, 24, 213
Frame Technology Group, 144
framed sextant, 169
frames, 176
framework knitting machine, 155
France, 102, 163, 209, 211, 283, 284
Frankenstein (Shelley), 98
Franklin stove, 93
Franklin, Benjamin, 33, 93, 114, 160
Frasch Process, 93
fraud, 184
fraudulent inventions, 175
free energy generator, 48
free energy technology, 48

free love movement, 63, 208
free software, 239
Free Software Foundation, 239
free unlimited energy, 180, 247
freedom, 285
freezing point, 62, 86
freezing point of water, 86
French Academy, 125
French Academy of Sciences, 125
French fries, 19, 190
French scientists, 102
Freon, 17, 175
frequency meters, 263
frequency modulation, 42, 113
freshwater mussels, 155
Fresnel lens, 94
Freud, Sigmund, 211, 218
Friday, 204
friendly compact, 93
Frigidaire, 78
fringe science, 106
Frisbee, 181
frog nerve cells, 115
fruit, 174, 219
frying food, 30
Frying Pan (guitar), 46
fuel, 19, 21, 22, 48, 49, 64, 76, 77, 91, 99, 103, 107, 111, 117, 121-123, 138, 155, 168, 175, 194, 200, 201, 206, 219, 229, 232, 242, 256, 257, 261
fuel cell, 107, 111, 175
fuel economy, 194
fuel efficiency, 122, 138, 168, 206, 232, 256
fuel injectors, 121
fuel pump laminations, 121
fuels, 22, 200, 219, 229, 250
full face dive mask, 212
full pressure suit, 64
fully automated aircraft, 113
Fulton, Robert, 102
fumigation, 109
funding, 95, 247
funding and patent issues, 247
fungal organisms, 104
furnace, 30, 118, 234
furnaces, 30
furniture, 18, 46, 138, 140, 187, 194, 223, 277
furniture industry, 194
furniture maker, 46, 140
fused deposition modeling, 69
fused filament fabrication, 69
fusion, 47, 48, 90, 225, 239, 272
fusion technique, 47

futurist, 44, 95, 149, 161, 230, 246
futurist painter, 44
G. A. Morgan Hair Refining Company, 180
GaAs molecular layer epitaxy, 189
gadgets, 230
galantamine, 203
Galanthus nivalis, 203
galaxies, 124
Galileo, 227, 249
Galvani, Luigi, 257
galvanic battery cells, 73
galvanism, 98
galvanized metal tie, 135
galvanometer, 17, 60, 81
Game and Watch, 27, 272
Game Boy, 27, 272
game designer, 72, 95, 173, 200, 217, 223, 254, 272
game engines, 242
Game of Thrones (TV series), 169
Game of Thrones (TV series), 28
game theory, 188
games, 9, 26, 27, 43, 120, 155, 169, 177, 272
gaming, 124, 167, 212
gamma camera, 41
gang nail plate, 135
Ganymede, 97
garage door opener, 25, 86
garden hoses, 9, 26, 218
gardener, 19, 177
gardeners, 98
gardening, 61
Gardner-Serpollet Steam Car, 231
Garibaldi, Giuseppe, 174
gas, 15-20, 25, 42, 44, 48, 65, 69, 71, 73, 75, 76, 89, 99, 107, 108, 130, 142, 143, 146, 153, 159, 180, 182, 183, 190, 193, 194, 200, 209, 210, 241, 242, 247, 250, 257, 263, 264, 276
gas absorption refrigerator, 182, 209
gas discharge lamp, 65
gas helium-neon laser, 130
gas laser, 48
gas lasers, 130
gas lighting, 108, 183
gas lighting system, 183
gas mask, 18, 180, 276
gas pipes, 264
gas pressured vessel, 42
gas production, 183
gas turbine, 25, 44, 241
gas, oil, and flame-based weapons, 159

gas-filled canisters, 159
gas-filled incandescent light bulb, 153
gas-permeable material, 99
gases, 31, 67, 78, 86, 123, 138, 143, 212
gaskets, 27, 154
gasoline, 9, 20, 21, 49, 86, 99, 141, 175, 187, 211, 250, 276, 277
gasoline powered engines, 211
gasoline tanks, 187, 277
GASPEC, 45
Gastornis, 209
gastronomy, 385
Gatling gun, 99
Gayheart, Rebecca, 385
GE, 27, 80, 387
gear mechanism, 204
gearless anchor, 39
gears, 73, 77, 186
GEET, 48, 200
Geet Fuel Processor, 48, 200
GEET Plasma Motor, 200
Geiger counter, 28, 99
gemstone industry, 112
gemstones, 70, 256
gene, 31, 53, 60, 83, 98, 129, 221, 236
gene targeting, 31, 60, 83, 236
gene-splicing technology, 53
genealogist, 385
genealogy, 3, 292, 385
general anaesthetic, 230
General Electric, 27, 80, 180, 248
General Motors, 22, 44, 78
General Motors Corporation, 78
general theory of relativity, 80
generator, 21, 26, 48, 74, 86, 106, 107, 117, 131, 138, 142, 200, 229, 250, 263
generator technology, 263
generators, 18, 50, 53, 92, 121, 131, 197, 208, 242, 248, 263
genes, 83, 98, 226, 236
genetic diseases, 236
genetic engineering, 53
genetic modification, 53, 236
genetic sample, 131
genetically engineered moss, 219
genetically modified (transgenic) mouse, 129
genetically modified organism, 129
genetically modified organisms, 53
geneticist, 60, 83, 129, 131, 219, 236
Genko's Forest Belt, 100

genome, 83
genomics, 238
genotype technique, 60
genus, 43, 158, 209
geochemist, 48
geodesic dome, 28
geodeticist, 239
geographer, 46, 117, 143, 148, 159, 196, 211
geological engineer, 44
geological findings, 116
geologist, 72, 73, 116, 117, 122, 148, 159, 218, 220, 222, 227, 236, 238
geology, 88, 158, 227
geometer, 118
geometric optics, 125
geometrist, 107, 122
geometry, 117, 142, 221, 235
geophysicist, 45
Georgia, 205
geostationary satellite, 30, 65
geostationary satellites, 65
GERD, 49
germ-killer, 90
germanium, 112
Germans, 159
Germany, 70, 112, 285
germination, 113
Gestetner Automatic Cyclostyle Mimeograph, 101
Gestetner Cyclograph Company, 101
Getting, Ivan A., 202
GFP, 251
ghost, 3, 287, 290
ghost stories, 3, 287, 290
GI, 243
giant crossbow, 73
Giemsa stain, 101
Giesl ejector, 102
Giffard Dirigible Airship, 102
gift cards, 202
Gilhoolie, 95
Gill Studios, Inc., 102
Gillette Company, 102
Gillette razors, 102
gimbal, 51
Ginkgo Bioworks, 144
ginseng, 230
girders, 66, 177
Girdler Sulfide Process, 99
glass, 18, 21, 23, 29, 32, 34, 40, 41, 44, 65, 76, 79, 83, 86, 88, 89, 92-95, 103, 111, 118, 130, 136, 143, 147, 155, 158-160, 169, 173, 183, 198, 229, 234,

236, 237, 266
glass disks, 44, 183
glass electrode, 111
glass fiber laser amplifier, 237
glass harmonica, 93
glass keys, 65
glass lenses, 155
glass mosaics, 160
glass plates, 79
glass tube, 86, 94, 229
glass-bottom boat, 147
glass-maker, 159
glass-making, 147
glassmaker, 236
glider, 18, 62, 146, 157, 178, 179
glider model, 62
glider recreation, 179
glider science, 179
global arms race, 166
Global Environmental Energy Technology, 200
global film industry, 255
Global Positioning System, 79, 101, 202
global unity, 275
Glock 19, 103
Glock firearms company, 103
Gloucestershire, UK, 237
glue, 135, 144
glycolic acid, 83
GM, 44
GMC, 78
GMO, 53, 129
GMS, 104
GNU operating system, 26, 239
goats, 256
God, 219, 289
goddess-worship, 167
godfather of cryptocurrency, 63
godfather of synthetic biology, 144
God's businessman, 156
Goebel Lamp, 103
gold, 115, 227, 385
gold miner, 227
Golden Gate Bridge, 230
goldsmith, 78, 109
golf, 18, 22, 147, 181, 290
golf ball maker, 181
golf club maker, 181
golf clubs, 18, 147
golf course designer, 181
golf educator, 181
golfer, 181
Golgi's method, 104
Gomori methenamine silver stain, 104

Gomori trichrome stain, 31, 104
Goodyear Blimp, 276
GoPro, 27, 267
GoPro, Inc., 267
Gore-Tex, 105
Gothic horror novel, 98
government, 23, 64, 79, 83, 89, 101, 148, 162, 174, 183, 184, 187, 202, 207, 208, 218, 219, 226, 260, 262, 282, 284, 285, 289, 291
government buildings, 89
government bureaucrat, 262
governments, 98
governor, 93, 113, 239, 385
GPS, 9, 24, 68, 79, 81, 101, 115, 119, 130, 163, 166, 202
GPS navigation systems, 81
GPS technology, 115, 130
Gracenote, 134
Graf Zeppelin (airship), 276
grain, 68, 93, 113, 261
grain milling, 68
grain trader, 93
grains, 261
Gram staining method, 31, 106
gramicidin S, 99
grammarian, 159
Gramme Armature, 106
Gramme Dynamo, 106
Gramme Machine, 106
Gramme Ring, 106
Grand Canyon National Park, 26
Grand Funk Railroad, 161, 162
grandfather, 262
grandfather of television, 74
Grant, Ulysses S., 174
granyonyi stakan, 237
grapes, 83
graphene, 100, 190, 191
graphic artist, 237, 385
graphic designer, 237, 385
graphic user interface, 51
graphics, 123, 124, 167, 187, 212
graphics cards, 124, 167, 212
graphite, 205
graphophone, 47
Grätzel Cell, 106, 195
grave, 230
gravel, 171
graves, 283, 385
Graves, Robert, 385
gravicembalo col piano e forte, 69
gravitation, 133, 142, 188
gravitational singularity theorem, 116

Gravitor, 55
gravity, 55, 89, 106, 122
gray water, 47
Grayway Rotating Drain, 47
grease-spot photometer, 57
Great Britain, 100, 278
Great Mother, 167
great outdoors, 174
Greathead Shield, 106
green energy, 200
green fluorescent protein, 31, 251
green material, 77
green physics, 48
green screens, 134
Gregorian telescope, 107
Gregory's series for the arctangent function, 107
Gretsch Company, 107
Gretsch drums, 16, 107
Gretsch guitars, 16, 107
Griffith, Andy, 385
grilling food, 30
grizzly bear attacks, 125
grocery store checkout scanners, 48
ground effect vehicles, 40
ground receiver, 57
Grove Battery, 107
Grove Cell, 107
growth, 52, 94, 127, 135, 151, 155, 183, 187
growth of affordable housing, 135
Guanella balun, 107
guard duty, 172
Guardian Angels, 24, 236
Guarneri, Bartolomeo G., 25
GUI, 51
guided missile systems, 113
Guidonian musical notation, 41
Guillaume, Maxime, 265
Guillotine, 108, 161
guitar, 16, 25, 46, 63, 86, 105, 107, 157, 202, 203, 220
guitar accessories, 220
guitar picks, 202
guitar-like instrument, 63
guitarist, 63, 157, 203, 385
guitars, 46, 87, 220
Güldner and Lüdeke Maschinenfabrik Magdeburg, 108
Güldner-Motoren, 108
gulonolactone oxidase, 219
gum disease, 34, 267
gum pain, 34
gum stimulators, 34
gun, 24, 50, 53, 66-68, 99, 142, 154, 167, 168, 170, 176, 182, 200, 213, 219, 227, 232, 241, 288, 385
gun designer, 213
gun manufacturer, 66, 213, 227
gun rights advocate, 385
gun technology, 142
gunpowder, 24, 142, 173, 227, 229, 230, 256
guns, 9, 23, 56, 183
gunsmith, 53, 56, 89, 92, 99, 168, 195, 219
Gurney Stove, 108
Guyana, 206
gyms, 119
gynecologist, 177, 220, 221, 273
gyrocar, 22, 232
gyrocar design, 232
gyrocompass, 41
gyrocopter, 22, 64
gyroplane, 64
gyroscope, 27, 51, 92, 113, 147, 161, 238, 264
gyroscope stabilizer, 147
gyroscopes, 232
Haber-Bosch Process, 111
hafnium, 119
hair brush handles, 202
hair combs, 202
hair curler, 187
hair growth measuring device, 187
hair-straightening cream, 180
hairdresser, 187, 260
Hale Telescope, 231
half-track, 139
Hall Braille writer, 112
Hall effect, 195
Hall, Charles M., 118
Hall-Héroult Process, 112, 118
ham, 28, 195, 253, 271
ham radio, 28, 253, 271
hammer-striking, 92
Hamming codes, 113
Hammond Castle, 113
Hammond Radio Research Laboratory, 113
Hampton Roads Conference, 289
Hampton Roads, Virginia, 83
hand mirrors, 202
hand power, 264
hand spinners, 114
hand-blown glass apparatus, 158
hand-held metal detector, 89
hand-loom weaver, 114
hand-sewing, 124
handball, 82
handgun, 66

handheld cellular phone, 67
handheld electronic game system, 273
handheld game console, 272
handheld mobile phone, 67
handheld tools, 74
handlebars, 240
handling, 206
handmade paper, 162
hands, 183, 248, 381
handwriting industry, 261
hang glider, 18, 146, 157
hanging, 131, 238
hanging bed, 131
Hannoversche Flugzeugwerke, 130
Hannoversche Flugzeugwerke limited liability company, 130
happiness, 260, 288
happy accidents, 34
happy accidents, and inventions, 34
hard disk drive, 133
hard disk drives, 9, 18, 242
hard rubber, 201
Harding, William G., 385
harmonic drive gear, 20, 184
harmonica, 16, 58, 93
harp guitar, 157
harp lute, 157
harp lyre, 157
harpischord, 65
harpoon gun, 50, 167
harps, 163
harpsichord maker, 69
Hartford, Edward V., 251
Harvard College, 115
Harvard University, 23, 115, 169, 292
Hasselblad 500EL/70 camera, 115
Hasselblad modular V-System, 115
hats, 180
haunted, 8, 287, 290
haunted houses, 290
Hawking Radiation, 116
Hawking, Stephen, 249
hay, 19, 224, 226, 282
hay baler, 19, 224
Hayden Valley, 116
haymaking machine, 19
haymaking machines, 226
hazardous material workers, 276
HDD, 133
HDPE, 187, 276
heading, 161, 260, 265
headlights, 94, 121
headphones, 234
healing, 31, 100, 109, 173, 200, 219, 220, 243
healing properties, 220
health, 3, 26, 29, 31, 75, 99, 100, 109, 114, 119, 123, 129, 139, 140, 143, 151, 152, 157, 166, 181, 186, 191, 196, 200, 203, 204, 216, 218, 219, 221, 222, 260, 288, 290, 385, 389
health and fitness, 3, 181, 385
health and fitness industry, 181
health and medicine, 166
health benefits, 26, 196, 218
health conditions, 203
health field, 151, 152
health food, 139
health hazard, 75
health issues, 204, 222
health risks, 109, 157
health spa, 140
healthcare, 145
hearing aids, 231, 249
hearing impaired, 47
hearing-impaired industry, 256
heart, 31, 32, 45, 46, 74, 81, 82, 101, 121, 130, 145, 151, 158, 201, 207, 212, 265, 381, 385
heart blockages, 82
heart defibrillator, 46
heart failure, 82
heart irregularities, 121
heart issues, 82
heart resuscitation, 46
heart rhythm abnormalities, 121
heart surgery, 46
heart-lung machine, 31, 101, 158
hearts, 34
heat, 22, 23, 27, 34, 40, 41, 50, 68, 72, 100, 119, 125, 128, 182, 191, 209, 210, 260, 261
heat energy, 40
heat pumps, 41
Heat-Indicator and Fire Alarm, 72
heat-proofing material, 125
heated air, 241
heater fan, 147
heating, 19, 22, 72, 83, 104, 128, 219, 226, 234, 257
heating oil, 219
heaven, 290
heavier-than-air flight, 157
heavier-than-air glider, 178
heavier-than-air vehicle, 92
heavy construction equipment, 156
heavy fighter aircraft, 207
heavy lifting, 91
heavy water, 99

Heimlich Maneuver, 30, 116, 265
Heimlich, Henry J., 265
helicopter, 22, 23, 49, 64, 73, 91, 92, 160, 176, 205, 234, 273
helicopter designer, 176, 273
helicopter-like device, 160
helicopters, 113, 122, 168, 176, 232, 242, 273
helium-neon, 48, 130
helium-neon laser, 130
hell, 116, 117, 174, 287, 289
Hell Printer, 117
Hellschreiber teleprinter system, 116
Helmholtz pitch notation system, 117
Helmholtz resonator, 117
HEMA, 265
hematology, 68, 101
hemodialysis machine, 145
Hemp Car, 91
hemp fiber, 91
hemp-based fuel, 91
Hendershot Generator, 117
Hendrix, Jimi, 161, 162
hepatitis, 30, 120
hepatitis A, 30, 120
hepatitis B, 30, 120
herbs, 61, 230
Hereditary Keeper of Raglan Castle, 237
hero, 32, 117, 118, 157, 205, 287
Heron's Engine, 118
Héroult furnace, 118
Héroult, Paul, 112
Hertz, 119
Hertz, Heinrich, 261
Hewlett-Packard Company, 119, 199
high altitude fighter aircraft, 207
high density data storage, 74
high output powers, 118
high performance fuel, 123
high quality paper, 249
high resolution capabilities, 100, 215
high resolution microscope, 276
high resolution spectroscopy, 85
high security-risk environments, 89
high voltage wires, 68
high wheelers, 240
high-altitude rockets, 249
high-compression automobile engine, 141
high-cut riding chaps, 45
high-density polyethylene, 187, 276
high-energy circular colliders, 195
high-energy physics research, 103

high-power diode laser, 98
high-pressure mercury-vapor lamp, 100
high-pressure steam engines, 186
high-resolution imaging, 50, 100, 215
high-resolution laser spectroscopy, 130
high-rise skyscrapers, 141
high-speed aircraft, 133
high-speed computing, 143
high-speed turbine industry, 154
hikers, 67, 263
hiking expeditions, 82
hills, 113, 268
Hindenburg (airship), 276
Hinge, 15, 172
hinged bed, 183
Hippomobile, 155
histochemist, 104
histological staining method, 102
histologist, 102, 104
histologists, 261
histology, 101
historian, 5, 41, 49, 134, 157, 159, 161, 188, 225, 228, 232, 380-382, 385
historic Scottish records, 181
history, 3, 6, 11, 15, 24, 32, 33, 35, 40, 81, 86, 91, 103, 105, 117, 134, 148, 174, 179, 208, 210, 223, 230, 264, 268, 273, 282-287, 289, 291, 292, 381, 382, 385
history museum docent, 385
HMS Beagle, 72, 114
hobbyists, 79
hockey, 22, 69
hoists, 210
hole punch, 237
hologram, 29, 74
holograms, 130
holographic devices, 97
holography, 31, 74, 97
Holter Monitor, 31, 121
home appliance industry, 42
home appliances, 79, 101, 120, 202, 218
home construction, 135
home decor, 138
home furnishings, 138, 233
home goods, 138
home video games, 43
homeland security, 99
homeopathic doctor, 146
homeowners, 100

homes, 77, 86, 114, 124, 156, 172, 271
honey, 9, 24, 213
honey production, 213
Hooke's Law, 122
Hopkins, Johns, 103
hops, 113
horizontal and vertical scans, 105
horizontal directional aerial, 168
horizontal transport, 239
hormesis stones, 109
hormones, 219
Hornby Model Railways, 122
hornless bronc saddle, 45
horologist, 115, 148, 158, 238
horror, 98
horse, 73, 77, 163, 166, 191, 261, 264
horse colossus, 73
horse power, 163, 191, 264
horse waste, 261
horses, 93, 256, 261, 290
horticulturist, 53
hoses, 9, 26, 155, 194, 218
hospital, 29, 93, 122
hospitals, 99
host fighter aircraft, 207
hot air balloon pioneer, 178
hot air balloon technology, 178
hot air balloons, 178, 241
hot spots, 51
Hot Tamales, 52
hot-wire galvanometer, 60
hotel, 26, 133, 140, 202
hotel room keys, 202
Hotz Box, 123
Hotz MIDI Translator, 123
Hotz MIDI vest, 123
Hou Feng Di Dong Yi, 117
house, 18, 95, 108, 124, 129, 152, 156, 166, 237, 287, 291
House of Representatives, 291
House, Royal E., 129
house-builder, 108
household, 42, 86, 113, 121, 122, 168, 174, 232, 242
household appliances, 121
household devices, 113
household fans, 122, 242
houses, 135, 290
housewares, 187, 277
Houten, Coenradd J. van, 124
Hovercraft, 23, 66, 147
HP, 27, 119, 199
HPV vaccine, 31, 94, 276
HTML, 26, 49, 80

HTTP, 26, 49
Hubble Space Telescope, 29, 124, 239
human aura, 219
human body, 219
human brain, 48, 287
human cancer, 30, 120
human diseases, 83
human eye, 140
human flight, 109
human hair, 194
human health, 99, 123
human memory, 122
human papillomavirus vaccine, 94, 276
human Swiss Army knives, 35
humanitarian, 108, 124, 162
humanity, 32, 228, 251
humans, 33, 34, 144, 172, 183, 230, 262
humidity, 57, 73, 115, 177, 227
humor, 3, 289, 385
humorist, 180
hunt, 180, 282
hunter, 64
hunters, 50, 66
hunting, 34, 168, 227
Huygens' Principle, 125
HVAC equipment, 272
Hyatt Filter, 126
hydraulic accumulator, 42
hydraulic brakes, 201
hydraulic cocoa press, 124
hydraulic engineer, 238
hydraulic interrupter gear, 67
hydraulic press, 203
hydraulic presses, 186
hydraulic shock absorbers, 123
hydraulic systems, 42
hydraulis, 69
hydro-mechanical astronomical clock tower, 238
hydrocarbons, 49, 89
hydrochloric acid, 256
hydroelectric dams, 86, 92
hydrofoil ship, 92
hydrofoils, 40
hydrogen, 75, 76, 87, 89, 102, 107, 111, 143, 153, 212, 225, 226, 246, 250, 276
hydrogen bomb, 225, 226, 246
hydrogen chloride, 212
hydrogen gas, 75, 76
hydrogen peroxide, 87
hydrogen welding, 153
hydrogen-filled aircraft, 102

hydrogen-oxygen fuel cell, 111
hydrogenated product, 190
hydrogenation of fats, 190
hydrogenation process, 190
hydrographer, 46
hydroplane, 20, 92
hydrostatic balance, 142
hydrostatic bed, 42
hydroxy ethyl methacrylate, 265
hydroxyl group, 83
hygrometer, 30, 73, 122, 227
hyperabrupt variable capacitance diodes, 189
hyperbaric welding, 142
hyperboloid design, 233
hypersonic rocket-powered aircraft, 20, 77
hypersonic speed, 165
hypertext, 49, 156, 187
hypertext markup language, 49
hypertext preprocessing, 156
hypertext transfer protocol, 49
hypnosis, 31, 173, 174
hypnotherapy, 173
hypnotism, 54, 173
hypnotist, 53, 102
hypochlorite, 49
hypodermic syringe, 31, 182
hyposulfite of soda, 118
hypotheses, 12, 107
Hz, 119
I/O devices, 188
iambic pentameter, 156
IBM, 27, 51, 57, 121, 133, 233
IBM Model 350, 133
iBOT, 21, 138
IC, 120
ICBM, 185
ICBMs, 63
ice, 22, 39, 57, 66, 67, 69, 138, 166, 169, 228, 275
ice calorimeter, 57
ice climbing, 39
ice hockey, 69
Ice Resurfacer, 22, 275
ice rinks, 275
ice screw, 39
ice shavings, 275
ice surface, 275
icebreaker, 20, 166
ichthyologist, 228
iconoscope, 21, 277
Ictíneo I (submarine), 179
Ictíneo II (submarine), 179
identification badges, 202
identity cards, 180

ideographic writing system, 50
ideoscope, 146
IF, 12, 13, 16, 17, 19, 20, 22, 24, 27-29, 32, 35, 63, 82, 115, 201, 204, 226, 273, 389
iGEM, 144
igniting fires, 263
ignition, 21, 92, 141
IKEA, 18, 138
Ilizarov Apparatus, 31, 127
Ilizarov Surgery, 31, 127
Ilizarov Technique, 127
ill, 276
Illinois, 112, 213, 282
Illinois School for the Blind, 112
illness, 94, 116
illumination of optics, 236
illumination technologies, 263
illusion, 123
illusion of motion, 123
illusionist, 123
illustrator, 76, 107, 135, 167, 213, 240, 265
Ilon Wheel, 127
images, 6, 11, 44, 53, 61, 62, 75, 80, 94, 123, 124, 140, 246
imaging, 28, 50, 53, 71, 92, 100, 119, 123, 186, 215, 222, 246
imaging device, 92
imaging instrument, 123
imaging methodology, 246
immediate auscultation, 151
immersion microscopy, 41
immune system, 219, 228, 256
immune system regulation, 219
immunologist, 67, 94, 119, 172, 276
impaired vision, 89
implantable pacemaker, 82
Impressionistic Period, 177, 217
improved handwriting, 261
in photography, 158
in situ hybridization, 31, 98
in vitro fertilization, 63
in-vitro fertilization, 80
inaccuracies, 12
inanimate objects, 143
inboard boat motor, 244
inboard motor engine, 95
incandescent electric lights, 107
incandescent lamp, 76, 159
incandescent light, 80, 103, 104, 121, 153, 243, 263, 271
incandescent light bulb, 80, 103, 104, 121, 153, 243
incandescent light technology, 271
incandescent lights, 248

Incas, 139
incident radiation, 118
incinerator technology, 78
inclined plane, 220
income, 273
independent filmmaker, 240
independent thinkers, 55
Index Thomisticus, 57
India, 237, 243, 281, 284
indigo, 44
individuals, 11, 12, 68, 77, 89, 100, 117, 140, 161, 162, 179, 181, 184, 194, 235, 263
individuation, 31, 135
Induced Pulse Transient system, 45
inductance, 118, 189
induction, 59, 86, 125, 128, 189, 246, 247
induction balance, 125
induction coil, 59
induction heat, 128
induction motor, 246
industrial, 21, 60, 62, 67, 77, 81, 92, 98, 99, 106, 109, 114, 121, 122, 135, 141, 144, 153, 155, 156, 166, 168, 170, 173, 177, 186, 203, 204, 206, 222, 224, 228, 232, 238, 262, 265, 291, 292
industrial applications, 98, 204
industrial assembly lines, 238
industrial chemist, 99, 109, 153
industrial designer, 67, 141, 177
industrial machine tools, 156
industrial manufacturers, 106
industrial plants, 92
industrial products, 206
industrial pumps, 121
Industrial Revolution, 114, 155, 170, 186, 222, 228, 262, 265
industrial solenoids, 121
industrialist, 42, 49, 60, 66, 71, 91, 101, 107, 124, 125, 163, 170, 173, 175, 190, 191, 200, 201, 206, 208, 210, 213, 218, 221, 231, 234, 238, 240, 241, 263
industrialization, 73
industrialization of the West, 73
industries, 42, 44, 51, 53, 61, 66, 76, 78, 83, 91, 92, 94, 99, 100, 112, 120, 132, 133, 135, 141, 156, 171, 174, 175, 177, 180, 184, 190, 191, 194, 197, 205, 207, 215, 218, 223, 231, 234, 236, 238, 247, 248, 251, 256, 261

industry, 19, 43, 44, 46, 49, 50, 53, 54, 63, 66, 73, 75, 78, 80, 82, 85, 87, 89, 91, 92, 94, 97, 111-113, 124, 128, 130, 138, 139, 147, 154, 159-162, 169, 172, 175, 177, 181, 183, 184, 187, 190, 194-196, 201, 208-210, 214, 221, 226, 228, 229, 231-235, 239-243, 246, 253, 255, 257, 260-262, 265-267, 271, 275, 281, 291
infant, 101
infants, 188
infections, 65, 81, 90, 193, 209, 267
infectious disease management, 98
infectious diseases, 145
infertility, 80
infertility issues, 80
inflammation, 65, 193, 209
inflatable rubber tire, 175
inflight refueling system, 231
influential scientists, 81
information, 11, 12, 33, 45, 54, 76, 109, 114, 134, 146, 152, 177, 185, 187, 202, 207, 208, 233, 234, 261, 267, 282, 291, 292, 385
Infoseek, 144
infrared heat lamps, 50
infrared light, 203
infrared radiation, 30, 118, 262
infrared video camera, 249
injection polio vaccine, 226
injectors, 121
ink, 27, 34, 89, 109, 160, 205, 210, 261, 385
ink cartridges, 89
ink tank, 210
inkblots, 222
inkjet printer, 235
inkwells, 261
innkeeper, 266
INPUT, 45, 82, 85, 144, 163, 203
input device, 82, 144, 163
insects, 155
insemination, 80
insoluble silver salts, 118
inspection systems, 92
instant noodles, 19, 41
Institute for Plastination, 112
instrument, 21, 24, 40, 41, 44, 46, 52, 54, 57, 63, 69, 74, 76, 85-87, 94, 98-100, 102, 103, 105, 107, 108, 111, 115, 117, 118, 123-125, 142, 143, 145, 157, 158, 163, 173, 179, 183,

186, 195, 203, 208, 211, 215,
220, 221, 223, 225-229, 238,
246-248, 256, 261, 262, 272,
385
instrument maker, 40, 74, 86, 105,
163, 173, 203, 208, 226, 228,
261
insulation, 16, 22, 155, 172, 194,
218, 219
insulin, 31, 44, 49, 138, 226
insulin pump, 31, 138
integral calculus, 250
integrated circuit, 120, 143
integrated circuitry, 155
integrated circuits, 137, 244
Intel, 27, 171, 179, 191
Intel 4004, 171
Intel Corporation, 179, 191
intellectual, 6, 94, 122, 145, 190,
211, 285
intelligence, 14, 49, 172, 177, 222
intelligence gathering, 172
intelligence level, 222
interchangeable gears, 186
intercity railway line, 240
intercontinental ballistic missile, 103,
145, 185
intercontinental ballistic missiles, 63,
249
interferometer, 25, 85, 206
Intergalactic Computer Network,
157
internal cleansers, 130
internal combustion engine, 44, 45,
67, 74, 76, 155, 170, 180, 189,
240, 276
internal combustion engines, 48, 49,
139, 179
internal combustion powered
engines, 231
internal combustion technologies,
241
International Business Machines
Corporation, 121
International Harvester, 20, 171,
180
international language, 275
Internet, 15, 51, 55, 62, 121, 130,
137, 143, 144, 152, 176, 181,
194, 199, 242, 250
Internet browsers, 242
Internet data transmission, 130
Internet protocol, 15, 62, 137, 250
Internet time standard, 176
interplanetary probes, 141, 249
interstate highway, 19, 81

interstate highway system, 19, 81
intra-aortic balloon pump, 31, 145
intrauterine device, 29, 220
introverted and extroverted
personality types, 135
invention suppression, 117
inventiveness, 6, 11, 14
inventor, 8, 9, 11, 13, 14, 33, 35,
39-57, 59-83, 85-109,
111-125, 127-135, 137-149,
151-163, 165-191, 193-197,
199-215, 217-251, 253-257,
259-268, 271-273, 275-277,
281-286, 290-293, 385
inverted-T rail, 241
investor, 40, 46, 52, 74, 132, 133,
180, 183, 255, 267
investors, 290
invisible containment barrier, 204
invisible plasma, 143
Io, 97
iodized silver plate, 71
ion implantation method, 189
ionization method, 87
ionized gas, 69
ionizing charged particles, 265
ionizing radiation, 99, 222
Ionocraft, 231
iOS, 26, 248
IP, 15, 62, 137, 250
iPad tablet, 27, 132
iPhone, 27, 132
iPod, 27, 132
IQ test, 49
Ireland, 285-287
iris diaphragm, 122
Irish, 46, 59, 65, 66, 100, 149, 201,
248, 282
Irish king, 100
iron, 22, 31, 53, 57, 66, 75, 78, 80,
82, 87, 131, 177, 188, 219,
226, 232, 242, 263, 264
iron catalyst, 87
iron core transformer, 75
iron homeostasis, 219
iron lung, 31, 78, 82, 232
iron ore, 80
iron plow, 131
iron wire, 66, 177
ironclad warship, 83
ironmonger, 217, 220
ironmongery, 39
ISH, 98
Israel-Palestine Philatelic Society,
234
Israeli postmaster, 173

Istanbul Observatory, 169
Italian socialist revolutionary, 174
Italy, 13
IUD, 29, 220
IVF, 31, 80
ivory turner, 91
J. S. Kemp Manufacturing Company, 140
J. S. Pemberton and Company, 205
Jacobi's law, 129
Jamaica, 227
James Webb Space Telescope, 29, 262
jansky, 130
Jansky antenna, 130
jars, 95
Jarvik 7, 130, 265
Jarvik Research, Inc., 130
Jarvik, Robert K., 265
jazz, 58, 63, 228
jazz musician, 63
jeans, 9, 16, 18, 73, 241
Jedlik Dynamo, 131
Jefferies, Richard, 385
Jefferson Airplane, 46, 220
Jefferson Davis Historical Gold Medal, 385
Jefferson, Thomas, 93
Jellinek, Mercedes, 48
jellyfish, 251
Jenkins Television Corporation, 131
Jesus, 260, 288, 290, 389
jet, 20, 49, 65, 77, 108, 133, 138, 154, 161, 163, 184, 251, 252
jet aircraft, 133, 163, 252
jet engines, 184
jet fuel, 49
jet motor, 138
jet stream, 65
jets, 113, 138
jeweler, 78
jewelry, 109, 147, 385
jewelry designer, 385
jewelry maker, 109, 385
jewelry-making, 147
Jews, 284
Jo-blocks, 132
João V, King, 109
Jobs, Steve, 262, 267
Johansson Gauges, 132
John Deere equipment, 22, 141
John Hancock Center, 141
Johns Hopkins Hospital, 122
Johns Hopkins University, 23, 103, 122, 219, 292
Johnson, Lyndon B., 262
Jolly Air Thermometer, 133
Jolly Balance, 133
Jones, Elvin, 107
Joukowsky Airfoil, 276
journalist, 68, 77, 100, 180, 235, 268
Judaism, 284
Judd, Ashley, 385
Judd, Naomi, 385
Judd, Wynonna, 385
judge, 107
judges, 283
judo, 196, 239
Judson Pneumatic Street Railway, 134
Judson, Whitcomb L., 243
jujitsu, 239
jujutsu, 196
JumpSport, Inc., 213
jurist, 232
justice of the peace, 180, 237
K and F Manufacturing Corporation, 87
Kalashnikov, 137
kaleidoscope, 55
karaoke, 17, 128, 187
karaoke machine, 17, 128, 187
Kay Computers, 139
kayakers, 53
Kaypro Corporation, 139
Kégresse Track, 139
Kelvin absolute temperature scale, 248
Kelvin, Lord, 249
Kennard Novelty Company, 140
Kennedy, John F., 262
kenotron, 90
Kentucky, 385
Kentucky Colonel, 385
Keough, Riley, 385
Kepler, Johannes, 249
Keplerian Telescope, 140
Kepler's Laws, 141
kerosene, 22, 158, 162
kerosene lamp, 22, 162
kerosene stove, 22
ketones, 83
Kettering Foundation, 141
keyboard, 15, 16, 44, 52, 65, 69, 74, 149, 173, 233, 238
keyboard composition, 69
keyboard construction, 69
keyboard design, 69
keyboard instrument, 52, 69, 238
keyboard instruments, 65
keyboard maker, 52

keyboard music, 69
keyboard performance, 69
keyless combination lock, 26, 272
Kiddicraft Self-Locking Building Bricks, 27, 199
Kill Devil Hills, 268
Kinder Surprise, 88
Kindermehl, 188
kinescope, 21, 277
kinetic energy, 92
Kinetograph, 75
Kinetophone, 75
Kinetoscope, 75, 80
King João V of Portugal, 109
King's College, 100
Kipp's Gas Generator, 143
Kirlian photographs, 143
Kirlian photography, 143
kirza leather, 209
kitchen appliance, 238
kitchen device, 95
kitchens, 89, 212
Kitty Hawk, NC, 268
Kitty Litter, 161
Klischograph, 117
knapsack parachute, 146
Knickerbocker Rules, 61
knife handles, 202
knife industry, 82
knife maker, 82, 210
Knight, Thomas A., 128
knighthood, 86
knockout mice, 83, 236
known scientific laws, 144
Koch's Postulates, 145
Kodak camera, 79
Korean water clock, 272
Kosmos, 145
Kossel, Albrecht, 175
Kremlin, 182
Kroll Process, 147
KS magnetic steel, 121
Ku Klux Klan, 288
Kukuruznik, 210
Kurzweil Computer Products, Inc., 149
Kurzweil K250 synthesizer, 149
Kwannon (goddess), 273
Kyanization, 149
Kyrle Grate, 201
La Fourche Truffault, 251
labor, 114, 135, 210
laboratories, 135, 141, 143, 152, 207, 256, 265
laboratory bottle, 83
laboratory filtration device, 229

laboratory hematology, 68
laboratory mice, 83
laboratory research, 154
lace gasser, 113
lactic acid bacteria, 172
ladder, 131, 240
lake, 95, 161, 162, 205
Lakehurst, NJ, 276
Lakhovsky Coil, 152
lamp, 16, 21, 22, 27, 28, 41, 65, 73, 76, 100, 104, 108, 119, 148, 159, 162, 238, 240, 263, 271
lamps, 50, 119, 174
land, 18, 55, 62, 66, 77, 93, 113, 127, 140, 146, 152, 156, 198, 236, 261
Land Camera, 152
land reclamation, 62
land surveyor, 261
land vehicle, 146
land vehicles, 113
land-based vehicles, 127
Land-Wheelwright Laboratories, 152
landscape photographer, 385
Langmuir, Irving, 257
Lansdowne, Lord, 266
lap desk, 131
lap steel guitar, 46
laptop, 87, 139, 233
laptop computer, 139
laptop computers, 87
laryngeal mask, 54
laser, 15, 30-32, 40, 45, 48, 51, 87, 97, 98, 105, 112, 118, 121, 130, 166, 186, 189, 203, 213, 237, 240
laser beam, 97
laser engraving, 186
laser microscopy, 186
laser printer, 15, 240
laser printers, 112
laser printing, 51, 118, 186
laser shows, 186
laser spectroscopy, 130
lasers, 9, 30, 48, 81, 118, 121, 130, 147, 155, 203
LASERTRACE, 45
lathe, 21, 170, 186
latitude, 14
Laufmaschine, 77
laughing gas, 73, 212
launch vehicles, 63
laundry, 49
laundry detergent, 49
law, 3, 6, 19, 68, 99, 103, 122, 129, 173, 180, 196, 202, 219, 257,

260, 276, 288
law enforcement, 68, 99, 103, 219, 257, 276
law enforcement agencies, 68, 103, 219, 257, 276
Law of Attraction, 3, 260, 288
Law of Elasticity, 122
lawn equipment, 74
lawnmower, 25, 56
laws of geometric optics, 125
laws of motion, 188, 292
Laws of Thermodynamics, 175
lawsuits, 103, 268
lawyer, 93, 104, 148, 160, 225
LC, 114, 169
LCD screen, 87
Le Prieur Rocket, 212
lead condensing chambers, 222
lead dioxide, 128
lead gasoline, 21, 141
lead shot, 233
lead-acid battery, 209
learning disabilities, 50
leather, 15, 16, 53, 73, 160, 190, 194, 209, 218
leather alternative, 209
leather diving suit, 73
leather hide-splitting, 53
leather worker, 53, 160
Lebedenko Tank, 154, 176
lecturer, 44, 51, 106-108, 112, 121, 122, 124, 131, 134, 139, 142, 151, 157, 158, 165, 167, 178, 184, 187, 190, 195, 211, 213, 217, 219, 222, 223, 240, 242, 243, 246, 255, 259, 264, 265
lectures, 259
LED, 26, 28, 49, 51-53, 55, 61-63, 67, 68, 72, 73, 80, 83, 89, 98, 112, 116, 121-123, 131-133, 144, 145, 148, 149, 157, 160-163, 169, 170, 173, 178, 179, 181, 185, 186, 189, 196, 202, 205, 207, 212, 214, 222, 237, 239, 245, 246, 248, 263, 276
LED bulb technologies, 263
LED technology, 189, 245, 246
Led Zeppelin, 248
LEDs, 147
Lee, Robert E., 174
Left-wing, 174
legal issues, 131
Legge, William, 264
leggings, 233
Lego, 27, 64, 199

legs, 46, 78
Lemuelson Foundation, 155
LENR, 90
lens, 28, 39, 40, 88, 89, 94, 99, 122, 125, 155, 167, 243, 249, 265
lens developments, 155
lens grinding and polishing, 125
lenses, 16, 44, 72, 75, 88, 99, 108, 122, 155, 167, 236, 265, 272
lensless photography, 97
Leo Gerstenzang Infant Novelty Company, 101
leprosy, 277
LET, 32, 87, 93, 184, 289
LeTourneau University, 156
letter-typing telegraph machine, 129
Levacar Mach 1, 147
Levi Strauss and Company, 241
levitate, 106
levitation, 106
lexicographer, 385
Leyden Jar, 184, 190
liberals, 385
libraries, 114, 169, 207, 281
library, 6, 29, 39, 93, 114, 166, 169, 281, 285
Library of Congress, 6, 29, 39, 114, 169
Library of Congress classification system, 114, 169
Lichtleiter, 53
lids, 187
life after death, 3
life cycles, 155
life energy, 218
lifeboat, 20, 167
lifeboats, 62
lifestyle, 152
lift cable, 196
light, 9, 16, 17, 23, 25, 26, 28, 44, 52, 53, 62, 64, 65, 69, 71, 73, 80, 81, 85-87, 94, 100, 103, 104, 108-110, 116, 119, 121, 122, 124, 125, 130, 138, 148, 153, 157, 160, 180, 183, 189, 191, 203, 206, 223, 239, 243, 245, 251, 262, 263, 271
light beam, 223
light beams, 94, 189
light bulb, 26, 80, 103, 104, 121, 153, 243
light bulbs, 9, 26, 121
light diffraction, 122
light dimmer switch, 121
light emitting diode, 116, 245

light meter, 28, 263
light researcher, 94
light socket, 16, 124
light switch, 85
light technology, 183, 271
light therapy, 28, 119
light wave, 125
light waves, 206
light-emitting diode, 160, 223
light-emitting transistor, 87
light-tension wheel, 62
lighter design, 76
lighter-than-air balloon, 109
lighters, 76, 263
Lightguide, 53
lighthouses, 71, 94, 108
lighting, 92, 100, 104, 108, 121, 128, 141, 142, 160, 170, 183, 186, 191, 201, 219, 263, 271, 285
lighting fuel, 219
lighting lenses, 108
lighting rod, 160
lighting systems, 92
lightning rod, 25, 93, 114
lightning strikes, 114
Lightning switch, 85
lightweight internal combustion engines, 49
lignite, 49
limbs, 145, 148, 226
limelight, 108
limitless energy, 117, 175, 247
Lincoln Logs, 27, 267
Lincoln, Abraham, 174
Lincoln's War, 184, 205, 213, 288
Lindbergh, Charles A., 117
linear homeoscope, 146
linear ramp generators, 50
linens, 202
linguist, 39, 122, 157, 159
linguistics, 255, 275
linoleum, 16, 260
linseed oil, 260
Lint-Pick-up, 172
Linux operating system, 26, 250
Lippmann Capillary Electrometer, 158
Lippmann Plate, 158
liquefaction of gases, 86
liquid, 15, 49, 65, 87, 89, 94, 103, 109, 116, 152, 155, 173, 183, 195, 201, 207, 230, 249, 272
liquid broth, 207
liquid crystal display, 15, 87, 201, 272

liquid crystal display devices, 272
liquid crystal display thermometer, 201
liquid crystal displays, 116, 152
liquid crystal industry, 87
liquid crystal thermometer, 201
liquid density-measuring glass flask, 173
liquid fuel rocket, 103
liquid fuel-power, 155
liquid helium, 195
liquid hydrocarbons, 49
liquid propellant engine rocket, 249
liquid rubber, 94
liquids, 39, 67, 83, 99
literature, 190, 281, 290, 291
lithium battery, 25, 253
lithium-ion battery, 251
lithographer, 189, 210
lithotomy, 31, 243
litterateur, 385
live cell imaging, 100, 215
Livens Projector, 159
livestock, 100
living cells, 276
living organisms, 158
Ljungström Regenerative Air Preheater, 159
Ljungström Turbine, 159
loading coils, 213
loading docks, 271
location, 24, 57, 68, 72, 79, 98, 101, 202
Locher Rack Railway System, 159
lock designs, 271, 272
lock maker, 210
lock making pioneer, 272
lock manufacturer, 272
lock technology, 272
locking gear, 159
lockstitch technology, 124
locomotion, 53, 184, 281, 286
locomotive designer, 102
locomotive engine design, 60, 159
locomotive engine performance, 102
locomotive engines, 238
locomotive technologies, 241
locomotives, 43, 102, 160, 183, 186, 221, 240, 244
log-stacking machine, 156
logarithms, 186, 197
logging, 91
logic, 74, 235, 255, 381
logical relationships, 255
logician, 143, 251, 255
login information, 152

London, England, 101, 173
London, UK, 66
Long Term Evolution, 203
long-playing (LP) phonograph record, 104
long-range heavy bomber aircraft, 207
longbow, 187
longitude, 115
Lord, 249, 266, 293
Los Alamos Laboratory, 195
Lotus Elan, 119
Lotus Elan +2, 119
Lotus Europa, 119
loudspeaker, 17, 112, 256
loudspeakers, 18, 242, 249
Louisiana, 267
love, 3, 287, 381, 382
Loveless, Patty, 385
low carbon emissions, 91
low cost, 90, 112, 116, 137, 162
low cost clean energy, 90
Low Engineering Company, 161
low income earners, 273
low threshold current, 118
low-energy nuclear reaction, 90
low-power blue laser, 186
LSD, 30, 120
LTE, 203
lubricants, 22, 219
lubrication, 153
lubricators, 171
Ludwig Drum Company, 161, 162
Ludwig drums, 16, 161, 162
Luftschiffbau Zeppelin, 276
lumber, 244
luminescence, 65
lunar days, 140
lunar rover, 140
lungs, 81, 151
Lunokhod 1, 140
Luppis, Giovanni, 264
luthier, 40, 63, 238
luxury vehicle manufacturer, 197
lycopodium powder, 189
Lycra, 16, 232
Lyon Whitewalls, 163
Lyon, Inc., 163
lyricist, 385
lysergic acid diethylamide, 120
M.K. Steel, 177
M-3 Half Track, 139
Macadam Road surface, 171
Macadamization, 171
MacBook, 27, 132
Mach 1, 147, 165

Mach 5, 165
Mach Cone, 165
Mach Number, 165
Mach Operating System, 26, 248
Mach Wave, 165
Mach, Ernst, 165, 211
machine gun, 24, 67, 99, 170, 200
machine guns, 56
machine manufacturer, 200
machine shops, 242
machine tool maker, 186
machine tooling, 184
machine tools, 120, 156
machinery, 20, 32, 44, 127, 141, 195, 224, 235, 248
machines, 56, 92, 121, 124, 133, 142, 146, 155, 180, 193, 195, 205, 217, 218, 226, 256, 271, 292
machines shops, 256
machinist, 117, 124, 221, 224, 235, 286
Mach's Principle, 165
Macintosh, 165
Macintosh computer, 132
Mack, 165
Mackintosh, 165
macOS, 26, 248
macromolecules, 87
magazine founder, 219
magic, 25
Magic Cube, 223
Magic Marker, 188, 222
magician, 102, 123, 240
Magna Carta, 93, 287
magnesium, 262
magnet, 109, 185
magnet-filled metal globes, 109
magnetic automobile clutch, 217
magnetic drum memory, 246
magnetic field, 196, 242, 246, 267
magnetic field system, 267
magnetic fields, 71, 117
magnetic healer, 200
magnetic healing, 31, 200, 219
magnetic properties, 177
magnetic recording, 17, 207, 211
magnetic recording tape, 17, 207
magnetic resistance, 121
magnetic resonance imaging machine, 71
magnetic sound recording machine, 211
magnetic speedometer, 263
magnetic steel, 20, 121, 177
magnetic stripe card, 202

magnetic wireless receiver, 168
magnetics, 141, 142, 195
magnetism, 196, 285
magneto, 208
magnifying glass, 23, 40
magnifying glasses, 94
mah-jongg, 119
mail, 208, 229
mainstream history, 117, 268
mainstream medicine, 152, 220
mainstream science, 33, 48, 55, 90, 106, 175, 180, 218, 219
mainstream scientists, 143
MAKS project, 161
Maksutov telescope, 166
malaria, 30, 101, 104, 145
male, 5-9, 11, 13, 14, 29, 32-35, 37, 53, 169, 281, 293
male inventors, 5-7, 9, 11, 13, 14, 32-34, 37, 53, 293
malleable cast iron, 53
malt kiln, 113
Malyutin, Sergei V., 277
mammalian brain circuits, 189
Manby Mortar, 167
manganin, 263
Manhattan Project, 154, 195, 210, 244, 246
manifold writer, 91
Manmade (Seabrook), 33
manmade alloy, 263
manmade DNA sequence, 144
manmade fiber, 232, 243
manmade gemstones, 256
manmade language, 28, 275
manmade mechanical limbs, 148
manmade organs, 145
manmade polymer, 144
manmade sweetener, 19, 85, 219
manned aircraft, 157
manned flight, 178, 268
manometer, 108, 125
mantle lamp, 263
Mantoux Test, 167
manufacturer, 44, 56, 60, 66, 78, 101, 140, 161, 163, 190, 194, 195, 197, 200, 213, 219, 226, 227, 234, 247, 260, 264, 272, 286
manufacturing, 13, 52, 67, 69, 76, 87, 97, 112, 114, 124, 126, 132, 133, 140, 154, 166, 171, 175, 184, 186, 195, 235, 256, 257, 263, 272, 291
manufacturing companies, 256
manufacturing costs, 175

manufacturing equipment, 124
manufacturing industries, 132
manufacturing industry, 97, 154
manure spreader, 19, 140, 195, 224
man's world, 32
mapping, 79, 98, 101, 172, 202, 251
Marconi, Guglielmo, 223
margarine, 19, 172
margarine producers, 172
marine, 20, 30, 44, 68, 79, 90, 101, 105, 111, 113-115, 147, 187, 202, 208, 220, 228, 232, 242, 265, 277
marine biologist, 228
marine buoys, 187, 277
marine chronometer, 115
marine concerns, 79, 101, 202
marine economy, 220
marine life, 68, 208
marine life preservation, 68
marine lightning conductor technology, 114
marine lightning rods, 30, 114
marine longitude, 115
marine military industry, 265
marine propulsion, 44, 202
marine researcher, 220
marine science, 147
marine screw propeller, 20, 90
marine sports, 68
marine-craft, 113
Marineland, 22, 57
Marineland, FL, 57
maritime compass, 142
marker protein, 251
marriage, 290
Mars, 141, 247
Marsh Rack Railway System, 168
marshmallows, 52
martial arts, 196, 239
Marvin, Lee, 385
Masculine Principle, 6
maser, 45, 130, 213
mask fabrication of fiber Bragg gratings, 237
masking tape, 9, 18, 19, 78
Mason jar, 29, 169
mass analyzer, 166
mass energy, 211
mass of an object, 165
mass production, 44, 210, 264
mass production system, 210
mass spectrometry, 87, 166, 167
Massachusetts, 23, 39, 113, 169, 186, 222
massage therapy, 200

master gene regulators, 129
Mastermind, 27, 147, 173
match, 17, 128, 203
match heads, 203
matches, 9, 17, 76, 219, 263
mater, 35, 167
material scientist, 194
mathematical contributions, 188
mathematical illustration, 255
mathematical innovation, 186
mathematical symbol, 197
mathematician, 39-43, 45, 51, 54, 55, 60, 62-64, 66, 73, 80, 91, 93-95, 97, 103, 107, 111, 113, 116-118, 120, 122, 123, 125, 133, 140, 142, 143, 146-148, 165, 169, 171, 176, 178, 184, 186, 188, 197, 200, 202, 210, 220, 221, 223, 224, 228, 232, 233, 238, 239, 248, 250, 253, 255, 256, 261, 276
mathematics, 47, 188, 235
Matryoshka Doll, 167, 277
matter, 19, 62, 69, 86, 103, 122, 142
Maude (TV series), 154
mauve, 206
Maxim Gun, 170
Maxim, Hiram S., 153
maximum power theorem, 129
mayonnaise, 195
mayor, 107
MBTI, 135
McCallum, Barney, 212
McCormick Harvesting Machine Company, 171
McCoy Displacement Lubricator, 171
McGavocks, 134, 287
McGraw, Tim, 385
McMillan, Edwin, 255
ME, 289, 381, 382
meadows, 94
mean time, 90, 204
measles, 30, 120
measurable force interaction, 100, 215
measure intelligence, 49
measurement, 21, 85, 154
measuring, 24, 41, 62, 69, 70, 73, 90, 100, 103, 111, 130, 132, 136, 145, 156, 170, 173, 187, 190, 226, 227, 231, 238, 248, 263, 285
measuring instruments, 69, 170
mecanum wheel, 127

Meccano, 27, 122
mechanic, 64, 68, 103, 115, 120, 124, 133, 180, 226, 229, 268
mechanical calculator, 203, 228
mechanical clock, 269
mechanical complications, 175
mechanical compressors, 182, 209
mechanical curiosity, 232
mechanical desktop calculator, 272
mechanical device, 105, 124, 186, 197
mechanical energy, 67, 86, 106, 131, 197
mechanical engineer, 42, 45, 50, 51, 65, 69, 75, 82, 104, 105, 107, 108, 120, 123, 124, 131, 132, 135, 138-140, 142, 146-148, 154, 157, 159, 161, 163, 170, 171, 175, 176, 178, 184, 191, 196, 199, 202, 204, 210, 217, 221, 226, 230-232, 238, 240-243, 246, 250, 251, 256, 259, 261, 264, 267, 272, 273, 276
mechanical engineering, 33, 261
mechanical pencil, 25, 205
mechanical performance, 206
mechanical property analysis, 100, 215
mechanical puzzle toy, 223
mechanical reaping machine, 171
mechanical scientist, 108
mechanical systems, 92
mechanical technology, 228
mechanical valves, 138
mechanician, 133
mechanics, 25, 67, 116, 117, 185, 291
Mechanics Made Easy, 122
media industry, 229
media personality, 186
media storage industry, 207
mediate auscultation, 151
medical applications, 220
medical center, 140
medical contributions, 260
medical devices, 161, 169
medical diagnosis, 124
medical diagnostics, 172
medical equipment, 18, 53, 87, 112, 121, 218, 242
medical establishment, 152, 157
medical field, 147, 200
medical imaging, 186, 222
medical instruments, 121, 272
medical issues, 129

medical operation, 127
medical operations, 122, 158, 225
medical researcher, 45, 49
medical researchers, 207
medical school, 29
medical science, 87, 90, 145, 146
medical science researcher, 146
medical screening tool, 215
medical sedative, 47
medical surgeries, 203
medical terminologies, 151
medication, 143
medication scale, 143
medications, 16, 31, 135
medicinal benefits, 196
medicine, 31, 51, 64, 74, 88, 92,
  97, 98, 100, 109, 112, 118,
  119, 124, 125, 130, 131, 133,
  139, 141, 148, 152, 155, 158,
  166, 174, 182, 184, 185, 187,
  190, 191, 195, 201, 207, 215,
  220, 226, 235, 246
medicine man, 109
medicines, 17, 120
medieval music theorist, 41
medieval scholars, 142
Méhauté, Alain Le, 125
Melissococcus plutonius, 213
Mellon Institute of Industrial
  Research, 60, 173
melted plastic, 69
membrane potential, 187, 226
membrane science, 41
Memjet Printer, 235
memorial, 186
memory, 56, 74, 122, 137, 163,
  169, 174, 188, 203, 217, 233,
  244-246, 272
memory cells, 74, 137, 244
memory impairment, 203
memory recovery, 174
memory stick, 245
memory sticks, 233
memory storage, 188, 217
men, 3, 6, 9, 11, 13, 15, 32-35, 97,
  244, 283, 285, 291
meningococcus, 30, 120
meniscus corrector lens system, 167
mental and physical health benefits,
  218
mental disorders, 49, 62
mental health, 186, 204, 222
mental health issues, 204, 222
mental illness, 94
merchant seaman, 105
mercury, 27, 32, 71, 86, 100, 119,
  133, 149, 200
mercury air pump, 133
mercury fumes, 71
Mercury-in-glass thermometer, 86
mercury-vapor lamp, 27, 100, 119
Mesmerism, 173, 219
mesmerize, 174
meso-porous oxides, 195
messages, 46, 107, 170
messaging, 114, 185, 194
Mestral, George de, 34
metal, 9, 18, 23, 26, 28, 34, 47, 52,
  54, 72, 77, 89, 92, 95, 102,
  105, 107, 109, 113, 125, 127,
  130, 134, 135, 137, 139, 143,
  147, 153, 154, 163, 169, 170,
  186, 217, 221, 229, 231, 233,
  234, 236, 238, 244, 253, 263,
  271, 276
metal arc welding, 153, 236
metal blades, 238
metal cartridge-styled ammunition,
  92
metal detector, 23, 47, 89, 125
metal detector technology, 125
metal detectors, 9, 23, 89
metal engine parts, 113
metal filament bulb, 263
metal filament bulbs, 263
metal fragments, 233
metal monoplane, 231
metal oxide semiconductor field-
  effect transistor, 137, 244
metal processing, 238
metal production, 221
metal strengthening, 186
metal suit, 125
metal works, 234
metallic influences, 142
metallic organ, 65
metallurgical binding technique, 142
metallurgical engineer, 153
metallurgical method, 201
metallurgist, 53, 70, 105, 121, 142,
  147, 159, 177, 201, 218, 234,
  238, 272
metallurgy, 147, 148
metals, 48, 70, 195, 236
metalsmith, 271, 272
metalworker, 89, 109, 169
Metaphysical Society, 53
metaphysics, 143, 148, 235
meteorological phenomena, 224
meteorological sciences, 178
meteorological technology, 211
meteorologist, 57, 62, 72, 98, 117,

143, 151, 177, 200, 211, 224, 228, 257, 265
meteorology, 79, 97, 101, 118, 119, 202, 235
meter tape, 208
methamphetamine, 30, 185
methane, 19, 49, 257
metrics, 166
metrology, 132, 133, 142
metronome, 16, 89, 266
metropolises, 66, 177
Meyer Water Fuel Cell, 175
mice, 83, 236
Michelin Guide, 175
Michigan, 147, 198
Mickey Mouse, 76, 128
micro CT, 31, 87
microbiologist, 62, 81, 106, 119, 144, 155, 207, 245, 266, 267
microbiologists, 207, 245
microbiome research, 172
microchip, 45, 54, 143, 233
microclimate, 100
microcredit, 23, 273
microelectronics, 100, 190, 215
microfiber, 16, 194
microfinance, 23, 273
micromachining, 231
microneurosurgery, 31, 272
microorganisms, 60, 83, 207
microphone, 16, 17, 49, 50, 80, 125, 128, 187, 211, 231
microphones, 9, 16, 234, 246, 249
microplanet, 250
microplate, 31, 245
microprocessor, 15, 85, 120, 143, 171, 232
microscope, 26, 28, 31, 49, 50, 88, 100, 122, 130, 155, 158, 215, 222, 223, 276
microscope technology, 155
microscopes, 39, 53, 152, 155, 236, 272, 277
microscopic organisms, 155
microscopic particles, 68
microscopic tissue structures, 261
microscopist, 155
microscopy, 41, 100, 176, 186, 215, 276
Microsoft, 26, 27, 40, 98, 99
Microsoft Corporation, 40, 98
microtiter plate, 245
microwave frequencies, 56
microwave oven, 15, 238
microwave ovens, 119
microwave radio beacons, 168

microwave receivers, 147
microwave signal amplification, 145
microwave signals, 52
microwave technologist, 56
microwaveable plastic, 75
microwaves, 168
MIG fighter jet project, 161
Mikoyan and Gurevich, 176
Mikoyan, Artem, 108
militaries, 138
military, 3, 22, 27, 34, 44, 55, 56, 60, 63, 66, 67, 70, 73, 74, 76, 77, 79, 81-83, 91-93, 97-99, 101-103, 125, 132, 133, 137, 139, 145-147, 154, 155, 157-159, 161, 163, 166, 167, 174, 181, 182, 186, 201, 202, 208, 210-213, 220, 225-227, 231, 233, 237, 239, 241, 249, 250, 256, 260-262, 265, 266, 273, 276, 287, 385
military aircraft, 249
military and civilian helicopters, 273
military applications, 55
military attache, 212
military command, 174
military doctor, 201
military engineer, 70, 137, 139, 155, 166, 186, 211
military footwear, 210
military forces, 133, 276
military gun, 241
military history, 3
military man, 93
military officer, 44, 60, 77, 99, 103, 137, 146, 158, 166, 167, 181, 208, 211, 212, 220, 225, 239, 260, 262, 265, 266, 385
military official, 147, 161, 163
military operations, 82
military pilot, 157, 231, 256
military satellites, 63
military space station, 63
military suit, 125
military tank, 154, 211
military tank designer, 154
military tank technology, 154
military technology, 145
military transportation, 276
military units, 233
milk, 19, 124, 172, 188, 206, 207, 256
milk chocolate, 19, 206
milk chocolate bar, 206
milk yield, 256
milking machine, 19, 86, 154

Milky Way, 124
mill design, 68
milling, 68, 113
mills, 68, 176, 259
Mimas, 118
MIMO, 203
mine collapses, 244
mine inspector, 50
mine-laying ships, 166
miner, 116, 227
mineral, 236
mineralogist, 238
mineralogy, 235
minerals, 109, 260
mines, 57, 73, 116, 233, 244
MingKwai typewriter, 157
mini space shuttle, 161
miniature RC cars, 113
miniature toolkit, 82
Minié bullet, 213
mining, 22, 55, 73, 79, 93, 99, 101, 112, 116, 157, 159, 190, 202, 210, 238, 240, 250
mining engineer, 93, 116, 210, 250
mining industry, 73, 240
mining-related inventions, 116
Minox, 27, 275
Mir, 141
miracle, 260, 290, 381, 389
miracle motor, 117
mirror, 25, 167, 209, 239
mirrors, 44, 89, 94, 148, 167, 202
misandrist war cries, 35
missiles, 63, 184, 212, 249
missing persons, 131
mists, 180
MIT Computer Science and Artificial Intelligence Laboratory, 177
MIT Registry of Standard Biological Parts, 144
mitochondrial disease, 104
mix nets, 63
mixing boards, 234
MLE, 189
mobile intercontinental ballistic missile, 185
mobile phone, 67
mobile phones, 116
mobile receiver, 125
mobile refrigeration, 17, 133
model hobby toys, 113
model steam engines, 171
model steam rail locomotive, 90
Model T, 91
modern bike, 78
modern city life, 138

modern computer, 80, 146, 157, 200, 233
modern computer science, 146
modern electronic dishwasher, 123
modern era, 81, 111, 168, 205, 256
modern father of safe sex, 94
modern fire alarm, 253
modern health authorities, 216
modern humans, 33, 34
modern hypnotism, 54, 173
modern lifestyle, 152
modern science, 121
modern society, 13, 87, 144
modern world, 5, 6, 11, 32, 34, 37, 87, 97, 131, 133, 183, 235, 244, 248
modular space station, 141
moisture displacement, 153
molder, 201
molecular biologist, 129, 182, 236, 238, 244
molecular biology, 226
molecular geneticist, 60
molecular research, 100, 215
molecular typing, 60
molecular weights, 87
molecules, 60, 64, 87, 226
Molniya, 145, 161
Molton Sea Ark Atomic Reconstruction Technology, 48
Monday, 204
money, 23, 180, 199, 235
money transactions, 180
monk, 41, 42, 229, 269
Monobloc Pump, 169
monoglyceride, 49
monoplane, 231
Monopoly, 72
Monty Python's Flying Circus (TV series), 28, 63, 65, 102, 127, 134, 200
mood, 143
Moog Music, 179
Moog synthesizer, 179
moon, 24, 89, 115, 116, 140, 141, 160
moon crater, 89, 160
moons, 97, 118
mops, 194, 275
Moray, John E., 179
Morgan, John H., 180
Morse code, 26, 46, 100, 181
Morse, Samuel F. B., 135
mortar, 24, 103, 167, 241
Mosby, John S., 385
Moscow State University, 23, 160

Moscow, Russia, 234
MOSFET, 137, 244
Mosin-Nagant rifle, 181
moss, 31, 219
moss bioreactor, 31, 219
mother, 167, 188, 287, 385
mother bird feeding her chicks, 188
mothers, 290
motion, 18, 24, 40, 41, 44, 51, 69, 74, 75, 80, 94, 115, 123, 132, 140, 147, 162, 175, 188, 209, 211, 213, 239, 240, 245, 261, 262, 283, 286, 292
motion blur, 245
motion picture, 41, 75, 80, 123, 132, 162, 209, 213, 239, 286
motion picture cameras, 80
motion picture industry explosion, 209
motion picture projector, 123
motion pictures, 44, 74, 75, 283
motion-picture camera, 94
motor, 20, 22, 50, 64, 72, 73, 75, 76, 83, 86, 91, 95, 96, 108, 118, 123, 129, 131, 138, 147, 155, 175, 187, 188, 194, 197-200, 214, 216, 232, 239, 241, 244, 246, 251, 274, 277, 281
motor capacitor, 50
motor oil, 187, 277
motorboat, 20, 156
motorcycle, 18, 20, 21, 86, 161, 232, 248
motorcycle engines, 86
motorcycles, 9, 21, 74, 78, 229, 238
motorized vehicles, 135
Mount Rainier National Park, 26, 182
Mount Washington, 169
mountain, 39, 290, 385
mountaineer, 39, 239
mountains, 35, 169, 385, 388
mouse, 15, 51, 76, 82, 128, 129, 144, 163
mouth washes, 22, 219
movable cannon, 103
movie film projector, 162
movie projector, 23, 131
movie projectors, 94
moving image, 105
Moviola, 28, 231
mowers, 74
MP3, 15, 54, 254
MP3 player, 15, 254
MPEG-1 Audio Layer 3, 54

MRI, 31, 71, 195
MRI machines, 195
MS, 26, 99
MSAART, 48
mud, 66, 138, 266
mudpacks, 109
mules, 290
multi-barrel, lever-actuated, machine gun, 200
multi-beam light interference, 85
Multi-purpose Reusable Aerospace System, 161
multi-robot system, 172
multi-track recording, 17, 203
Multi-wave Oscillator, 152
multiparty computation, 63
multiplane camera, 76, 128
multiple-fire-tube boiler, 230
multiplex switchboard, 214
multiplexing, 214
multiplication, 16, 197
multiplying, 66
multipurpose product, 153
multiuniverse, 116
multi-instrument musician, 385
mumps, 30, 120
muon, 265
Murphy Bed, 26, 183
Murphy, William L., 34
muscle fatigue reduction, 261
muscle tissue, 104
muscles, 29, 98, 112, 226
Museum of Ornamental Art, 66
music, 9, 15-17, 28, 41, 54, 58, 69, 123, 128, 143, 148, 173, 179, 187, 197, 203, 207, 228, 230, 235, 256, 266, 385
music arranger, 230
music files, 54
music industry, 207, 256
music producer, 123, 385
music system, 128, 187
music technologist, 123
music therapy, 28, 143
music writing, 41
musical genres, 228
musical groups, 183
musical instrument amplifiers, 246
musical instrument designer, 46, 57, 87, 123, 179, 220
musical instrument designer and maker, 87, 179
musical instrument maker, 74, 173, 203, 226, 228
musical instruments, 87, 107, 163, 202, 235, 246, 259

musical keyboard synthesizers, 149
musical notation, 41, 54
musical products, 179
musical works, 230
musician, 16, 40, 46, 47, 51, 57, 63, 65, 69, 78, 87, 89, 93, 100, 118, 123, 127, 128, 143, 146, 149, 153, 157, 161, 173, 179, 195, 201, 203, 206, 209, 226, 228, 230, 238, 241, 268, 385
musicians, 183, 197, 248, 286, 385
musicianship, 47
musicologist, 140
muskets, 195
mustard, 195
mutated genes, 236
mutual insurance company, 93
muzzle-loading musket, 109
Myers-Briggs Type Indicator, 135
myopathy, 104
myopia, 96
mysteries, 3, 385
mysteries and enigmas, 3
mysticism, 385
nail-making, 53
nails, 135
naming system, 158
nanoscale, 100, 215
nanotechnology, 100, 191
Naperian Logarithm system, 186
naphthalene, 83
Napier's Analogies, 186
Napier's Bones, 186
NASA, 82, 95, 202, 262, 264
NASA space missions, 82
natal sex, 33
national barriers, 275
National Investigations Committee On Aerial Phenomena, 56
national parks, 9, 26, 182
national security, 79, 101, 201, 202
National Security Act, 201
National String Instrument Corporation, 46
natural diffusion process, 190
natural energies, 152
natural environment, 262
natural foods, 152
natural gas, 250, 257
natural gas industry, 257
natural health, 3, 129
natural health concept, 129
natural history, 3, 24, 148, 385
natural philosopher, 107, 259
natural radiation, 109, 152
natural radiation hormesis, 109

natural scientist, 117
natural selection, 72
natural tooth cleaners, 34
naturalist, 50, 57, 72, 107, 117, 122, 140, 158, 182, 215, 218, 228, 230, 262
nature, 11, 12, 100, 119, 156, 185, 202, 288, 385
nature reserve, 100
naturopathic medicine, 139
Nautilus (U.S. submarine), 220
naval architect, 82, 105, 155, 166
naval battle, 83
naval engineer, 166
naval officer, 46, 55, 68, 116, 166, 212
navigation, 41, 51, 66, 68, 79, 81, 97, 101, 102, 111, 115, 202, 220, 230, 284
navigation instrument, 41, 111
navigation technology, 68
navy, 105, 114, 285
nearsightedness, 96
necklaces, 109
Neelsen, Friedrich, 277
negative electrodes, 69
neologist, 131, 385
neon gas, 48, 65
neon light, 65
neoprene, 16, 53, 61
neoprene wetsuit, 53
nephew, 205
nephoscope, 211
nerve fibers, 49
nerve physiology, 115
nervous system, 47, 49, 104, 185, 205, 272
nervous system strengthening, 205
Nesmith, Michael, 107
Nestlé Company, 188
Nestle, Heinrich, 188
Netherlands, 281, 287
network router, 26, 176
network technology, 51
Network Time Protocol, 26, 176
networking technology, 174
neural network simulator, 26, 176
neural networking, 137, 244
neurobiologist, 103, 188
neuroengineer, 188
neurohistologist, 49
neurological diseases, 129
neurologist, 47-49, 56, 62, 94, 102, 168, 188, 204, 212
neurology, 115, 117, 281
neurons, 104

neuropathologist, 49, 102
neuroscience, 189
neuroscientist, 48, 62, 188
neurosurgeon, 272
neutrally buoyant underwater diving gear, 59
neutron, 62
New England, 285
New Hampshire, 22
New Jersey, 57, 75, 93, 266, 276
New Orleans, LA, 267
New Testament, 288
New World, 54
New York, 24, 42, 61, 100, 184, 247, 281-286, 289-292
New York City, 184
New York Knickerbockers Base Ball Club, 61
Newcomen engine, 188
news, 11
newspaper, 27, 180, 183, 204, 233
newspaper founder, 204
newspaper publisher, 183, 233
Newton, Isaac, 249
Newtonian telescope, 28, 188
NeXT, 132, 142, 260
NeXTSTEP Operating System, 26, 248
NICAP, 56
Niépce, Claude, 189
night, 28, 81, 109, 119, 160, 175
Night Hawk Minerals, 109
night vision, 28, 81, 119, 160
night vision equipment, 28, 119
night vision technology, 81
Nikonos underwater camera, 68
Nintendo, 27, 272
Nintendo Entertainment System, 27, 272
Nipkow Disk, 189
nitrates of cerium, 262
nitric oxide, 212
nitrobenzene, 111
nitrocellulose, 75, 201, 256
nitrogen, 111, 157, 212
nitrogen dioxide, 212
nitrogen fertilizer, 111
nitrogen-enriched fertilizer, 157
nitroglycerine, 75
nitrous oxide, 29, 31, 73, 212
Nivalin, 31, 203
Nobel Prize, 13, 39, 40, 44-46, 49-51, 53-55, 60, 62, 64, 71, 80, 81, 83, 87, 88, 90, 97, 98, 100, 103, 104, 111, 119, 124, 143, 145, 147, 153, 154, 156, 158, 168, 172, 182, 186, 187, 190, 191, 195, 204, 213, 219, 222, 223, 226, 228, 229, 236, 243, 251, 256, 260, 265, 273, 276, 277
Nobel Prize nominee, 46, 228, 229
Nobel Prizes, 190
nobility, 249
nobleman, 105, 142
non-destructive imaging, 100, 215
non-explosive match, 128
non-lethal handheld weapon, 68
non-linear optics, 74
Non-linear Systems, 139
non-recoiling harpoon gun, 50
non-responsive key, 52, 69
non-volatile memory, 137, 169, 244, 272
non-volatile memory cells, 137, 244
nondairy creamers, 19, 190
nonelectric gas mantle lamp, 263
Nordenfelt machine-gun, 200
Norman, Duane, 107
Normandie (ocean liner), 273
North, 20, 77, 85, 243, 268, 288, 291, 385
North Carolina, 268, 385
Northerners, 288
Northwest, 116
Norway, 13
novelist, 169, 222, 257
NPO Molniya, 161
NTP, 176
nuclear fuel, 64
nuclear medicine pioneer, 246
nuclear medicine scans, 119
nuclear physicist, 56, 141, 148, 246
nuclear physics, 62
nuclear power, 25, 62, 86, 99, 148, 234
nuclear power plant, 148
nuclear power plants, 86, 99
nuclear program, 148, 225
nuclear reaction, 90
nuclear reactor, 40, 88, 142, 148
nuclear reactor maintenance and repair, 142
nuclear reactors, 64, 99
nuclear science, 88
nuclear scientist, 88, 225
nuclear submarine, 220
nuclear technology, 225
nuclear-powered subs, 166
nucleic acid, 88, 98
nucleic acid sequences, 98
nucleic acids, 245

nucleotide level, 226
nurse, 29
Nutella, 19, 88
nutrients, 256
nutrition, 3, 119, 228, 243, 385
nutritionist, 139
nuts, 127
Nvidia Corporation, 124, 167, 212
nylon, 61, 144, 194, 243
nylon replacement, 243
Oberon, 118
obese, 216
obesity, 216, 243
objective psychology, 47
oblique wing design, 133, 256
obsessive-compulsive disorder, 30, 204
obsolete data, 12
obstetrician, 98, 177, 221, 273
occupation, 63, 169, 182, 188
OCD, 30
ocean, 27, 68, 72, 273
ocean ecology, 68
ocean exploration pioneer, 68
ocean liner, 273
ocean voyage, 72
oceanographer, 166
oceanography, 24, 156
OCR, 17, 92, 104, 149, 245
Odhner arithmometer, 193
Odic Force, 219
Odin (god), 219
odometer, 232
Odor-o-no, 183
office equipment, 202
office products, 237
offices, 87, 206, 286
offshore drilling platform, 156
Ogle Carburetor, 193
Ohain, Hans von, 108, 265
Ohio, 141
ohmmeters, 263
oil, 25, 41, 44, 51, 67, 89, 91, 108, 109, 123, 130, 142, 147, 153, 159, 162, 174, 175, 187, 190, 219, 221, 227, 244, 247, 260, 277, 291
oil consumption, 175
oil driller, 227
oil fields, 162
oil industry, 44, 130, 162, 190
oil lamp, 41, 108
oil lamps, 174
oil pipeline, 244
oil pipelines, 190
oil refineries, 162

oil refinery technology, 190
oil rig maintenance and repair, 142
oil shale, 89
oil tanker, 190
oil tycoon, 190, 221
oil worker, 244
oil yield, 89
oils, 86, 172, 230
Old South, 289
Old Sparky, 238
Old Testament, 32
Old Tom, 181
Oldsmobile, 22, 78, 194
OLED, 28, 116, 245
Oleomargarine, 172
Olivier Salad, 195
omni-font OCR software, 149
omnidirectional wheel, 127
omniwheel, 127
One Day at a Time (TV series), 28, 154
one-cylinder engine, 155
one-hand bareback rigging, 45
OneID, 144
oneiromancy, 235
online bank, 183
online city guide, 183
online content, 157
online media, 28, 229
onomastician, 385
onomastics, 3, 385
onscreen cursor, 82
open hearth furnace, 234
open-heart bypass surgery, 101
open-source scripting language, 156
open-source software, 239
operant conditioning, 24, 236
operant conditioning chamber, 236
operating system, 26, 99, 221, 223, 239, 248, 250
operating systems, 9, 26, 132, 242
operator services, 132
ophthalmologist, 88, 96, 117, 143, 275
ophthalmology, 88
ophthalmoscope, 31, 117
opinions, 11, 12
optical character recognition, 17, 92, 104, 149, 245
optical characteristics, 236
optical communications technology, 189
optical computer mouse, 144, 163
optical data communications, 121
optical device, 123
optical disks, 233

optical fiber technology, 189
optical industries, 236
optical lenses, 167
optical logic, 74
optical mirrors, 167
optical physicist, 45, 122
optical scientist, 118
optical space observatory, 239
optical surfaces, 167
optics, 30, 40, 62, 70, 74, 94, 100, 107, 111, 112, 117, 122, 125, 130, 140, 141, 143, 148, 158, 166, 169, 185, 188, 191, 195, 211, 223, 229, 237, 239, 240
optics designer, 40
optics engineer, 107
optics researcher, 111, 169, 211, 229, 239
optics scientist, 125, 130, 148, 158, 166, 211, 223
optics specialist, 240
Optophonic Piano, 44
OPV, 225
oral contraceptive pill, 63, 208, 221
oral polio vaccine, 225
Orbitrap, 166
orchestral percussion instruments, 163
order, 14, 26, 109, 135, 151, 158, 171, 172, 183, 205, 218, 220
ore deposits, 45
ore separator, 118
ores, 147
organ, 16, 29, 65, 69, 72, 115, 157
organ perfusion pump, 157
organ transplant science, 115
organic chemist, 49, 83, 89, 111, 119, 218, 229
organic chemistry, 144, 157
organic farming, 62
organic light emitting diode, 116, 245
organic molecules, 87
organic objects, 156
organic synthesis, 265
organisms, 44, 53, 104, 155, 158, 220
organist, 54, 157, 385
organizations, 12, 13, 221, 385
organized medicine, 221
organs, 29, 112, 145, 158
orgone accumulator, 218
orgone energy, 218
orientation, 51
ornaments, 202
ornithological observations, 178

ornithologist, 115, 158
ornithopter, 73, 205
orrery, 18, 105, 259
orsted, the, 196
Ørsted's Law, 196
orthopedic surgeon, 127
orthopedist, 44
orthostatic riders, 138
oscillating engines, 183
oscilloscope, 21
oscilloscopes, 184
osmosis, 26, 133, 190
osteotomy, 127
Otis Elevator Company, 196
Otis elevator safety brake system, 196
Otto Flugmaschinenfabrik, 197
Ouija Board, 27, 52, 95, 140
outboard motor, 20, 83, 251
outdated research, 12
outdoor furniture, 187, 277
outdoor painting, 177
outdoors, 174, 217
outdoorsmen, 263
outer space, 152
output torque, 184
over-dubbing, 17, 203
overnight passenger travel, 213
ovulation, 273
oxide, 29, 31, 73, 99, 137, 207, 212, 244, 257
oximeter, 31, 227
oxygen, 65, 99, 101, 107, 111, 138, 212, 260
oxygen concentration, 65
oxygen gas, 107
ozone layer, 85
pacemaker, 31, 82
Pacinotti Armature, 199
pack ice, 166
packaging, 54, 109, 112, 144, 196, 218
packaging film, 54
packset, 120
paddle game, 212
paddlewheel steamboat, 244
padlock, 26, 210
padlocks, 272
pain, 34, 172, 174, 183, 230, 249
pain management, 174
pain reduction, 172, 249
paint industry, 231
paint spray, 276
paint tubes, 28, 217
painted glass disks, 44
painted wooden figures, 167, 277

painter, 28, 44, 45, 50, 51, 61, 71, 73, 85, 144, 147, 167, 177, 181, 199, 213, 217, 385
paintings, 181
paintmaker, 51, 104
paints, 28, 147, 217
palace, 114, 213
paleontologist, 122
paleontology, 24, 156
Palmer College of Chiropractic, 200
palpitations, 121
pancakes, 88
pandemic, 30, 120
pandemic flu, 30, 120
panning technique, 50
pans, 128, 221
pantelegraph, 62
Pantone, Paul, 48
Pap smear, 201
Pap test, 31, 201
Papanicolaou Test, 201
paper, 9, 27-29, 43, 60, 61, 75, 81, 87-89, 91, 95, 99, 109, 118, 144, 147, 149, 162, 174, 212, 237, 243, 249, 262, 285
paper clip, 28, 87
paper hole punch, 237
paper making, 174
paper production, 249
paper towels, 81
papermaker, 178
papillomavirus, 94, 276
parachute, 20, 73, 77, 146
parachute design, 146
parachute science, 146
parachutes, 184, 229
paraffin wax, 22
paralysis, 189
paralytic polio, 226
paranormal, 3, 106, 140, 143, 219, 385
paranormal investigator, 106
paranormal parlor game, 140
paranormal researchers, 143
parasites, 101, 104
parchment paper, 29, 88, 89, 212
Paris, France, 211
Parisian Nouveau Réalisme, 144
Parkes Process, 201
Parkes, Alexander, 126
Parkesine, 201
Parkinson's disease, 189
parliamentarian, 82
parlor game, 52, 95, 140
parquet carpet, 49
particle accelerator, 154, 255

particle accelerators, 195
particle collector, 265
particle physics, 40, 64, 195
particle physics research, 195
particles, 57, 63, 64, 68, 103, 106, 147, 193, 195, 243, 265
Parton, Dolly, 385
party, 174, 193
party telephone line apparatus, 193
Pascal, Blaise, 266
Pascaline, 203
Pascal's Law, 202
Pascal's Principle, 202
Pascal's Theorem, 203
Passarola Airship, 109
passenger carriage, 232
passenger-carrying blimp, 102
passengers, 169
Pasteur-Chamberland filter, 62
pastries, 19, 190
patch clamp technique, 31, 187, 226
patent applications, 131, 235
patent attorney, 61
patent authority, 153
patent issues, 247, 268
patents, 11, 12, 67, 68, 80, 87, 94, 97, 107, 113, 115, 128, 131, 133, 155, 156, 182, 186, 222, 223, 235, 238, 248, 263, 267, 272, 281-284, 291, 292
paternity issues, 131
pathogenic bacteria-borne ailments, 145
pathologist, 43, 67, 104, 172, 225, 260, 267
pathology, 98, 158
patients, 47-49, 53, 54, 62, 78, 82, 99, 130, 139, 232, 243
patriotism, 158
patriots, 205
patrol, 172, 236
Pavlov's dog, 204
payment processing, 114, 185
PayPal, 27, 183
PCB, 81
PCR, 182
peace, 24, 61, 180, 190, 237
peanut butter, 19, 61, 80, 139, 190, 241
peanut butter making machine, 241
peanut paste, 80
peanuts, 61
Pearl Musical Instrument Company, 272
peas, 195
Pecqueur, Onésiphore, 135

pedal radio, 250
pedals, 77, 82, 250
pediatrician, 56, 193, 228
pediatrics, 141, 228, 243
pedology, 61
Peeps, 52
Pemberton, John S., 205
Pemberton's French Wine Coca, 205
pen, 25, 29, 34, 50, 73, 89, 91, 160, 188, 202, 205, 210, 222, 232, 256, 261, 385
pen maker, 91
pencil, 25, 160, 205
pendulum clock, 125
pendulums, 122
penetrating oil, 153
penicillin, 31, 90
Pennsylvania, 23, 93, 284
penny farthings (bicycles), 240
pens, 9, 24, 50, 248, 261
pentode, 16, 246
pepper, 26, 195
peppers, 195
peptides, 87
perception, 117
percussion accessories, 161, 162
percussion cap, 109
percussion instruments, 161-163, 272
percussion lock firearm mechanism, 92
percussion lock gun mechanism, 53
percussion pill, 109
percussionist, 161, 385
perforated film rolls, 94
performance monitoring, 41
performance outerwear, 105
performing musician, 161
perfume, 17, 206
perimeter patrol, 172
Periodic Law, 19, 173
Periodic Table, 9, 19, 173
periscopes, 236
perishable foods, 133
Perkin Reaction, 206
permaculture, 62
permanent artificial heart, 130
permanent wave machine, 187
perpetual motion device, 175
persisting currents, 195
personal computing, 132, 171
Personal Home Page, 156
pest control, 168
pesticides, 22, 87, 219
PET, 29, 31, 98, 119, 161, 204, 246
pet fence, 29, 204

pet industry, 161
pet owners, 98
pet training, 204
Peter the Great, 236
Peter, Daniel, 188
Peter's Chocolate, 207
Petlyakov Pe-8, 207
Petri Dish, 29, 207
petroleum industry, 162
petroleum jelly, 75
Pettit, Robert, 200
Petty-Fitzmaurice, Henry, 266
pewter plate, 189
PFGE, 60
phage therapy, 81
phagocyte, 172
phagocytosis, 172
Phantoscope, 131
pharmaceuticals, 111, 166, 206, 220
pharmacist, 46, 80, 83, 93, 120, 143, 144, 148, 157, 162, 169, 172, 182, 185, 188, 196, 205, 230
pharmacologist, 44, 106, 182, 238
pharmacology, 100, 143, 158, 187, 215, 226, 243, 259
phase-contrast microscope, 31
phase-contrast microscopy, 276
phased array antenna, 55
phenakistiscope, 209
phenol, 22, 49, 75
phenolphthalein, 44
philanthropist, 40, 55, 60, 68, 81, 91, 98, 103, 122, 173, 180, 181, 190, 217, 221, 239, 255
philatelist, 234
philology, 235
philosopher, 43, 54, 57, 60, 73, 80, 94, 95, 97, 102, 104, 105, 107, 117, 122, 135, 140, 142, 143, 148, 152, 157, 159, 165, 169, 188, 202, 211, 212, 218, 232, 235, 238, 259
philosopher of the Arabs, 143
philosophy, 3, 117, 235
phobias, 30, 174, 204
phone payment cards, 180
phonograph, 49, 75, 80, 104, 283
phonograph record disk, 49
phosgene, 75
phosphate ore, 154
phosphor, 128
phosphorus, 203
phossy jaw, 203
photo booth, 28, 134
photo editing, 134

photo process, 79
photochromolithography, 107
photocopiers, 101
photocopying technology, 61
photoelectret, 185
photoelectric device, 241
photoelectric effect, 81
Photoelectric Image Scanning Tube, 117
photoelectric musical instrument, 183
photoelectric sensor, 189
photoelectric solids, 185
photoelectric technology, 81
photoelement, 241
photoepitaxy, 189
photogastrograph, 167
photogram, 28, 262
photogramer, 262
photograph development, 243
photograph produced by X-rays, 60
photographer, 41, 53, 60, 68, 75, 94, 107, 115, 118, 134, 144, 146, 147, 152, 158, 162, 171, 173, 184, 187, 189, 190, 210, 213, 218, 228, 229, 234, 236, 243, 249, 262, 385
photographers, 115, 263
photographic chemistry, 210
photographic film, 201
photographic image, 189
photographic lens diaphragm, 249
photographic paper, 43, 262
photographic process, 118
photographic reproduction process, 189
photographs, 94, 116, 143
photography, 3, 9, 27, 51, 52, 71, 79, 94, 97, 118, 134, 143, 152, 158, 163, 171, 184, 185, 189, 210, 213, 235, 272, 385
photography accessories, 272
photography filters, 152
photography science, 210
Photomaton, 134
photomechanical printing, 210
photonics, 237
photons, 53
photophone, 47
photos, 6, 24, 75, 115, 143
photovoltaic cell, 46
photovoltaic effect, 46
PHP, 26, 156
phylogenetic trees, 158
phylum, 158
physical biochemist, 236

physical chemist, 49, 111, 112, 156
physical processes, 144
physical science, 100, 215
physician, 42, 43, 48, 53, 60, 63, 67, 69-71, 82, 88, 89, 94, 98, 101, 102, 104, 106, 108, 109, 111, 112, 114, 116, 117, 119, 127, 129, 139, 143-146, 151, 158, 167, 169, 173, 175, 183, 185, 188, 200, 201, 205, 207, 209, 220, 221, 225, 226, 232, 235, 236, 241, 243, 245, 260
physicians, 151, 208, 267
physicist, 39-42, 44-49, 51-57, 59-65, 68, 69, 71, 73-75, 77, 79, 80, 85-88, 90, 92, 94, 97-105, 107, 112, 116-119, 121, 122, 125, 129, 130, 133, 138, 140-142, 145, 147, 148, 151-155, 158-161, 165-167, 169-171, 178, 184-191, 193-196, 199, 200, 202, 203, 206-211, 213-215, 218, 220-223, 225, 227, 229, 231, 233, 234, 236-241, 243, 244, 246, 248-250, 253, 255-257, 259, 261-266, 272, 275-277
physicists, 142, 255
physics, 18, 40, 47, 48, 51, 62, 64, 99, 103, 117, 133, 190, 195, 230, 235, 249, 265, 275
physiognomy, 235
physiologist, 47, 49, 52, 63, 80, 81, 88, 98, 117, 161, 172, 204, 212, 227, 260
physiology, 115, 158, 190, 243
physiophonic phenomenon, 174
pianforte, 65
pianist, 47, 108, 230, 385
piano, 16, 44, 47, 52, 69, 108, 202, 238
Piano Concerto No. 5 in E-flat Major (Beethoven), 47
piano keys, 202
Piano Sonata No. 14 in C-sharp Minor (Beethoven), 47
piano-maker, 108
pianoforte, 52, 69, 238
pianos, 69, 163
pickleball, 22, 212
pickles, 195
picture frames, 89
pictures and sound, 131
pies, 19, 190
pigment, 44, 51, 104, 210
pigs, 256

pike, 187
pike and bow, 187
Pilatus Railway, 22, 159
pile driving, 186
pill, the, 63, 208
pillow, 288
pills, 29, 119
pilot, 48, 64, 67, 77, 92, 109, 130, 134, 153, 157, 197, 210, 211, 231, 234, 256, 268
pin diode, 189
pin photo diode, 189
pin tumbler lock, 26, 271
pinch-to-zoom technology, 88, 120
pine resin, 260
ping pong, 212
pinhole camera, 40
Pink Floyd, 161, 162
pins, 127, 146
pinscreen animation technique, 40
pinwheel calculating machine, 193
pipe organ, 16, 69
pipe wrench, 30
pipeline construction, 190
pipeline maintenance and repair, 142
piperine, 196
Pirogoff Angle, 208
Pirogoff's Aponeurosis, 208
Pirogov's Triangle, 208
Pirquet, Clemens von, 228
pistols, 56
piston valves, 60
pit props, 244
Pitney Postal Machine Company, 208
Pitney-Bowes, 208
Pitney-Bowes Postage Meter Company, 208
Pixar, 132
pixel response time, 245
pizza, 19, 190
planchette, 52
plane, 20, 77, 86, 115, 118, 125, 146, 147, 205, 220, 241, 268
plane engines, 86
planetarium, 18, 89, 105, 259
planetarium projection hall, 259
planetary system, 106
planets, 18, 51, 89, 105, 259
planets' relative positions, 105
Planophore (model aircraft), 205
plant cells, 122
plant food, 52
plant growth, 52
plant material, 226
plant physiologist, 52
plantation, 134, 287

plants, 52, 53, 86, 92, 99, 135, 147, 154, 156, 158, 201
plasma, 28, 69, 143, 200, 249
plasma display, 249
Plasmodium, 101
plasmoid fusion technology, 48
plasmoids, 48
plaster casting process, 208
plastic, 9, 16, 18, 25, 31, 43, 55, 69, 75, 76, 88, 89, 91, 95, 112, 116, 146, 187, 199, 201, 202, 204, 223, 243, 264-266, 277
plastic bags, 204
plastic bottles and containers, 187, 277
plastic explosive, 55
plastic interlocking toy, 199
plastic mattress, 116
plastic pen bodies, 202
plastic pipe materials, 264
plastic surgery, 31, 243
plastics, 43, 46, 109, 111, 144, 147, 155, 194, 202, 218, 228
plastics chemistry, 144
plastics engineer, 228
plastination, 29, 112
plastination exhibitions, 112
Plateau, Joseph, 239
plating warships, 67
platters, 133
Playmobil toys, 46
playwright, 127, 153, 211
plectrum, 52, 69
plinkers, 168
plow, 19, 93, 99, 131, 196, 217, 283
plowing process, 93
plows, 93, 226, 259
plugs, 175, 214
plumber wrench, 132
plumbing, 16, 187, 218, 277
plumbing pipes, 187, 277
Plunkett, Roy J., 34
Pluto, 250
Pluto Platter, 181
plutonium, 99
PMD, 138
PMMA, 265
pneumatic tire, 78, 175, 229, 248
pneumatic tire valve, 229
pneumatic tires, 78
pneumococcus, 30, 120
pnip transistor, 189
Po-series aircraft, 210
pocket calculator, 143
pocket openings, 73

pocket watches, 250
pockets, 73
poem, 156
poet, 39, 73, 89, 95, 114, 117, 148, 153, 156, 157, 159, 190, 238, 244, 275, 385
poetry, 3, 156, 235, 289
point-writing system, 44
pointing stick, 15, 230
Poitevin process, 210
Poland, 162
Polaris missile program, 166
polarizing filters, 152
polarizing technology, 152
Polaroid Corporation, 152
Polaroid Land Camera, 152
Polhem's Lock, 210
police, 24, 52, 66, 257
police specialist, 257
polio, 31, 78, 82, 225, 226, 232
polio vaccine, 31, 225, 226
political activist, 162
political commentator, 236
political parties, 385
political theorist, 39, 212
politician, 39, 49, 56, 79, 89, 96, 107, 108, 122, 131, 137, 154, 162, 166, 167, 173, 180, 195, 199, 210, 212, 232, 233, 236, 238, 239, 248, 266
politics, 3, 148, 385
pollutants, 87
pollution, 139, 179, 207
poltergeists, 290
polyalphabetic code system, 40
polyamide-6, 144
polybutadiene, 154
polycaprolactam, 144
polycarbonate, 75
polyester, 194
polyethylene bottle, 178
polygraph machine, 131
polymath, 39, 43, 60, 62, 73, 89, 93, 97, 117, 118, 122, 130, 143, 148, 159, 169, 188, 202, 211, 228, 232, 233, 238, 250, 385
polymer, 46, 51, 104, 105, 144, 174
polymer industry, 46
polymerase chain reaction, 19, 182
polymerizing, 218
polymers, 144, 202
polymethyl methacrylate, 265
polystyrene foam, 172, 182
polytetrafluoroethylene, 105
polyvinyl acetate, 144

polyvinyl chloride, 15, 46, 144, 218
Pons, Stanley, 90
pop music, 228
Popper, Karl, 211
Popper-Lynkeus, Josef, 211
poppet valves, 60
population, 68, 111, 131
population explosion, 111
population research, 131
porcelain basin, 114
Porro prism, 211
portable cardiac monitoring device, 121
portable crane, 156
portable defibrillator, 201
portable electronics industry, 253
portable hydraulic riveter, 251
portable infusion pump, 138
portable kidney dialysis machine, 31, 138
portable military weapon, 67
portable mortar, 103
portable offshore drilling platform, 156
portable photo studio, 134
portable refrigeration, 133
portable revolving bridge, 73
portable voltmeter, 263
portable workbench, 119
Portland cement, 42
portraitist, 272
Portugal, 109
positron emission tomography, 31, 119, 246
Post-it note, 95
postage machinery, 235
postage meter, 29, 208
postage printing machine, 208
postage stamp, 29, 119
postal administrator, 119
postal efficiency, 119
poster artist, 385
postmark information, 208
postmaster, 93, 173
potassium chlorate, 92
potatoes, 61, 195
potter, 262
poultry, 256
powdered milk, 206
power, 17, 20, 21, 25, 44, 56, 62, 74, 77, 79, 81, 83, 85, 86, 92, 98-101, 109, 112, 115, 117, 129, 145, 147, 148, 154-156, 159, 163, 175, 179, 182, 186, 191, 196, 202, 204, 209, 211, 218, 231, 234, 238-241,

244-247, 263, 264
power distribution systems, 239
power efficiency, 245
power generation, 44
power grids, 79, 101, 202
power lines, 196
power plant, 20, 74, 148, 241
power plant builder, 74
power plant industry, 241
power plants, 86, 99, 147, 154, 156
power stations, 86, 148, 159
power-outage emergencies, 115
powered mobility device, 138
powers, 32, 91, 118
praseodymium fluoride glass fiber laser amplifier, 237
praxinoscope, 123
pre-Christian Rome, 181
pre-Darwinian evolution, 122
pre-prepared gases, 143
precession of the equinoxes, 120
precipitation producing technique, 257
Precise Path Robotics, 134
precision clock, 115
precision dimensions, 132
precision medical instruments, 272
prefabricated concrete homes, 156
pregnancy, 220
prehistoric bird, 209
prehistoric times, 167
prehistory, 33, 180
presenter, 65, 200
preservationist, 385
preservatives, 172
preserving human tissue, 112
preserving wood, 149
president of the United States, 57, 81, 166
President of the United States of America, 81
presidential history, 3
Presley, Elvis, 385
Presley, Lisa M., 385
presses, 186
pressure, 44, 57, 64, 65, 69, 95, 100, 108, 109, 111, 115, 116, 128, 142, 145, 147, 177, 186, 202, 207, 221, 238, 250, 256
pressure suit, 64
pressure-variable key, 69
pressurized ink cartridges, 89
priming powder, 109
Primus Stove, 158
Princeton University, 23, 57, 75, 266

print media, 28, 229
printed circuit board, 81
printer, 15, 93, 102, 117, 125, 210, 233, 235, 240, 268
printing method, 107
printing paper, 162
printing press, 34, 73, 109, 272
printing surface, 69
printing technology, 210, 235, 240
printing telegraph, 124, 269
printing type telegraph, 125
printmaker, 45
prisons, 98
Pritchett bullet, 24, 213
probiotic, 172
probiotics, 56, 172
processing and sorting machines, 217
proctologist, 129
prodigy, 202, 385
producer, 50, 55, 57, 63, 65, 73, 75, 76, 80, 94, 102, 107, 123-125, 128, 132, 139, 154, 162, 163, 167, 175, 177, 188, 201, 212, 221, 230-232, 235, 239, 241, 267, 272, 385
product assembly, 172
product information, 114, 185
product testing, 41
production designer, 240
productivity, 114, 233
products, 12, 13, 19, 22, 27, 35, 60, 72, 78, 79, 81, 101, 108, 115, 124, 149, 151, 154, 167, 168, 179, 187, 188, 194, 196, 202, 206, 207, 212, 219, 233, 234, 237, 249, 263, 267, 272, 277
professional ridicule, 268
program designer, 242
programmable computer, 277
programmable electronic, 91
Prohibition, 205
Project 1794, 95
Project Xanadu, 187
projectiles, 53, 229
projecting electrotachyscope, 41
projective test, 222
propane gas, 182, 209
propeller, 20, 53, 64, 65, 82, 83, 90, 121, 122, 141, 168, 220, 232, 242
propeller blades, 53
propeller-driven guided, aerial missile, 141
propellers, 67, 191, 284
props, 122, 134, 168, 232, 244
propylene, 187, 276

prosateur, 385
prosthetics, 148, 185, 189
prostitution, 287
protection, 6, 62, 68, 79, 100, 101, 125, 142, 153, 172, 202, 219, 226, 260, 282, 291
protection of forests, 62
protective mask, 276
protein, 31, 219, 251
protein synthesis, 219
proteins, 80, 87, 220
proton, 62
protons, 64
prototyping, 69
protozoa, 155
protozoans, 43
pseudoscience, 175, 218, 220
psychedelic drug, 120
psychiatrist, 47-49, 62, 94, 102, 135, 218, 222
psychoanalysis, 30, 94
psychoanalyst, 30, 94, 135, 218, 222
psychoanalysts, 120
psychologist, 30, 47-49, 94, 99, 117, 135, 157, 204, 235, 260
psychology, 30, 47, 49, 135, 143, 211, 218, 235, 286
psychology pioneer, 135
psychotherapist, 135
psychotherapy, 94
PTFE, 105
public health, 114
public hospital, 29, 93
public safety, 79, 97, 101, 202
public transport, 73, 113, 159, 234
public transport vehicle, 73, 113, 159, 234
public works project, 81
publisher, 6, 11, 12, 73, 174, 175, 183, 233, 268, 385
publishers, 233, 292
publishing designer, 385
pull chain light socket, 16, 124
pulleys, 73
Pullman Sleeper, 213
Pullman, IL, 213
pulping, 249
pulse clock, 227
pulsed-field gel electrophoresis, 60
pulsilogium, 227
pumping, 116, 267
pumps, 41, 68, 121, 156, 186
Punch, 27, 28, 109, 113, 121, 146, 237
punch card readers, 113
punch cards, 121, 146
punch lock, 109
Pupinization, 213
Pupin's Coils, 213
puppet animation, 240
pure metals, 195
pure water, 152
purified gas, 183
purifying water, 62
Puritan, 115
puzzle video game, 200
PVA, 144
PVC, 15, 46, 144, 218
pycnometer, 29, 173
Pyréolophore, 189
pyrocollodion, 173
pyrometer, 184, 262
Q-Tips Baby Gays, 101
QR code, 23, 114, 185
quad skate, 209
quadrant, 27, 142, 169, 286
quadric surface, 233
Quadricycle, 91
quality of life, 139, 179
quanta energy, 209
quantum compass, 68
quantum electrodynamics, 45
quantum electronics, 45, 213
quantum energy, 211
quantum engineer, 68
quantum fluid, 195
quantum navigation, 68
quantum physics, 51
quantum sensors, 68
quantum theory, 25, 81, 188, 209
quantum theory of light, 81
quantum well, 118, 121, 147
quantum well laser, 118, 121
quantum well laser diode, 121
quartet, 17, 47
quartz, 46, 72, 89
quartz lenses, 72
quartz spectrophotometer, 46
queen bee excluder, 24, 213
Quetelet Index, 216
quick release rear hub, 60
Quick Response code, 114, 185
quill maker, 91
quinine, 206
quiz, 3, 289
R. G. LeTourneau Company, 156
R.A. Moog Company, 179
rabbi, 104
rabbit ear TV antenna, 260
rabbit ears, 175
RABCO, 217
Rabens, Neil W., 91

rabies, 209
Rabinow Reading Machine, 217
race car driver, 64, 227, 231
racehorse saddle, 166
racing cyclist, 60
rack railway, 22, 50, 159, 168, 221, 242
rack railway design, 221
racks, 67
radar, 30, 51, 74, 92, 119, 145, 168, 246, 261
radar electronics, 51
radar technology, 30, 145
radial fan, 225
radial tire, 227
radiant energy, 179
radiant matter, 69
radiation, 28, 30, 41, 69, 72, 85, 99, 100, 109, 116, 118, 152, 153, 168, 171, 222, 261, 262
radiation hormesis, 109
radiator, 108, 226
radio astronomy, 130, 218
radio broadcast, 88
radio chemist, 156
radio communication, 54
radio engineer, 51, 124, 130, 179, 218
radio industry, 207
radio personality, 203, 236
radio pioneer, 88
radio receivers, 223
radio science, 52
radio signals, 65, 90, 204, 253, 271
radio technologist, 235
radio technology, 183, 223
radio telegraphy, 167, 168
radio telephones, 223
radio telescope, 130, 218
radio transmissions, 260
radio transmitters, 223
radio wave control, 113
radio wave pioneer, 113
radio wave receiving beam antenna, 130
radio waves, 52, 54, 72, 119, 130, 168
radio-collar-receiver, 204
radio-control, 113
radio-controlled technologies, 113
radio-transmitting wire, 204
radio-wave emission strength, 130
radioactive elements, 99
radioactive stones, 100, 109
radioactive tracer, 119
radioactivity, 41, 195

radioactivity measurements, 41
radiocarbon dating, 24, 156
radiometer, 69
radiophase, 45
radios, 46, 235, 246
Radioskop, 249
radiosonde, 57, 177
radiotelephony, 168
rail air brake, 263
rail locomotive, 90, 221, 250
rail locomotive builder, 221
rail locomotives, 240, 244
rail track, 159
rail transport line, 240
rail travel, 159
railroad car, 26, 153, 159, 213
railroad car toilet, 153
railroad car wheels, 159
railroad engineer, 160
railroad industry, 160, 241
railroad locomotives, 102
railroad spike, 57, 241
railroad technology, 230
railroad track, 169
railroad tracks, 26
railroadman, 171
railway, 22, 50, 62, 73, 106, 122, 130, 134, 155, 159, 168, 169, 208, 221, 234, 239, 240, 242, 243, 263, 267, 281
railway air signal system, 263
railway equipment, 62
railway line, 159, 240
railway locomotive, 73
railway padding, 155
railway pioneer, 240, 242
railway station, 122
railwayman, 221
railways, 79, 101, 106, 122, 202, 248
rain, 23, 228, 257, 272
rain gauge, 23, 272
rainwater catcher, 131
ramp, 50, 220
ranch hand, 385
ranchers, 98, 256
ranching, 91, 100
ranching concerns, 100
Rand International Limited, 55
range finders, 235
RAR file format, 26, 222
rarefied gas, 69
rationalists, 285
Ravensdale, Cassidy, 6
RC, 113
RD-170, 103

reading aids, 94
reading machine, 149
real McCoy, 171
reality, 74, 117, 121
reaping machine, 19, 171
reaping machines, 226
reasoning, 203
rebar, 9, 16, 66, 152, 177, 178
receiver, 57, 78, 124, 125, 168, 179, 204, 222
rechargeable batteries, 201
rechargeable battery, 251
rechargeable electric battery, 209
recipe, 19, 128, 171, 172, 205, 260
recipes, 61, 143, 195
recoilless rifle, 24, 184
recombinant DNA, 53
reconnaissance aircraft, 207
reconnaissance missions, 172
Reconstruction, 48, 174, 289
record player, 49, 80
record players, 235
record producer, 123, 230
recording, 9, 16, 17, 49, 50, 74, 77, 121, 123, 128, 143, 149, 155, 162, 179, 200, 203, 207, 211, 218, 220, 231, 256, 260, 385
recording artist, 123, 149, 220
recording drum, 200
recording engineer, 123
recording equipment, 231
recording industry, 49, 207, 256
recording material, 162
recording studio mixing engineer, 385
recovery time, 172, 249
recreation, 79, 101, 179, 202, 291
recreational vehicles, 182, 209
recycling of fresh water, 113
red blood cells, 155
red dye, 206
red phosphorus, 203
reeds, 74
reel-to-reel technology, 207
reference, 3, 40, 90, 225, 282, 285
reflected light, 189
reflecting telescope, 111
reflectron, 167
reflexologist, 47
reflexology, 47
refractometer, 39
refrigerant, 17, 34, 175
refrigeration, 17, 41, 42, 50, 70, 86, 109, 111, 133, 182, 209
refrigeration technology, 133
refrigeration unit, 42

refrigerator, 15, 17, 27, 42, 70, 86, 175, 182, 209
refrigerators, 121, 182, 209
refueling system, 231
regeneration of damaged cells, 172, 249
Regent Street Polytechnic Institution, 62
regulations, 6, 61, 266
regulator, 50, 110, 212, 250
rehabilitation therapies, 189
Reiche, E. C., 140
Reichstein Process, 32, 219
reinforced concrete, 66, 152, 177
reinforcing bar, 66, 152, 177
relay towers, 63
relays, 21, 121
reliability, 42, 137, 203
religion, 3, 290, 385
religion and spirituality, 3
religious practice, 167
religious services, 128
Remington Model 700, 219
Remington Typewriter, 233
remote control, 28, 166, 210, 259
remote control aircraft, 259
remote controls, 112, 119
remote-control, 113
removing water from food, 50
Renaissance Man, 122, 385
renewable fuel, 91
REO automobile, 194
REO Motor Car Company, 194
REO Speedwagon, 161, 162
repeating pistol, 66
replaceable blade, 102
reproduction biologist, 63
Republican Party, 174
repulsion motor, 75
research, 11, 12, 34, 54, 57, 60-63, 68, 69, 81, 83, 85, 90, 98-100, 103, 106, 113, 116, 119, 122, 130, 131, 135, 139, 140, 145-147, 154-157, 160, 166, 168, 171-173, 175, 176, 185, 188, 189, 191, 195, 201, 203, 206, 208, 210, 212, 213, 215, 219, 222, 223, 226, 229, 232, 236, 244-246, 250, 251, 255, 256, 261, 264, 381
research chemist, 139
research engineer, 68, 69, 229
researcher, 40, 45, 46, 49, 51-55, 60, 62, 74, 80, 85-88, 92, 94, 98-100, 103, 104, 106, 108, 109, 111, 112, 114, 115,

117-119, 121, 123, 129-131,
    135, 139, 146, 153, 169, 172,
    173, 175, 178, 186, 190, 195,
    196, 202, 204, 208-212, 217,
    219-222, 226, 228, 229, 239,
    243, 249, 260, 262, 267, 276
reservoir pen, 91
residential refrigerator, 42
resin, 75, 91, 218, 260
resin (plastic) composites, 91
resins, 243
resistance, 64, 121, 189, 263, 268
resistors, 64
respirator, 31, 78, 82, 122, 232
restaurants, 176
restoring surfaces, 249
retail displays, 272
retinopathy, 48
retractable landing gear, 205
reusability, 261
reverb, 17, 203
reverse opening rodeo chute, 45
reversible steam turbine, 154
revolution, 60, 97, 98, 114, 135,
    155, 170, 186, 222, 228, 262,
    265
Revolutionary period, 3
Revolutionary War, 93, 195, 223,
    240, 385
Revolutionary War officer, 240
Revolutionary War soldiers, 385
revolving book stand, 131
revolving clothes rack, 131
revolving cylinder, 66
revolving door, 16, 138
RF, 145
RFIQ, 19, 128
RFIQin, 128
rhinoplasty, 32, 243
Rhode Island, 55
Rhode Island College, 55
rhyming patterns, 156
rhythm guitarist, 385
rhythm mandolinist, 385
Rhythmicon, 248
rib cage, 116
Rich, Buddy, 161, 162
Richter Magnitude Scale, 109, 220
Rickenbacher Guitar Company, 46,
    220
Rickenbacker guitars, 46, 220
riding chaps, 45
Rife Machine, 220
rifleman's rifle, 266
rifles, 56, 213
Riggenbach Counter-Pressure Brake,
    221
Riggenbach Rack Railway System,
    221
rigid airship, 229, 276
rigid endoscopes, 122
ring armature design, 199
ring armature motor, 214
ring binder, 28, 237
rings of Saturn, 125
rivers, 106, 176
riveted jeans, 16, 73
RK-1, 146
Roach, Max, 161, 162
road, 25, 70, 152, 171, 190, 206,
    248
road builder, 171
road building, 171, 190
road building system, 171
road grip, 206
road hauling, 248
road stability, 171
Roadster (Tesla), 251
roadway material, 236
roadways, 119, 171
roasting foods, 119
Robert Fulton of air navigation, 102
Roberval Balance, 221
Robervallian Lines, 221
robot, 25, 73, 172
robotic, 32, 73, 138, 140, 141, 184,
    226
robotic arm prosthesis, 32, 138
robotic exoskeleton, 226
robotic soldier, 73
robotic technology, 184
robotics, 51, 121, 124, 134, 155,
    167, 169, 172, 189, 212
robotics engineer, 172
robots, 9, 25, 92, 113, 127, 172,
    226
rock groups, 248
rock music, 58
Rockefeller Foundation, 221
Rockefeller University, 221
rocker skate, 209
rocket, 20, 45, 55, 63, 67, 77, 103,
    138, 141, 145, 161, 185, 211,
    212, 229, 249
rocket complex, 45
rocket designer, 185, 211, 229
rocket engineer, 45, 63
rocket engines, 103
rocket mail, 229
rocket motor design, 138
rocket propulsion, 103
rocket scientist, 77, 103, 145

rocket technologist, 141
rocket-driven motorcycle, 161
rocketry, 67, 145
rockets, 113, 229, 249
Rocky Mountains, 35, 385, 388
rod and crank engine, 244
rod lens system, 122
rodeo chute, 45
Roland, 180
roll film, 79
rolled aluminum, 187
roller bearing, 126
roller chain, 238
roller conveyor, 224
roller skate designs, 209
roller skates, 173
roller skating club, 209
roller skating rink, 209
Rollerbladed, 18
Rollerblades, 173
Rolling Stones, 248
Romain, Pierre, 223
Roman alphabet, 124
Romantic Period (music), 47
Rome, 181
Röntgenograms, 222
roofing, 16, 218
room, 26, 40, 44, 90, 98, 202
room temperature continuous wave operating diode laser, 40, 98
rope line, 167
ropes, 243
Rorschach Test, 31, 222
rotary motion, 69
rotary steam engine, 263
rotating cardboard disk, 209, 239
rotating magnetic field, 246
rotating propellers, 67
rotational mechanical energy, 86
rotational motion, 51
rotators, 131
Rothamsted Experimental Station, 154
rotorcraft, 64
Round Script, 237
Rover bicycle, 240
rover technology, 140
Rowland Institute of Science, 152
Royal Canadian Air Force, 95
Royal National Lifeboat Institution, 167
Royal Navy, 105, 114
Royal Polytechnic Institution, 62
Royal Society for the Prevention of Cruelty to Animals, 29
Royal Society of London, 190

Royal Swedish Academy of Sciences, 158, 160
Royal Teleochordan stop, 65
Rozière Balloon, 223
RSPCA, 29, 104
RT-21 Temp 2S, 185
rubber, 9, 13, 15, 21, 22, 25, 27, 61, 67, 78, 94, 104, 105, 139, 146, 154, 169, 175, 190, 201, 205, 206, 230, 248
rubber band, 25, 146, 205, 206
rubber ice cube tray, 22, 67
rubber surgical tubing, 67
rubber tires, 9, 21, 248
rubella, 30, 120, 209
rubidium, 57
rubies, 256
Rubik's Cube, 27, 223
Rucker, Edmund W., 385
rudder, 109, 264
rugby, 22, 59, 82
Rugby School, 82
Rugby World Union Tournament, 82
Rugby, UK, 82
Rummikub, 27, 119
rummy, 119
Rundgren, Todd, 161, 162
Rundschrift, 237
running machine, 77
running rails, 221
runway, 130
Russia, 23, 146, 160, 161, 182, 234, 286
Russian Archimedes, 148
Russian culture, 167
Russian emperor, 236
Russian heavy bomber, 207
Russian nesting dolls, 167, 277
Russian oil industry, 190
Russian Salad, 19, 195
Russian soldiers, 99
Russian space program, 141
rust, 138, 152, 153
rust protection, 153
Rutgers University, 23, 93, 224
Ruth Cinder Spreader, 224
RVs, 182, 209
Sabbath, 161, 162
saccharin, 19, 85, 219
saddle, 45, 166, 288
safes, 271
safety, 17, 18, 24, 27, 51, 73, 79, 97, 101, 102, 119, 128, 180, 183, 196, 202, 203, 206, 209, 213, 236, 240, 260, 267

safety bicycle, 240
safety formula, 183
safety hood, 180
safety match, 17, 128, 203
sailors, 115
sailplane, 18, 62
salad, 19, 195
sales and retail industry, 97
sales offers and discounts, 114, 185
salesman, 68, 103, 134, 161, 204, 231, 261
salicyl alcohol, 206
salmon, 226
salt, 71, 195
salts, 70, 118
salvage diver, 101
Salyut, 63
Sambo, 21, 196, 239
samozashchita bez oruzhiya, 196, 239
Samsung, 27, 58
Samsung Group, 58
San Francisco, CA, 230, 241
sand, 138, 152, 161
sandblasting, 28, 34, 249
sandpaper, 78
Sanford and Son (TV series), 154
Sanger Sequencing, 226
sanitation, 114, 257
sapphires, 256
satellite communications, 30, 130, 145, 183
satellite communications service, 183
satellite technology, 65
satellite TV, 28, 124, 253, 260, 271
satellites, 63, 65, 68, 92, 113, 119
saturated cadmium cell, 263
Saturday, 28, 175, 204
Saturday Night Live, 28, 175
Saturday Night Live (TV comedy show), 28
Saturday Night Live (TV show), 28, 175
Saturn, 55, 125
Saturn V rocket, 55
Savage Model 99, 227
Savage Tire Company, 227
sawdust, 260
sawmill, 68
saxophone, 25, 228
SCA, 201
scale pans, 221
scales, 142, 221, 256
scanned light beams, 189
scanners, 48, 53, 92, 130, 234, 267
scanning probe microscopy, 100, 215
scanning technique, 71

scanning technology, 189
scanning tunneling microscope, 49, 222
scenographer, 167
Schaefer, Vincent J., 257
Schick Test, 32, 228
Schiff Test, 32, 228
Schjeldahl Company, 228
Schlenk Flask, 32, 229
Schmidt Camera, 24, 229
Schmidt Telescope, 229
Schmitt Trigger, 17, 229
scholar, 117, 120, 143, 157, 158, 169, 195, 232, 259, 264, 266, 385
scholars, 142
school, 13, 24, 29, 41, 55, 82, 112, 115, 144, 160, 208, 213, 223, 255, 264, 286
School of Beekeeping, 213
Schrader Valve, 229
Schrader, George, 229
science, 3, 24, 29, 33, 40, 41, 48, 51-53, 55, 56, 60, 61, 65, 66, 68, 87, 88, 90, 93, 97, 98, 100, 106, 109, 112, 115, 116, 118, 120-122, 129, 133, 141-148, 152, 154, 156-158, 163, 168, 170, 172, 175, 177, 179, 180, 184, 186, 188-191, 195, 196, 210, 215, 218, 219, 223, 229, 230, 232, 237, 249, 255-257, 261, 286, 291, 382, 385
science awards, 190
science competition, 144
science fiction, 52, 53, 65, 116
science fiction stories, 52
science fiction writer, 65, 116
science laboratories, 256
science of electromagnetism, 196
science teachers, 223
scientific advisor, 246
scientific applications, 88
scientific community, 90, 117, 247
scientific fields, 133, 148, 160, 182, 238, 259
scientific instrument designer, 143
scientific instrument maker, 86, 105, 208, 261
scientific instruments, 18, 46, 120, 142, 166, 169, 190, 193, 224, 227, 242
scientific principle, 173
scientific research, 98, 130, 145, 166, 185
scientific synthesists, 248

scientist, 39-41, 43-47, 49, 51-56,
    59, 61-63, 65, 67, 68, 72-74,
    76-80, 82, 83, 85-91, 93-97,
    99, 100, 102-109, 112, 113,
    116-122, 124, 125, 127-133,
    135, 137, 138, 140-149,
    151-161, 163, 165-180,
    183-191, 194-196, 199-204,
    206-215, 218-226, 228-232,
    234-239, 241, 243-246,
    248-251, 253, 255-257,
    259-267, 272, 276, 277, 284
scientists, 12, 62, 69, 81, 88, 100,
    102, 109, 143, 145, 211, 247,
    249, 261, 275, 290, 385
sclera, 88
scleral contacts, 88
Scorpion, 156
Scotch Masking Tape, 78
Scotch tape, 78
Scotland, 13, 24, 181, 282, 286,
    290, 291
Scott, George C., 385
Scrabble, 27, 58
screenwriter, 50, 55, 57, 63, 65, 76,
    102, 127, 154, 175, 188, 221,
    231, 240, 385
screw, 20, 21, 39, 41, 73, 82, 83,
    89, 90, 148, 169, 170, 186,
    267
screw anchor, 89
screw elevator, 148
screw propeller, 20, 82, 83, 90
screw threads, 186
screw-cutting lathe, 186
screw-on metal band, 169
screws, 135
Scriabin, Alexander N., 183
scrolling script screen, 229
SCUBA, 22, 53, 59, 73, 212
scuba diver, 212
scuba divers, 53
scuba gear, 22, 73, 212
scuba instructor, 59
sculptor, 44, 45, 73, 141, 144, 223,
    385
SDM, 236
sea, 5, 6, 20, 27, 48, 76, 100, 109,
    146, 147, 156, 179, 205, 207,
    208, 287-290, 385, 388, 389
sea depths, 208
sea plane, 20, 146, 147, 205
sea plane technology, 147
sea planes, 146
Sea Raven Press, 5, 6, 100, 109,
    287-290, 385, 388, 389

Seabrook, Lochlainn, 3, 35, 287,
    385, 389
sealants, 155
search and rescue, 24, 91, 172
searchlights, 238
Sears Tower, 141
seaside estate, 113
seasonings, 206
seatbelts, 9, 21
seawall, 152
SeaWorld, 22, 176
secession, 93, 184, 289
Second Earl of Dartmouth, 264
secret message, 60
secret of life, 152
secret writing, 60
secretary of state, 166
secure communication, 152
Secure Network Programming, 152
Secure Socket Layers, 152
Secure Sockets, 152
secure voting systems, 63
seed, 19, 226
seed drill, 19
seed drills, 226
seeds, 61, 128, 264
Seguin, Camille, 230
Segway Human Transporter, 138
Segway Personal Transporter, 138
Segway PT, 138
Seinfeld, 230
Seinfeld (TV series), 28, 73
seismograph, 24, 98, 117, 200
seismologist, 98, 109, 220
seismology, 79, 101, 202
seismometer, 117, 200
Self Contained Underwater Breathing
    Apparatus (SCUBA), 212
self defense, 68
self-contained propulsion unit, 83
self-defense, 196, 239
self-igniting firearm, 142
self-inductance, 118
self-powered carriage, 148
self-powered weapon, 170
self-propelled torpedo, 163, 264
self-propelling carriage, 232
self-regulating clock, 217
self-sealing lid, 169
self-taught, 53, 76, 82, 86, 93, 115,
    122, 133, 148, 153, 160, 179
Semantography, 50
semaphore, 26, 63
semaphore visual telegraph, 63
semiconducting device, 45, 54, 233
semiconductor, 55, 74, 85, 112,

118, 121, 137, 147, 186, 189, 244, 246
semiconductor chip, 85
semiconductor device, 121
semiconductor inductance, 189
semiconductor injection laser, 112, 189
semiconductor laser, 118
semiconductor material, 147
semiconductor memory cells, 74
semiconductor RAM technology, 246
semiconductors, 70, 155
semiotician, 50
semisolid fat, 190
Semtex explosive, 22, 55
sensitized paper, 118
sensory fluid dynamics, 117
separate condenser, 261
Sequoia National Park, 26
serologists, 245
serotherapy, 43
Serrin, Victor, 110
Serrurier Truss, 231
servant, 66, 180, 256
services, 9, 29, 79, 101, 128, 132, 134, 202, 273, 283, 285
session player, 385
set theory, 255
setting angles, 132
Setun, 56
sewage systems, 267
sewing machine, 29, 76, 124, 126, 235, 274, 291
sewing machines, 180
sexual intercourse, 273
Shakespeare, William, 156
sheep, 256
sheet vinyl flooring, 260
shelf life, 172, 190, 204
Shelley, Mary, 98
shells, 103, 166, 233, 241
shelving, 18, 89, 223
shelving system, 223
shielded metal arc welding, 153, 236
Shimomura, Osamu, 251
ship, 20, 25, 30, 63, 83, 92, 105, 113, 114, 142, 148, 159, 166, 167, 202, 220, 241, 265, 273
ship building engineer, 273
ship camouflage, 265
ship design architect, 273
ship designer, 241
ship hull designs, 273
ship maintenance and repair, 142
ship propeller, 220

ship-builder, 105
shipping, 171, 241
shipping industry, 241
ships, 40, 82, 114, 121, 122, 125, 146, 147, 166, 168, 232, 238, 265
ships' hulls, 105
ship's silhouette, 265
shock absorber, 205, 251
shock absorbers, 20, 123
shock therapy, 49, 62
shock wave, 165
shock waves, 256
shock-absorbing qualities, 175
shoes, 16, 21, 187, 243, 277
shopping cards, 180
shopping online, 152
shops, 180, 242, 256
short wave technology, 168
short-wave, 57
short-wave telemetric radio, 57
shotguns, 56
shower curtains, 16
shrapnel, 24, 233
Shrapnel Shell Ammunition, 233
Shredded Wheat, 206
shrink resistant, 194
Shukhov hyperboloid, 234
Shukhov Tower, 234
Shure Radio Company, 234
Shure, Inc., 234
shutter, 94, 115
SI unit, 118
side mirrors, 94
side-delivery rodeo chute, 45
sidereal time, 204
siding, 16, 218
Siemens, 234
Sierra Club, 182
Sigma, 27, 173, 272
Sigma Corporation, 173, 272
signal conditioning applications, 229
signal switching, 137, 244
Sikorsky helicopter, 23, 234
silage, 256
silent intruder alarm, 24, 183
silent movies, 249
silica glass, 89
silicon, 74, 121, 212, 231
silicon chip, 74
silicon tetrafluoride, 212
silicon tunnel diode, 121
siloxane-methacrylate, 99
silver, 18, 49, 71, 104, 118, 201, 234, 257, 261, 285
silver chromate, 104

silver nitrate, 261
silver oxide, 257
silver staining, 261
silversmith, 89
Sinclair C5, 235
sing, 128, 187
Singer Manufacturing Company, 291
Singer Sewing Machine, 29, 235
singer-songwriter, 123
single crystals, 70
single-lens reflex camera, 243
single-mode optical fiber, 237
sink and shower water, 47
sisal, 91
sisters, 385
Site-Directed Mutagenesis, 32, 236
six-shooter, 66
Skafandr, 64
Skaggs, Ricky, 385
skeptics, 194, 268
sketch artist, 213, 385
ski industry, 266
ski lift, 19, 266
Ski-doo, 52
skimmed milk, 172
skin, 14, 32, 90, 167
skin color, 14
skin infections, 90
skippers, 115
skydiver, 146
skyscraper, 18, 152
skyscrapers, 66, 138, 141, 177
slavery, 213, 288
SLBM, 166
slide rule, 16, 186, 197
slipstream, 65
sloping ramp, 220
slotted armature, 17, 263
Smakula Effect, 236
small applications, 74
small businesses, 273
small camera design, 275
small game hunting, 168
small loans, 273
smart card, 15, 180
smart phones, 121
smart TVs, 113
smartphones, 17, 113, 231
SMF, 237
Smithsonian Institution, 22, 118, 153, 236
smithsonite, 236
smoke, 9, 22, 72, 157, 276
smoke alarm, 72
smokeless candles, 174
smokeless explosive, 75

smokeless grate, 42
smokeless gunpowder, 24, 173, 227, 256
smokeless gunpowder alternative, 173
smoking, 157
smooth flowing ink, 261
smoothbore military gun, 241
snorkel, 22, 73
snorklers, 53
snow, 114, 138, 228
snowdrop plant, 203
snowdrops, 203
snowmobile, 20, 52
snowpack, 257
SNP, 152
soccer, 22, 59, 180, 185
social communications, 115
social media platform, 27, 183
social media-sharing functions, 134
social phenomenon, 215
social sciences, 158
social systems designer, 211
socialism, 282, 283
socialist, 174, 285
Society for the Prevention of Cruelty to Animals, 29, 104
sociobiology, 290, 292
sociologist, 49, 187, 215
Soennecken, 237
soft contact lens, 265
soft drink, 205
soft ground, 106
soft-landing, 141
software, 15, 123, 132, 149, 200, 221, 222, 239, 242, 248, 250, 267
software applications, 242
software developer, 200, 221, 222, 248, 250, 267
software operating systems, 132
soil, 56, 60-62, 87, 91, 93, 100, 106, 109, 140, 152
soil analyses, 140
soil decontamination, 87
soil erosion, 100
soil science, 61
soils, 93
solar cell, 23, 46, 106, 195, 241
solar panel, 23, 46
solar panels, 94
solar power, 81
solar radiation, 41
solar system, 18, 105, 250
solar technology, 141
soldier, 73, 101, 123, 134, 289

soldiers, 66, 93, 99, 102, 134, 184, 250, 289, 385
solicitor, 180
solid rubber tires, 248
solid-body electric guitar, 203
solid-state focusing optical fibers, 189
solids, 67, 78, 99, 185
solo sonata, 17, 47
Solventil, 71
solvents, 87, 219
sonar technology, 239
song cycle, 17, 47
songwriter, 63, 123, 127, 149, 203, 230, 385
sonic boom, 165
sonic oil drilling technique, 67
sonics, 67
sonnet, 28, 156
Sonnet 18, 156
Sons of Confederate Veterans, 385
Sony, 27, 127, 180
Sony Group Corporation, 127, 180
sore throat, 22
sore throat lozenges, 22
soul, 290
sound, 26, 50, 52, 74-76, 131, 133, 143, 163, 165, 173, 183, 204, 211, 218, 230, 231, 234, 248, 249, 255
sound dynamics, 163
sound engineer, 230
sound measuring devices, 231
sound pioneer, 249
sound telegraph, 174
sound-on-disk system, 255
sound-on-film technology, 249
sounds, 16, 25, 44, 69, 75, 145, 151, 157, 260
soup, 19, 77
South, 3, 28, 57, 58, 66, 77, 106, 119, 137, 174, 183, 184, 201, 241, 272, 288, 289, 292, 381, 385
South Africa, 292
South Park, 201, 241
South Park (TV series), 28
Southern Blot, 32, 238
Southern Cause, 289
Southern Confederacy, 288, 289
Southern states, 93
Southerners, 174, 213, 289
Soviet, 40, 99, 103, 140-142, 144, 145, 148, 161, 166, 207, 225, 226, 243
Soviet ANT bomber aircraft series, 207
Soviet atomic bomb, 141, 148
Soviet chemical weapons program, 144
Soviet missile deployments, 166
Soviet nuclear program, 148
Soviet rocket engines, 103
Soviet space program, 103
Soviet spaceflight program, 145
Soviet Union, 243
Soviet war effort, 207
soybeans, 61
Soyuz, 145
space, 14, 28, 29, 55, 56, 63, 69, 77, 79, 81, 82, 89, 94, 95, 101, 103, 113, 122, 124, 125, 131, 133, 140, 141, 145, 147, 152, 161, 170, 183, 202, 239, 262
space agencies, 95
space dog, 141
space engineering team, 140
space exploration, 55, 56, 81, 94, 140, 183
Space Exploration Technologies Corp, 183
space exploration vehicle, 140
space flight, 103
space industry, 113
space pen, 89
space science, 122, 145
Space Shuttle, 77, 161
space station, 63, 141
space technology, 183
space-based radio-navigation system, 79, 101, 202
spacecraft, 20, 45, 115, 145, 161, 244
spaceport, 45
spacewalk, 141
Spandex, 16, 232
Spanish War, 170
spanner, 30, 132, 242
spanners, 132
spark plugs, 175
spark stick, 263
spas, 119
SPCA, 104
speaker, 17, 49, 132
speakers, 128, 187, 196, 229
spear, 187
special theory of relativity, 81
special visual effects innovator, 128
special-purpose electronic computer, 42
speciality paper, 162
species, 34, 72, 99, 129, 158

spectral classification of stars, 249
spectrograph, 249
spectrometers, 40
spectrometry, 87, 166, 167
spectroscopic tool, 275
spectroscopy, 85, 130, 186
speech recognition computer
    software, 15, 149
speed, 40, 64, 65, 73, 79, 90, 91,
    101, 122, 125, 130, 133, 138,
    141, 143, 145, 154, 165, 168,
    174, 184, 201-203, 206, 207,
    232, 235, 241, 248, 251, 265
speed bomber aircraft, 207
speed of light, 125, 130
speed of sound, 165
speed-changing device, 184
speedometer, 21, 43, 47, 263
speleology, 212
Spellbound (film), 248
sperm, 80
spermatozoa, 155
Sperry Electric Mining Machine
    Company, 238
spices, 61
spin-flip Raman laser, 32, 203
spinal cord, 272
spindles, 114
spinel, 256
spinet, 238
spinning frame, 114
spinning jenny, 114
spinning jenny technology, 114
Spiral project, 161
spirit, 8
spiritual emanation, 143
Spiritualism, 219
Spiritualist, 52, 129, 158, 200
spiritualists, 143
spirituality, 3, 167, 287, 385
splitting atoms, 40
spodumene, 179
sport, 34, 59, 61, 69, 82, 146, 181,
    196, 212, 237, 239, 266
Sport Sambo, 196, 239
sports, 9, 16, 19, 22, 27, 53, 59, 61,
    66, 68, 73, 74, 79, 100, 101,
    105, 119, 180, 181, 185, 186,
    191, 196, 202, 239, 246, 291
sports arenas, 119
sports enthusiasts, 105
sports game, 180
sports gliding product, 181
sports industry, 181
sports scoreboards, 246
sports vehicles, 74

sportsman, 61, 68, 147, 180
sportswear, 233
spotlight lamp, 148
SPP-1 Underwater Pistol, 235
Sprague Electric Railway and Motor
    Company, 239
spray nozzle, 178
sprinting races, 52
Sputnik, 103, 141, 145, 249
Sputnik I, 103, 141, 145
spy industry, 275
square roots, 197
SSL, 152
SSL technology, 152
St. Augustine, FL, 57
stability, 95, 139, 171, 174, 206,
    213, 232, 263
stability problems, 95
stabilized bombsight, 231
stabilized HT circuits, 50
stable-resistance alloys, 263
stage actor, 47
stain resistant, 194
stain-resistant, 201
staining method, 31, 101, 102, 106
staining technique, 277
stamp collector, 234
Stampfer, Simon von, 209
stamping dies, 124
stamps, 9, 28, 208
Standard Oil Company, 291
Standard Time Zones, 153
Standard Triangle Language, 125
standardizing biological parts, 144
Stanford University, 23, 239
star catalog, 120
star maps, 118
star noise, 130
Star Trek, 221
Star Trek (TV series), 28
Starley and Sutton Company, 240
Starlink, 27, 183
Starry, Allen C., 261
stars, 249
starting blocks, 52
statesman, 93, 131, 264
static induction transistor, 189
statistician, 120, 215, 251
statistics, 13, 215, 255
statues, 32
steam boilers, 113
steam cranes, 248
steam engine, 20, 57, 93, 108, 118,
    160, 169, 170, 188, 210, 220,
    222, 228, 241, 261, 263, 286
steam engine design, 241

steam engine designer, 220
steam engine designs, 262
steam engine-powered locomotives, 160
steam engines, 53, 150, 157, 171, 183, 186, 230, 241
steam guns, 183
steam hammer, 186
steam jet, 108, 154
steam jet nozzle, 154
steam locomotive, 53, 60
steam locomotive engine design, 60
steam locomotives, 183, 186
steam plow, 196
steam power, 204, 240, 264
steam powered engines, 211, 231, 241
steam pressure, 128
steam pump, 201
steam rail locomotive, 90, 250
steam ship engines, 113
steam traction, 248
steam turbine, 60, 154, 159, 169, 202
steam turbine torpedo boat, 60
steam-driven plow, 93
steam-power, 155
steam-powered, 70, 83, 102, 169, 179, 180
steam-powered boat designs, 180
steam-powered locomotive, 169
steam-powered road vehicle, 70
steam-powered vessel, 83
steamboat, 90, 95, 244, 281
steel, 9, 20, 29, 46, 53, 61, 66, 91, 100, 106, 118, 121, 132, 135, 142, 177, 180, 191, 212, 242, 264
steel blocks, 132
steel friction wheel, 142
steel industry, 53
steel rods, 66, 177
steel sheet, 135
steep gradients, 221
steering, 70, 78, 201
steganography, 60
stellarator, 239
stem cells, 32, 172, 249
stem-cells, 129
stenographer, 131
Stephens, Alexander H., 385
Stephenson Safety Lamp, 240
Stephenson, Robert, 240
stereo microphone, 50
Stereobelt, 204
stereogonic telescope, 211

stereolithographic file format, 125
stereolithography process, 125
stereophonic sound, 50
stereoscope, 32, 55, 264
stereoscopic TV, 44
steroids, 135
stethoscope, 145, 151
stick game, 68
stick-guided artillery device, 67
sticky note, 95
Stihl, 18, 241
Stihl chainsaw, 241
stimuli, 52, 204
Stirling Engine, 241
STL, 125
stock price ticker system, 59
stock prices, 77
stocking frame, 155
stockmen, 256
Stodola Arm, 241
Stokes Mortar, 241
stomach, 65, 193, 209
stonemason, 42, 199
stop motion animation films, 40
stop-motion animation, 240
Stopette, 178
storage, 74, 76, 85, 125, 137, 187, 188, 195, 207, 217, 233, 244, 245, 277
store loyalty cards, 202
stores, 42, 76, 180, 183, 184, 267
stove, 15, 17, 22, 42, 67, 71, 93, 108, 158
stoves, 242
straight wing design, 133
strain wave gear, 184
stratified microbial communities, 266
street car, 73, 208
street lighting, 263, 271
street lighting industry, 271
street lights, 121, 231
street sweeping truck, 25, 55
streetcars, 238
streetlight illumination, 247
streptococcus, 30, 120
stress, 152, 174, 183
stress management, 152, 174
string galvanometer, 81
string trimmer, 44
stringed instrument maker, 40, 226
stringed instruments, 46, 220
strings, 105
strip camera, 211
stroboscope, 25, 123, 209, 239
stroboscopic effect, 239

stroke, 74, 108, 176
strong winds, 135
strongly correlated quantum fluid, 195
Strowger Automatic Telephone Exchange, 242
structural engineer, 141, 233
structural interiors, 265
structural system, 141
structures, 66, 106, 119, 152, 177, 222, 233, 261, 272
students, 140, 149
stunt performer, 46, 123
stunt pilot, 134
Styrofoam, 9, 16, 17, 172, 182
Su-series fighter aircraft series, 242
subatomic particle, 62
subconscious mind, 94
submarine, 25, 95, 142, 166, 179, 208, 220, 259
submarine maintenance and repair, 142
submarine technology, 179, 208
submarine-launched ballistic missile, 166
submarines, 113, 122, 125, 146, 168, 179, 232, 242, 265
submersible, 25, 207, 208
submersible technology, 208
subminiature camera, 275
substance abuse, 30
subtracting, 66
sudden cardiac arrest, 201
sugar, 19, 81, 126, 188, 205
sugar cane mill, 126
sugar packet, 19, 81
sugar syrup, 205
suitcase with wheels, 21, 147
Sukhoi Design Bureau, 243
Sukhoi Su-7, 242
sulfite pulping, 249
sulfur, 93, 94, 104, 212
sulfur dioxide, 212
sulfur industry, 94
sulfuric acid, 76, 89, 154, 212, 221, 256
sun, 27, 51, 71, 85, 105, 111, 119, 259
sun lamps, 119
Sundbäck, Gideon, 134
sundial, 131, 148, 272
sunglasses, 15, 152
sunlight, 106, 144, 152, 195
sunscreen, 18, 147
sunshine, 385
Sunshine Sisters, the, 385

super computer, 43
super scooper, 205
super-strengthened hull, 166
supercapacitors, 185
superconductivity, 195
supercritical airfoil, 264
superglue, 67
superheated ionized gas, 69
superheated water, 93
superphosphate, 154
supersonic aircraft, 133
supersonic flight, 138, 259
supersonic passenger jet, 20, 251
supersonic shock wave, 165
support beams, 149, 244
suppressed invention, 44, 194
suppressed inventions, 55, 91, 94, 106, 117, 147, 174, 179, 193, 200, 218, 220, 246
suppressed inventor, 201
supraglottic airway device, 54
Supreme Court, 104, 225
surface cleaning and restoration, 249
surface condenser, 113
surface topography, 100, 215
surfers, 53
surfing pools, 176
surgeon, 44, 46, 53, 78, 91, 96, 98, 101, 104, 108, 109, 116, 127, 145, 161, 205, 207, 208, 225, 230, 236, 243, 272
surgeons, 67, 183, 208, 225
surgeries, 130, 203
surgery, 30, 31, 46, 101, 127, 172, 208, 222, 225, 230, 243, 249
surgical instruments, 183
surgical kidney stone removal, 243
surgical method, 272
surgical operations, 186
surpressed inventions, 48
surrealist painter, 50
SURTRACE, 45
surveillance equipment, 231
surveillance operations, 172
surveying, 51, 79, 97, 101, 116, 202
surveying expedition, 116
surveyor, 44, 66, 89, 113, 116, 122, 228, 239, 261, 286
survival kits, 128
survival of the fittest, 72
survival situations, 263
suspension bridge, 25, 230
sustained flight, 268
swarm bots, 25, 172
swarming technology, 172
sweat, 183

Sweden, 179
Swedish patriotism, 158
Swedish Wheel, 127
Sweet'N Low, 19, 81
swim fins, 22, 93
swimmers, 53
swimming pools, 176
swimsuits, 233
Swiss Army Knife, 15, 17, 82
switchboard, 214
switches, 214, 250
Switzerland, 188, 285, 286
swiveling chair, 131
Sylvester, the, 244
symbols, 290
symphony, 17, 47
Symphony No. 5 in C Minor
 (Beethoven), 47
Symphony No. 9 in D Minor
 (Beethoven), 47
symptoms, 121
sync-pulse separators, 50
synchronized sound, 75, 255
synchronized transmission, 131
Synchronous Multiplex Railway
 Telegraph, 267
synchrotron, 32, 255
syngas, 89
synthesizer, 16, 149, 173, 179, 183
synthetic diamond, 112
synthetic fabric, 209
synthetic fiber, 61
synthetic gas, 89
synthetic gemstones, 70
synthetic indigo dye, 44
synthetic laundry detergent, 49
synthetic leather, 15, 218
synthetic liquid hydrocarbons, 49
synthetic material, 194
synthetic particles, 63
synthetic plastic, 43, 201
synthetic rubber, 61, 154, 190
synthetic thermoplastic polymer, 46
syntonic telegraphy, 167
syringe, 31, 32, 182, 203
System 2000, 235
T-34 medium tank, 146
table tennis, 202
table tennis balls, 202
table-glass, 236
tablet, 15, 26, 27, 132, 139, 233
tablets, 87, 121, 133, 196
tabulating machine, 121
Tabulating Machine Company, 121
tachymeter, 17, 211
taillights, 94

tailor, 73, 132, 241
talk show host, 236
talkies, 75
talking board, 140
tallow, 172
Tanaka Engineering Works, 120
tandem rotor helicopter, 91
tangents, 238
tank, 20, 73, 139, 146, 154, 176,
 181, 210-212, 261
tank design, 181
tank designer, 146, 154, 181, 211
tank models, 181
taped music, 128, 187
target practice, 168
tartaric acid, 206
Taser, 68
task lighting, 121
taxicabs, 56
taxidermist, 50
taximeter, 25, 56
taxonomy, 158
TCP, 15, 62, 137, 250
TCP/IP, 62, 137, 250
tea, 19
teacher, 18, 25, 42, 44, 45, 47, 49
tech pioneer, 133
tech products, 233
technician, 78, 272
technological explosion, 264
technological visionaries, 200
technologies, 33, 45, 48, 55, 59, 76,
 79, 90, 95, 101, 102, 112, 113,
 117, 121, 132-134, 137, 147,
 153, 156, 168, 181, 183, 186,
 189, 193, 194, 196, 199, 202,
 207, 208, 224, 230, 238, 241,
 244-246, 260, 263, 264, 267,
 275, 277
technologist, 56, 88, 123, 133, 141,
 149, 157, 229, 235, 245
technology company, 132, 183
technology innovator, 141
technology inventor, 125
Teflon, 19, 34, 210
telecommunications, 85, 87, 97,
 118, 119, 132, 143, 173, 206,
 214, 218
telecommunications engineer, 173
telecommunications industry, 214
telecommunications switching
 systems, 132
telecommunications system, 214
telegraph, 45, 59, 62, 63, 68, 74,
 80, 118, 124, 125, 129, 148,
 156, 168, 181, 183, 224, 228,

267, 269
telegraph engineer, 45
telegraph lines, 62
telegraph signals, 224
telegraph stock ticker tape, 59
telegraphic alphabet, 46
Telegraphone, 211
telegraphy, 52, 54, 72, 108, 124, 135, 167, 168, 181, 234
Telemareograph, 224
Telemeteograph, 224
telemeter, 17, 211
telemetric device, 177
telemetric radio, 57
telepathy, 48
telephone, 24, 47, 49, 74, 78, 90, 115, 132, 153, 174, 181, 193, 214, 218, 224, 242, 267, 282, 291, 387
telephone exchange, 193, 214, 242
telephone exchange apparatus, 193
telephone handset, 78
telephone industry, 242
telephone line, 193, 214
telephone lines, 214
telephone network, 214
telephone operator, 214
telephone patent, 47, 174
telephone signals, 224
telephone transmitter, 267
telephonic engineer, 115
telephonic technology, 115, 193, 224
telephony, 130, 155, 214, 218
telephony pioneer, 214
Teleprompter, 28, 229
telescope, 24, 26, 28, 29, 97, 107, 111, 118, 124, 125, 130, 140, 158, 160, 166, 169, 188, 211, 218, 229, 231, 239, 262
telescopes, 63, 92, 122, 231, 236
telescopic device, 158
telescopic objective, 211
Teletrofono, 174
television, 15, 28, 39, 44, 50, 51, 55, 60, 74, 87, 104, 105, 113, 119, 131, 132, 161, 175, 189, 210, 222, 228, 249, 260
television broadcasting station, 131
television engineer, 228
television industry, 113, 175, 260
television pioneer, 105, 222
television technologies, 132
television viewing, 210
TeleVista, 161
teliochordon, 65

temperance drink, 205
temperature, 40, 53, 57, 62, 72, 86, 89, 98, 111, 115, 153, 177, 191, 201, 248, 262
temperature changes, 153
temperature strip, 201
temperatures, 87, 90
tempered glass jar, 169
Tennessee, 287, 385
tennis, 22, 109, 202, 212, 266
tennis balls, 266
tennis elbow, 109
tennis nets, 266
tennis patent, 266
tennis pioneer, 266
tennis rackets, 266
tensile strength, 66, 105
tension regulator, 50
ternary computer, 56, 93
terrestrial magnetism, 117
terrestrial radioactive rocks and soil, 152
terrestrial stationary waves, 247
Tesla Coil, 152, 247
Tesla Motors, 251
Tesla of Križevci, 142
Tesla Tower, 247
Tesla, Nikola, 96, 131, 142, 179, 249
test tube, 245
testing, 12, 31, 41, 68, 89, 90, 172, 249
tetraethyl lead fuel, 175
tetrafluoroethylene gas, 210
Tetris, 27, 200
Tetris Company, 200
text, 6, 11, 61, 82, 88, 120, 149, 187, 194, 290
text-to-speech synthesizer, 149
textile engineer, 54
textile industry, 124, 243
textile lace gasser, 113
textile manufacturing, 114
textile producing machinery, 195
textile production technology, 155
textiles, 99, 111, 186, 201
textilist, 74
TFE, 210
thallium, 69
thanatologist, 172
thanatology, 143
thaumatrope, 123
The Avengers (TV series), 28, 188
The Boring Company, 183
The Dam Busters (film), 259
The Day the Earth Stood Still (film),

248
the East, 285
The Federalist, 166
The Federalist Papers, 166
The Jeffersons (TV series), 28, 154
The Lost Weekend (film), 248
The Mary Tyler Moore Show (TV series), 28, 55, 57
the North, 20, 77, 288
The Sea of Energy in Which the Earth Floats (Moray), 179
The Simpsons (TV series), 28, 107
The Sopranos (TV series), 28, 63
The South, 66, 381
The Tonight Show (TV show), 28, 40
The Truffault Fork, 251
The Twilight Zone (TV series), 28, 231
The Wall Street Journal, 77
the West, 40, 73, 120, 143, 178
The Wire (TV series), 28, 235
thealogy, 167
theater scene designer, 71
theologian, 57, 140, 169, 186, 188, 197, 202, 212, 228, 260
theology, 143, 158, 167, 235
theoretical inventor, 80
theoretical physicist, 116, 195
theoretical theorist, 246
theoretician, 138, 209
theories, 12, 81, 92, 138, 148, 157, 171, 178, 276
theory of relativity, 80, 81, 165
theory of sonics, 67
theory of spontaneous generation, 155
theory of the conservation of energy, 107
therapeutic creams, 153
therapy, 28, 49, 62, 81, 119, 139, 143, 200, 218
Theremin, 16, 248
thereminists, 248
thermal energy, 249
thermal engineer, 76
thermal imaging, 28, 119
thermal machine, 44
thermal paper, 75
thermionic tube, 90
thermionic valve, 90
Thermo King, 133
thermodynamicist, 218
thermodynamics, 117, 175, 180, 195
thermometer, 30, 32, 48, 63, 86, 97, 133, 169, 182, 201, 227

thermometers, 9, 30, 86, 119
thermonuclear bomb, 148
thermonuclear fusion, 239
thermoplastic, 46, 144, 201
thermoplastic polymer, 46
thermoplastic polymers, 144
Thermos, 15, 75
thermoscope, 227
thermosonic bonding, 68
thermostat, 67, 170
thermostatically controlled shock absorber, 205
thin-film transistors, 137, 244
thixotropic visco-elastic ink, 89
Thomas Edison of Finland, 249
Thomas Edison of Japan, 115
thorium oxides, 263
thread, 39, 124, 243
thread rolling machine, 124
threaded metal rods, 127
three laws of planetary motion, 24, 140
three-level maser, 130
Three-Line Rifle, Model 1891, 181
three-phase electric power system, 17, 77, 263
three-way automotive catalytic converter, 139, 179
three-way traffic light, 180
three-wheel clock, 93
threshing, 92, 261
threshing machines, 92
throat and lung diseases, 139, 179
throat infections, 90
thrust, 20, 55, 138, 175
thrust force, 138
thrust-bearing engine, 175
thumb drive, 245
thumb drives, 233
Thunder and Lightning Harris, 115
Thunderstorm Generator, 48
Thursday, 204
ticks, 43
tidal science, 143
Tilghman, Benjamin C., 34
tilting pad principle, 175
timber, 68, 229
time, 13, 14, 24, 26, 28, 29, 33, 34, 44, 47, 55, 72, 79, 81, 82, 85, 86, 90, 91, 93, 101, 103, 106-108, 114, 118, 120, 125, 131, 133-135, 147, 153, 154, 158, 160, 161, 167, 171, 172, 176, 179, 189, 195, 197, 200, 202, 204, 205, 208, 211, 217, 222, 227, 238, 241, 245-247,

249, 251, 256, 260-262, 266, 268, 271, 282, 283, 286, 290, 382, 389
time period of Man, 262
time-of-flight analyzers, 167
timepiece, 120
timing device, 269
Tinder, 15, 217
Tinkertoys, 27, 199
TiO2, 106, 195
tire designer, 227
tissue, 72, 102, 104, 112, 115, 172, 249, 261
tissues, 49, 98
Titania, 118
titanium, 105, 106, 147, 195
titanium dioxide, 106, 195
titanium ores, 147
TKS, 63
TN-LCD, 87
toast, 88
toaster, 25, 67
tobacco, 157
tobacco-based cigarette, 157
Tocqueville, Alexis de, 93
toilet, 30, 114, 153
Tokamak, 225
toll coin collection device, 193
Tompion Regulator, 250
tonic, 205
Tonight Show Band, the, 161, 162
tonka tree, 206
tool boxes, 242
tool chest, 132
tool maker, 186, 210
toolkit, 82
toolmaker, 195
tools, 74, 81, 85, 101, 120, 156
toothed rail, 221
top soil, 62
top soil protection, 62
top-down cosmology, 116
topographer, 211, 239
topology, 117
toroidal aircraft propeller, 168
toroidal props, 122, 168, 232
torpedo, 21, 60, 163, 166, 259, 264
torpedo boats, 166
torpedo nets, 259
torpedoes, 163, 227, 238
torque, 21, 184, 238
torretta butoscopica, 101
Toshiba, 27, 120
Toshiba Corporation, 120
touchscreen interface, 132
touchscreen technology, 132

towels, 81, 194
tower clocks, 120
towers, 63, 234
towns and cities, 183
toxic emissions, 44, 123, 139, 179
toxic gases, 123
toxicology, 243
toxins, 52, 276
toy, 22, 46, 64, 88, 91, 95, 102, 112, 122, 141, 181, 199, 200, 205, 217, 223, 267, 272
toy designer, 122, 141, 199, 217
toy inventor, 102, 122, 267
toy maker, 199, 272
toy model enthusiast, 46
toy model train set, 102, 112
toys, 27, 46, 85, 113, 155, 157, 187, 202, 205, 277
traction, 139, 175, 239, 248
tractor, 18, 20, 62, 70
trading cards, 202
traffic light, 180
train, 17, 20, 26, 43, 50, 86, 102, 112, 122, 134, 155, 156, 169, 171, 205, 221, 242, 246, 267
train braking systems, 155
train engines, 86
train locomotives, 43
train refrigeration, 50
train station departure boards, 246
train stations, 267
training aircraft, 207
tram, 208
trampoline, 24, 213
trampoline beds, 213
trampoline safety net enclosure, 213
tranquillizer gun, 24, 182
trans fat, 190
transducers, 263
transfer vehicles, 127
transfixion wires, 127
transformer, 50, 75, 86, 263
transformer technology, 263
transformers, 121, 247, 248, 263
transgenesis, 129
transgeneticist, 129
transgenics, 129
transistor, 25, 45, 54, 74, 87, 121, 137, 189, 233, 244, 246
transistor laser, 87, 121
transistor technology, 246
transit time, 189, 211
transit time negative resistance diodes, 189
translator, 114, 123, 157, 212
Transmission Control Protocol, 15,

62, 137, 250
transmission of power by electricity, 211
transmissions, 77, 260
transmitter, 60, 78, 80, 125, 152, 267
transmitting clandestine messages, 107
transparent cellophane tape, 78
transparent glass, 89, 266
transport, 63, 70, 73, 91, 113, 154, 159, 176, 207, 217, 234, 239-241
transport aircraft, 207
Transport-Supply Ship, 63
transportation, 44, 76, 79, 101, 112, 133, 147, 196, 202, 234, 276
transversal filter, 50
trap mass analyzer, 166
trauma, 183
travel, 64, 79, 81, 82, 101, 141, 159, 193, 202, 213
travel in space, 141
traveler, 182, 213
travelers, 67
traveling-wave tube, 145
travels, 20, 64, 165, 283
treads, 139
treasurer, 55
treatment of disease, 172, 249
tree, 19, 162, 206, 226
tree bark, 162
trees, 158
triathletes, 53
trichloromethane, 109
tricycle, 70, 251
triglyceride, 49
trigonometry, 30, 107, 120, 142, 186
trigonometry functions, 107
triode, 74, 90
triple barrel canon, 73
triquetrum, 142
Trojan Ballistics Suit of Armor, 125
trolley car, 22, 73, 208, 281
trolley cars, 231
trolley poles, 159, 234
trolley wheel, 267
trolley wheels, 159, 234
trolleybus, 22, 159, 234
truck brakes, 196
truck engines, 86
trucks, 17, 78, 113, 122, 209, 229, 231
Tructractor, 65
true believers, 247

Truffault-Hartford Suspension, 251
Trump, Donald J., 183
trumpet, 25, 65
truth, 90, 148, 281, 381, 382
Trypanosoma, 101
Tsar Bell, 182
Tsar Tank, 154, 176
Tsusensan, 230
tube, 16, 50, 55, 56, 69, 74, 78, 86, 90, 94, 117, 125, 130, 145, 151, 159, 172, 217, 222, 229, 230, 233, 245, 246, 249, 277
tubercle bacillus, 144
tuberculin skin test, 32, 167
tuberculosis, 30, 144, 145, 167, 277
tubes, 28, 45, 54, 60, 73, 81, 109, 200, 212, 217, 218
tubular condenser, 131
tubular water slide, 22, 223
Tuesday, 204
tungsten steel, 121
tuning fork, 65
tunnel construction, 183
tunneling shield, 56, 106
tunneling techniques, 170
tunnels, 106, 191
Tupolev Tu-144 Rossiya, 251
turbines, 25, 86, 122, 154, 168, 202, 232, 242
Turbinia (boat), 60
Turbo-cooker, 242
Turbococina, 19, 242
turbofan engine, 163
turbogenerator, 50
turbojet, 20, 108, 138, 194, 265
turbojet engine, 108, 194, 265
turbojet technology, 138
Turing Machine, 251
turret clocks, 120
tutor, 197, 211
TV, 9, 15, 27, 28, 40, 44, 55, 57, 63, 65, 73, 76, 82, 102, 107, 112, 113, 124, 125, 127, 131, 134, 139, 140, 143, 154, 168, 169, 175, 188, 189, 200, 201, 204, 207, 210, 221, 222, 230, 231, 235, 241, 249, 253, 259, 260, 271, 277, 385
TV images, 140
TV industry, 207
TV music, 230
TV producer, 76
TV remote controls, 112
TV scientist, 131
TV sets, 189
TV show, 76, 385

TV show host, 76
TV signals, 113, 253, 271
TV system, 249, 260, 277
TV technology, 124, 143, 249
twisted nematic liquid crystal display, 87
Twister, 27, 91, 217
Twitter, 27, 183
two-cycle diesel engine, 21, 141
two-cylinder steam engine, 210
two-dimensional plane, 118
two-electrode radio rectifier, 90
two-fluid electric cell, 107
two-piston engine, 70
two-stroke diesel engine, 108
two-stroke internal combustion engine, 74
TWT, 145
typewriter, 18, 157, 233, 282, 292
typhus, 30, 145
typos, 12
U.S. Air Corps, 117
U.S. army, 155
U.S. Atomic Energy Commission, 156
U.S. Bureau of the Budget, 262
U.S. Constitution, 17, 93, 166
U.S. Energy Information Administration, 202
U.S. Federal Court, 218
U.S. government, 64, 79, 83, 101, 187, 202, 218, 219, 226, 260, 285, 291
U.S. House of Representatives, 291
U.S. military, 102, 133, 220
U.S. military forces, 133
U.S. Space Shuttle, 77
U.S. Steel, 20, 61, 180
U.S. Thermo Control Company, 133
U.S.A., 62, 131, 180
U-joint, 122
U-tube manometer, 125
UAVs, 113
UCAVs, 113
UEFI standard, 223
UFO researcher, 55
UFOs and extraterrestrials, 3
UHF, 253, 271
ultracentrifugation, 32, 243
ultracentrifuge, 243
ultramarine blue, 144
ultramicrotome, 88
ultramicrotomists, 88
ultrasonic energy, 68
ultrasuede, 194
ultraviolet radiation, 85

Ulyanovsk Oblast, 100
UMBO shelving and furniture, 223
uncle, 211
undergarments, 233
underground cable system, 113
underground deposits, 93
undertaker, 242
underwater, 22, 24, 26, 59, 68, 89, 101, 106, 122, 123, 133, 142, 147, 160, 168, 186, 212, 232, 234, 235
underwater archaeology, 212
underwater breathing device, 68
underwater camera, 68
Underwater Club, 212
underwater communications, 186
underwater construction, 142
underwater dive mask, 89
underwater dive suit, 123
underwater diving gear, 59, 234
underwater exploration, 212
underwater observation chamber, 101
underwater salvaging operations, 142
underwater speleology, 212
underwater technologies, 101
underwater tunnel, 26, 106
underwater viewing, 160
underwater welding, 142
unicycle, 21, 82, 184
Unified Extensible Firmware Interface Standard, 26, 223
Union, 6, 23, 68, 82, 116, 134, 174, 184, 242, 243, 276, 289, 385
Union army, 116, 174, 242
Union military observer, 276
Union soldier, 134
Union soldiers, 134, 289
United Kingdom, 242
United States, 6, 21, 39, 57, 79, 81, 95, 101, 104, 131, 166, 202, 246, 281, 283, 284, 289, 291, 292
United States Air Force, 79, 95, 101, 202
United States Army, 95
United States government, 291
United States of America, 6, 21, 81, 131, 289
Universal Aether Telegraph Co., 183
universal bottle neck, 199
universal gravitation, 188
universal gravitational pull, 165
universal joint, 122
universal language, 50
universal life force, 218

Universal Product Code, 153
universe, 124
University of Chicago, 23, 221
University of Pennsylvania, 23
University of Virginia, 131
UNIX operating system, 26, 221, 248
unlimited power, 179
unmanned aerial vehicles, 113
unmanned automated earth landing, 161
unmanned combat aerial vehicles, 113
unmanned heavier-than-air airplane, 153
unmanned spacecraft, 145
unrefined foods, 34
unsaturated acids, 206
unstable soil, 56
UPC code, 23, 153, 234, 267
upholstery, 194
upper atmosphere, 85
upper classes, 213
Uranus, 118
urban development, 114
urea, 49
urinary catheter, 32, 93
urinary tract, 81, 90
urinary tract infections, 90
URL, 152
Ursus suit, 125
USA, 5, 6, 35, 93
usability, 139
USB drive, 15, 245
USB drives, 169, 233
uterus, 220
utopian, 95, 236
V-2 rocket, 55
V-thread, 39
vaccine, 30-32, 65, 94, 111, 193, 209, 225, 226, 276
vaccines, 9, 30, 120, 209
vaccinologist, 119
vacuum, 9, 16, 17, 45, 54, 65, 74, 75, 78, 90, 159, 233, 246
vacuum cleaner, 17, 78
vacuum design, 78
vacuum tube, 16, 74, 159, 233, 246
vacuum tubes, 45, 54
vacuums, 75
valve system, 65
valveless pulse engine, 20, 138
valves, 21, 60, 121, 138
Van Gieson's Stain, 32, 102
Van Houten Company, 124
Van Rysselberghe System, 224

Vanderbilt University, 23, 255
vape, 157
vaping, 157
vapor calorimeter, 57
variable aircraft wing system, 256
Variomatic transmission, 77
varying current, 118
vascular conditions, 272
Vassar College, 23, 181, 255
vegetable oils, 172
vegetable waste, 162
vegetables, 219
vegetarian, 139, 206
vehicle cameras, 92
vehicle headlights, 121
vehicle tires, 155
vehicles, 9, 19, 40, 44, 52, 63, 67, 74, 85, 95, 113, 123, 124, 127, 135, 142, 154, 167, 182, 206, 209, 212, 229, 259, 271, 276
veins, 67
Velcro, 34, 174
Velcro S.A., 174
Velocimeter, 47
velocipede, 77, 184
velocity, 51, 92, 125, 154, 165, 211
velours, 174
Velox, 43
vending machines, 271
Venn Diagram, 16, 255
Venn Diagrams, 255
ventilating chimney valve, 42
ventilation, 108
ventriloquist, 265
venture capitalist, 174
Venus, 141, 160
Veritiphone, 255
Verneuil Process, 256
Vernier Caliper, 256
Vernier Scale, 16, 256
Versorium, 102
vertical color filter technology, 173
vertical takeoff and landing aircraft, 94
vertical transport, 239
Vespa motor scooter, 72
veterinarian, 65, 182
Vezdekhod, 211
VHF, 253, 271
vibration dampening devices, 251
vibration-resistant, 177
vibratory transmitter, 60
Victoria and Albert Museum, 66
Victoria, Queen, 237, 266
Victorian alcoholic beverage, 205
Victorian Era, 71, 250, 257, 263

Victorian Period, 3, 42, 53, 206, 226
Victorian scientific instrument, 211
Victorian Spiritualist parlor game, 52
Victorian times, 54
Victorian tricycle, 251
video, 28, 29, 43, 76, 85, 88, 92,
    120, 124, 128, 134, 149, 155,
    167, 169, 177, 194, 197, 200,
    207, 212, 249, 254, 272
video camera, 249
video capabilities, 134
video editing, 124, 167, 212
video game console, 29, 43, 149
video game designer, 200, 254, 272
video game developer, 43, 272
video game producer, 177, 272
video games, 43, 169, 177
video recording, 155
video storage, 207
video technology, 128, 207
videographer, 123, 385
Viewpoints Research Institute, 139
Vinalon, 16, 243
vintage aesthetic, 17, 261
vintner, 61, 91
vinyl chloride, 218
vinyl polymer, 144
vinyl records, 218
Vinylon, 243
viola, 25, 40
violent individuals, 68
violin, 25, 40, 47, 226
Violin Concerto in D Major
    (Beethoven), 47
violins, 9, 25, 40, 65
Virginia, 83, 131, 155, 166, 281,
    385
Virginia House representative, 166
virologist, 225, 276
virologists, 245
virtual backgrounds, 134
viruses, 81
vision, 28, 81, 89, 100, 119, 160,
    215, 256
vision science, 100, 215
vision-impaired industry, 256
visionary, 95, 123
vison-impaired, 149
visually impaired, 45, 54, 76
Vitagraph Company, 50
Vital Assist, 145
vitalist phenomenon, 219
Vitamin C, 26, 219
Vitamin C, natural, 219
Vitamin C, synthetic, 219
vitamin D, 260, 290

VMX Inc., 170
vocalist, 385
VOCs, 260
voice, 20, 47, 76, 88, 194, 260, 265
voice actor, 47, 76, 265
voicemail, 15, 51, 133, 170, 212
voicemail technology, 133, 170
Voigtlander camera, 28
volcanologist, 200
volt, 257
Volta, Alessandro, 98
voltage, 17, 21, 48, 59, 68, 74, 86,
    139, 143, 247
voltage fluctuations, 48
voltaic pile, 98, 257
voltmeter, 17, 64, 139, 263
voltmeters, 263
volunteer fire department, 29
Von Neumann computer
    architecture, 188
Vonnegut Jr., Kurt, 257
Voskhod, 145
Vostok, 145
voting systems, 63
VTOL, 94
vulcanization, 104, 105, 201, 206
vulcanization process, 104, 201
vulcanizing, 94
VZ-9-AV Avrocar, 94, 95
waffles, 88
Walden Inversion, 32, 259
Walden Two (Skinner), 236
Wales, 282, 286, 292
walkie-talkie, 120
walking, 21, 147
wall plug, 15, 25, 89, 124
wall receptacle (electric), 124
Wallace, Alfred R., 72
Wallis, Barnes, 34
walls, 18, 44, 66, 177
Walt Disney Company, 76
Walt Disney World, 76, 95
war, 3, 65, 83, 91, 93, 99, 102, 133,
    134, 146, 155, 159, 166, 170,
    174, 182, 184, 195, 205, 207,
    212, 213, 223, 226, 227, 240,
    259, 261, 265, 276, 287-289,
    382, 385, 389
War Between the States, 83, 134,
    289
war effort, 91, 207
War for the Constitution, 83, 385
War of the Currents, 247
Wardenclyffe Tower, 247
warfare, 55, 159, 264
warhead, 67

warming food, 30
wars, 67, 74, 170, 187, 219, 283
Warthin-Starry Stain, 32, 261
washing line, 67
washing machines, 121, 271
Washington and Lee University, 23
Washington, George, 93
waste ash, 77
waste minimization, 41
wastewater products, 272
wastewater treatment, 26, 87
watch technology, 105, 207
watches, 87, 120-122, 196, 250
watchmaker, 103, 105, 204, 250
water, 19-23, 25, 26, 34, 39, 41, 47, 48, 50, 51, 53, 56, 60, 62, 64, 66, 68, 69, 75, 77, 82, 86, 89, 92, 93, 97, 99, 102, 104, 107, 109, 112-114, 116, 122, 126, 138, 147, 148, 152, 156, 160, 161, 163, 165, 168, 171, 175, 176, 194, 200, 201, 205, 210, 212, 218, 220, 223, 227, 230, 232, 234, 241, 250, 259, 263, 269, 272, 275
water clock, 69, 89, 148, 272
water control devices, 259
water current-powered barge, 148
Water Displacement 40th Formula, 153
water drainage, 171
Water from Cologne, 87
Water Fuel Cell, 175
water gauge, 23, 272
water injector, 102
water level, 41
water levels, 220
water meter, 227, 234, 263
water mills, 68
water molecules, 64
water navigation, 66
water organ, 69
water park, 22, 176
water pillar, 116
water pump, 20, 25, 86, 97
water pumps, 156
water purification, 126
water resistant, 194
water scooper, 205
water slides, 176
water sports, 53
water stones, 109
water supply pipes, 75
water turbine, 92
water weight, 69
water-absorbing clay, 161

water-resistant, 51, 104, 165, 201
water-resistant paint, 51, 104
water-walking skis, 147
waterbed, 42, 116
watercolorist, 385
watercraft, 83
Waterman, Lewis E., 34
watershed protection forest belt system, 100
Watson, James, 175
Watt Engine, 261
Watt, James, 150, 188
wattmeter, 248
wattmeters, 263
Watts, James, 222
wave drag, 264
wave theory, 125
wave theory of light, 125
wave-particle duality, 81
wavelength frequency modulation, 113
Wayne, Ronald G., 132, 267
WD-40, 153
wealth, 113, 260, 288
weapon of the century, 138
weapon technologies, 137
weaponry, 51, 153, 219
weapons designer, 181, 186
weapons engineer, 96, 158, 272
weapons maker, 170
weapons manufacturer, 195
weapons program, 144
weapons systems, 92, 145
weaponsmith, 272
weather, 57, 138, 177, 207, 218, 228, 257
weather balloons, 177
weather control, 228
weather modification, 257
Web browser, 49
Web designer, 385
Webb Ellis Cup, 82
Webcams, 92
Website address, 152
Website development, 156
Websites, 114, 185, 187
Wedgwood Company, 262
Wednesday, 219
Weed Eater, 44
weight, 59, 69, 97, 131, 138, 173, 186, 215, 216, 221, 234
weight balancing, 97
welding, 23, 47, 130, 142, 153, 186, 203, 236, 248, 264
welding equipment, 248
welding metal, 130

welding technique, 264
welding technologies, 153
well counter, 41
well plate, 245
Welsbach Mantle, 262
Welsh, 72, 107, 134, 266, 282, 286
West Virginia, 281, 385
Western Union, 23, 68
Western Union Telegraph Company, 68
Westinghouse Air Brake Company, 263
Weston Electric Light Company, 263
wet concrete, 66, 177
wet-cell battery, 107
whale hunters, 50
Wham-O, 181
wheat, 91, 206
Wheatstone bridge, 64
wheel, 9, 20, 51, 62, 70, 78, 82, 93, 127, 131, 142, 163, 168, 169, 201, 240, 267, 269
wheelchairs, 127
wheelless hovercraft vehicle, 147
wheellock, 24, 53, 142
wheellock gun mechanism, 53
wheels, 20, 21, 73, 147, 156, 159, 201, 206, 209, 221, 234, 242
whiskey, 27
whiskeys, 113
white light, 71
White Mountains, 169
white phosphorus, 203
Whitehead Torpedo, 264
Whittle, Frank, 108
Who, the, 46, 220
wholistic doctor, 243
wholistic health program, 139
wholistic land management, 62
wi-fi, 113, 130, 174, 196
wide angle lens, 28, 243
wife, 32, 123, 385
wild celery, 230
wilderness preservationist, 182
wildfires, 20, 205
wildlife, 3, 182, 385
wildlife preservationist, 182
Wilson Cloud Chamber, 265
Wilson, Woodrow, 385
Winchester Model 70, 266
Winchester Repeating Arms Company, 266
wind, 25, 33, 34, 40, 46, 64, 66, 68, 73, 74, 86, 122, 152, 168, 177, 191, 223, 227, 228, 232, 242, 259

wind energy, 68
wind gauge, 227
wind instrument, 74, 228
wind powered sawmill, 68
wind speed, 40, 73
wind strength, 46
wind tunnel instrumentation, 64
wind tunnels, 191
wind turbines, 25, 86, 122, 168, 232, 242
wind-up radio, 46
windbreak, 100
window, 16, 92, 218, 250
window frames, 218
windows, 18, 26, 34, 75, 99, 198
windpipe, 116
wine, 61, 205
wing shape, 264
winglet, 264
Winogradsky Column, 29, 266
WinRAR file archiver, 26, 222
wire, 25, 27, 28, 60, 66, 73, 113, 152, 155, 159, 172, 177, 196, 204, 211, 213, 230, 235, 238, 242, 253, 271
wire cable suspension bridge, 25, 230
wire coils, 213
wire gauze chimney, 73
wire insulation, 155
wire rope maker, 113
wire "fishbone" antenna, 28, 253, 271
wired connection, 174
wired LAN technology, 174
wireless, 32, 52, 54, 74, 85, 105, 117, 167, 168, 174, 196, 203, 207, 210, 247, 253, 271
wireless antenna, 253, 271
wireless communications, 168
wireless devices, 168
wireless LAN, 196
wireless light switch, 85
wireless radio, 105, 167
wireless radio telegraphy system, 167
wireless remote pioneer, 210
wireless system, 203
wireless technologies, 168
wireless technology, 203
wireless telecommunication, 117
Wireless Telegraph and Signal Company Limited, 168
wireless telegraphy, 52, 54
wireless telegraphy age, 54
wireless transmission station, 247
wireless TV remote control, 210

wireless-telegraph apparatus, 183
wisdom, 286
Witherspoon, Reese, 385
witness, 134
Witte, Olivier de, 125
WLAN, 196
Woden (god), 219
Woden's Day, 219
wolfsbane, 230
Womack, Lee Ann, 385
women, 3, 35, 173, 257, 282, 283, 285, 286, 288
women's lives, 201
wood carver, 167, 277
wood glue, 144
wood preservation, 153
wood products, 149
Wood Screw Pump, 267
wood stoves, 242
wood surfaces, 135
wood turner, 91
wooden beams, 135
wooden dock pilings, 149
wooden frames, 146
wooden joints, 135
wooden lathes, 146
wooden matches, 76
wooden mine support beams, 149
wooden pen box, 232
wooden railroad ties, 149
wooden railway track, 130
wooden trusses, 135
wooden tube, 151
woods, 267
woodturner, 277
Woog, Philippe-Guy E., 34
wool, 114
word association test, 31, 135
word processing, 51, 117, 155
worksites, 251
world, 3, 5, 6, 8, 11, 15, 20, 29, 32, 34, 35, 37, 48, 49, 51-54, 59, 62, 65, 67, 68, 76, 78, 80, 82, 87, 89-91, 95, 97-100, 102, 104, 113, 114, 117, 128, 130-134, 137, 139-141, 143, 146, 152, 157, 159, 161, 162, 166, 169, 170, 179, 180, 182, 183, 186, 187, 191, 195, 198, 205, 207, 208, 212, 213, 218, 219, 223, 227, 230, 231, 233, 235, 237, 243, 244, 247, 248, 251, 253, 257, 259-261, 263, 265-268, 271, 275, 277, 283, 285, 290, 380, 385
world economy, 183

world history, 3
world traveler, 213
World War I, 102, 159, 166, 170, 182, 212, 265
World War II, 65, 91, 99, 133, 146, 166, 182, 207, 227, 259, 261, 265
World Wide Web, 49, 59, 80, 157
wound healing, 219
wounds, 67
Wozniak, Steve, 132, 262
wrangler, 385
wrappers, 187
wrapping materials, 196
wrapping paper, 162
wrench, 30, 132, 242
wrestling, 196, 239
Wright Brothers, 130, 157, 227, 268
Wright Flyer, 268
Wright Flyer III, 268
Wright, Frank L., 267
Wright's Stain, 32, 267
wrinkle-resistant, 194
writer, 40, 41, 43, 45, 48, 52, 55, 59, 63, 65, 68, 73, 90, 91, 100, 107, 112, 116, 125, 134, 165, 200, 201, 213, 230, 236, 241, 242, 268, 380, 381, 385
writers, 11, 14, 233, 385
writing, 3, 11, 12, 28, 41, 44, 50, 54, 60, 90, 112, 160-162, 233, 248, 381, 382, 385
writing implement, 50, 160
writing paper, 28, 162
writing pens, 248
writing surface, 162
written records, 33
Wyoming, 5, 6, 35, 116
x (multiplication symbol), 197
X Corp, 183
X Holdings Corp, 183
X-10 PWM, 235
X-ray, 142
X-ray machines, 142
X-rays, 60, 69, 119, 124, 222
Xanthium strumarium, 174
xerographic copier, 61
xerographic process, 61
xerography, 61, 128, 185
xF Technologies, Inc., 59
Yablochkov Candle, 271
Yagi Antenna, 253, 271
Yagi-Uda Antenna, 253, 271
Yale Infallible Bank Lock, 272
Yale Lock Manufacturing Company, 272

Yale University, 23, 41, 208, 223
Yale, Elihu, 208
Yankee myth, 288
Yardbirds, the, 161, 162
Yazu Arithmometer, 272
Yellowstone National Park, 26, 116
yoga pants, 233
York Minster, 120
Yosemite National Park, 26, 182
yttrium, 262
Yuzpe Regimen, 32, 273
Z3, 277
Zamboni Pile, 25, 275
Zamboni, the, 275
Zeppelin, 229, 248, 276
Zeppelin airships, 276
Zeppelin, Ferdinand von, 229
zero emissions fuel source, 200
zero gravity, 89
Ziegler-Natta Catalyst, 187, 276
Ziehl-Neelsen Stain, 32, 277
Zike, 235
zinc, 76, 122, 201, 236
zinc alloy toy cars, 122
zinc spar, 236
Zip2 Corp, 183
zipper, 22, 134, 243
ZN stain, 277
zoetrope, 123
Zond, 145
zookeeper, 385
zoologist, 98, 111, 115, 158, 169, 172, 229, 238
zoology, 235
zoopraxiscope, 184
Zuma 3D Graphics, 123
ZX Spectrum, 235
ZX80, 235

Praise for Author-Historian-Artist
Lochlainn Seabrook

Comments from our readers around the world

★ "Lochlainn Seabrook is a genius writer!" — STEVEN WARD

★ "Best author ever." — EMILY (last name withheld)

★ "We get asked a lot what books we use and read. We don't do many modern historians, but we make an exception for some, and Lochlainn Seabrook is one of them. His works are completely well researched from original documents, and heavily footnoted and documented." — SOUTHERN HISTORICAL SOCIETY

★ "Looking forward to more Lochlainn Seabrook books, my favourite historian!" — ALBERTO IGLESIAS

★ "Lochlainn Seabrook is one of the finest authors on true history in this century. His books should be on every student's desk." — RONDA SAMMONS RENO

★ "All of Col. Seabrook's books are great. I have bought most of them and want to end up buying them all." — DAVID VAUGHN

★ "Lochlainn pulls together such arcane facts with relative ease, compiling these into ordinary prose that strike to the heart with substance, no fluff-speak. I am awestruck! Really. He is an inspiration to me. . . . He is truly a revolutionist. He dares to speak what others whisper; he writes with a boldness and an authoritative knowledge that is second to none." — JAY KRUIZENGA

★ "Mr. Lochlainn Seabrook is . . . the most well researched and heavily documented author I've ever read. His books are must haves. Everything he writes should be required reading! I assure you, you won't be disappointed. One simply cannot go wrong with his books. Mr. Seabrook is awesome! . . . I have never read any other author as well researched and footnoted as him. I've been in love with Mr. Seabrook for almost 5 years now. His quick wit and logic is enough reason to purchase his books. But the mere fact that he's so extensively researched is icing on the cake. Mr. Seabrook is my favorite, hands down." — LANI BURNETTE RINKEL

★ "My favorite book is the Bible. Lochlainn Seabrook wrote my second favorite book." — RICHARD FINGER

★ "I have a new favorite author and his name is Lochlainn Seabrook." — J. EWING

★ "Lochlainn Seabrook is an incredible writer and I love all of his books on the South. . . . His writing is brilliant. . . . I look forward to reading more of his masterpieces. Thank you." — JOEY (last name withheld)

★ "It's hard to choose just one of Lochlainn's books!" — ROSANNE STEELE

★ "Mr. Seabrook, thank you ever so much for blessing us with your most enlightening works." — LAURENCE DRURY

★ "I recommend anything written by Lochlainn Seabrook." — HOTRODMOB

★ "Awesome books . . . by a great writer of truth, Lochlainn. Thank you so much. Keep up the great work you do." — WILDBUNCH19INF

★ "I love Lochlainn Seabrook's style and approach. It's not the 'norm.' What a miracle his books are. . . . He is a literal life changing author! Amazing books!" — KEITH PARISH

★ "I adore Mr. Seabrook's style and I love his books. I love an author that does proper research, and still finds a way to engage the reader. Mr. Seabrook does an admirable job of both." — DONALD CAUL

★ "Lochlainn Seabrook's books are much more well researched and authoritative than those eminently celebrated as being the authorities on the subjects he writes on. You can always trust to find the truth in his writings. . . . He does not rewrite history, but instead shows it as it is." — GARY STIER

★ "I love all of Colonel Seabrook's books. They are informative and enlightening, and his warm Southern hospitality writing style makes you feel right at home." — KEITH CRAVEN

★ "Lochlainn Seabrook's work is an absolute treasure of scholarship and historic scope." — MARK WAYNE CUNNINGHAM

★ "Mr. Seabrook's command of . . . history is breathtaking. . . . He deserves great renown—check out his books!" — MARGARET SIMMONS

★ "I love Seabrook's writings. LOVE!!! . . . So grateful to know the truth! Keep writing Lochlainn!!!" — REBECCA DALRYMPLE

★ "Lochlainn Seabrook . . . [has] probably [written] the best book on mental science in existence by a living author. Along with Thomas Troward, Emmet Fox, and Jack Addington, Mr. Seabrook is one of the top four mental science authors of all time, since biblical times." - IAN BARTON STEWART

★ "Glad I discovered Mr. Seabrook! . . . He writes eye opening books! Unbelievable the facts he unearths - and he backs it all up with truth, notes, footnotes, and bibliography! . . . He always amazes me! His books always see the whole picture. His timelines and bibliographies are incredible. He always provides carefully reasoned arguments! He's the best. To me I think he's better than the late great Shelby Foote! America needs more like Lochlainn Seabrook. I can't wait to own all of his books on the war someday. Everyone who wants the Truth, who seeks the Truth and wants the full story, should read his books." — JOHN BULL BADER

★ "I love all of Colonel Seabrook's books!" — DEBBIE SIDLE

★ "Lochlainn Seabrook is well educated and versed in what he writes and I'm impressed with the delivery." — THOMAS L. WHITE

★ "Thank you Lochlainn Seabrook for your wonderful books! You are the real deal! You are an amazing author and I love your books!!" — SOPHIA MEOW CELLIST

★ "I really enjoy Mr. Seabrook's books! His knowledge is beyond belief!" — SANDRA FISH

★ "Love Lochlainn Seabrook. Awesome!!" — ROBIN HENDERSON ARISTIDES

★ "Kudos to Lochlainn Seabrook who is a very good and informative professional truthful historian. We need more like him!" — AMY VACHON

Patent drawing of America's first locomotive, "Tom Thumb," built by Peter Cooper, 1829-1830.

Photo of "Tom Thumb," 1927.

MEET THE AUTHOR

AMERICAN POLYMATH LOCHLAINN SEABROOK is a bestselling author, award-winning historian, and world acclaimed artist. A descendant of the families of Alexander Hamilton Stephens, John Singleton Mosby, Edmund Winchester Rucker, and William Giles Harding, the neo-Victorian scholar is a 7th generation Kentuckian, and one of the most prolific and widely read writers in the world today. Known by literary critics as the "new Shelby Foote," the "American Robert Graves," the "Southern Joseph Campbell," and the "Rocky Mountain Richard Jefferies," and by his fans as the "Voice of the Traditional South," he is a recipient of the United Daughters of the Confederacy's prestigious Jefferson Davis Historical Gold Medal, and is considered the foremost Southern interpreter of American Civil War history—or what he refers to as the War for the Constitution (1861-1865). A lifelong litterateur, the Sons of Confederate Veterans member has authored and edited books ranging in topics from ancient and modern history, politics, science, comparative religion, diet and nutrition, spirituality, astronomy, entertainment, military, biography, mysticism, photography, and Bible studies, to natural history, technology, paleography, music, humor, gastronomy, etymology, onomastics, mysteries, alternative health and fitness, wildlife, comparative mythology, genealogy, Christian history, and the paranormal; books that his readers describe as "game changers," "transformative," and "life altering."

One of America's most popular living historians, he is a 17th generation Southerner of Appalachian heritage who descends from dozens of patriotic Revolutionary War soldiers and Confederate soldiers from Kentucky, Tennessee, North Carolina, and Virginia. Also a history, wildlife, and nature preservationist, the well-respected scrivener began life as a child prodigy, later maturing into an archetypal Renaissance Man. Besides being cofounder and co-CEO of Sea Raven Press, an accomplished writer, author, historian, biographer, lexicographer, encyclopedist, neologist, publisher, editor, poet, creative, onomastician, etymologist, and Bible authority, the influential prosateur is also a Kentucky Colonel, eagle scout, entrepreneur, businessman, composer, screenwriter, nature, wildlife, and landscape photographer, videographer, and filmmaker, artist, artisan, painter, watercolorist, sculptor, ceramic artist, visual artist, sketch artist, pen and ink artist, graphic artist, graphic designer, book designer, book formatter, editorial designer, book cover designer, publishing designer, Web designer, poster artist, cartoonist, content creator, inventor, aquarist, genealogist, jewelry designer, jewelry maker, former history museum docent, and a former ranch hand, zookeeper, and wrangler. A contemporary songwriter (of some 3,000 songs in a dozen genres), he is also a pianist, organist, drummer, bass player, rhythm guitarist, rhythm mandolinist, percussionist, classical composer, film composer (currently his musical work has been featured in 11 movies), lyricist, band leader, multi-instrument musician, lead vocalist, backup vocalist, session player, music producer, and recording studio mixing engineer, who has worked and performed with some of Nashville's top musicians and singers.

Currently Seabrook is the multi-genre author and editor of over 100 adult and children's books (totaling some 30,000 pages and 15,000,000 words) that have earned him accolades from around the globe. His works, which have sold on every continent except Antarctica, have introduced hundreds of thousands to vital facts that have been left out of our mainstream books. He has been endorsed internationally by leading experts, museum curators, award-winning historians, chart-topping authors, celebrities, filmmakers, noted scientists, well regarded educators, TV show hosts and producers, renowned military artists, venerable heritage organizations, and distinguished academicians of all races, creeds, and colors. He currently holds two world records: He is the author of the most books (12) on American military officer Nathan Bedford Forrest, and he was the first to publicize and describe the 19th-Century platform reversal of America's two main political parties, namely that Civil War era Democrats (primarily in the South—the Confederacy) were Conservatives, while Civil War era Republicans (primarily in the North—the Union) were Liberals.

Of northern, western, and central European ancestry, he is the 6th great-grandson of the Earl of Oxford and a descendant of European royalty through his Kentucky father and West Virginia mother. A proud descendant of Appalachian coal miners, trainmen, mountain folk, and wilderness pioneers, his modern day cousins include: Johnny Cash, Elvis Presley, Lisa Marie Presley, Billy Ray and Miley Cyrus, Patty Loveless, Tim McGraw, Lee Ann Womack, Dolly Parton, Pat Boone, Naomi, Wynonna, and Ashley Judd, Ricky Skaggs, the Sunshine Sisters, Martha Carson, Chet Atkins, Patrick J. Buchanan, Cindy Crawford, Bertram Thomas Combs (Kentucky's 50th governor), Edith Bolling (second wife of President Woodrow Wilson), Andy Griffith, Riley Keough, George C. Scott, Robert Duvall, Reese Witherspoon, Lee Marvin, Rebecca Gayheart, and Tom Cruise.

A constitutionalist, avid outdoorsman, wilderness conservationist, and gun rights advocate, Seabrook is the author of the international blockbuster, *Everything You Were Taught About the Civil War is Wrong, Ask a Southerner!* He lives with his wife and family in the magnificent Rocky Mountains, heart of the American West, where you will find him hiking, filming, and writing.

For more information on Mr. Seabrook visit
LOCHLAINNSEABROOK.COM

386 ∞ MANMADE

Scan and shop our Webstore!

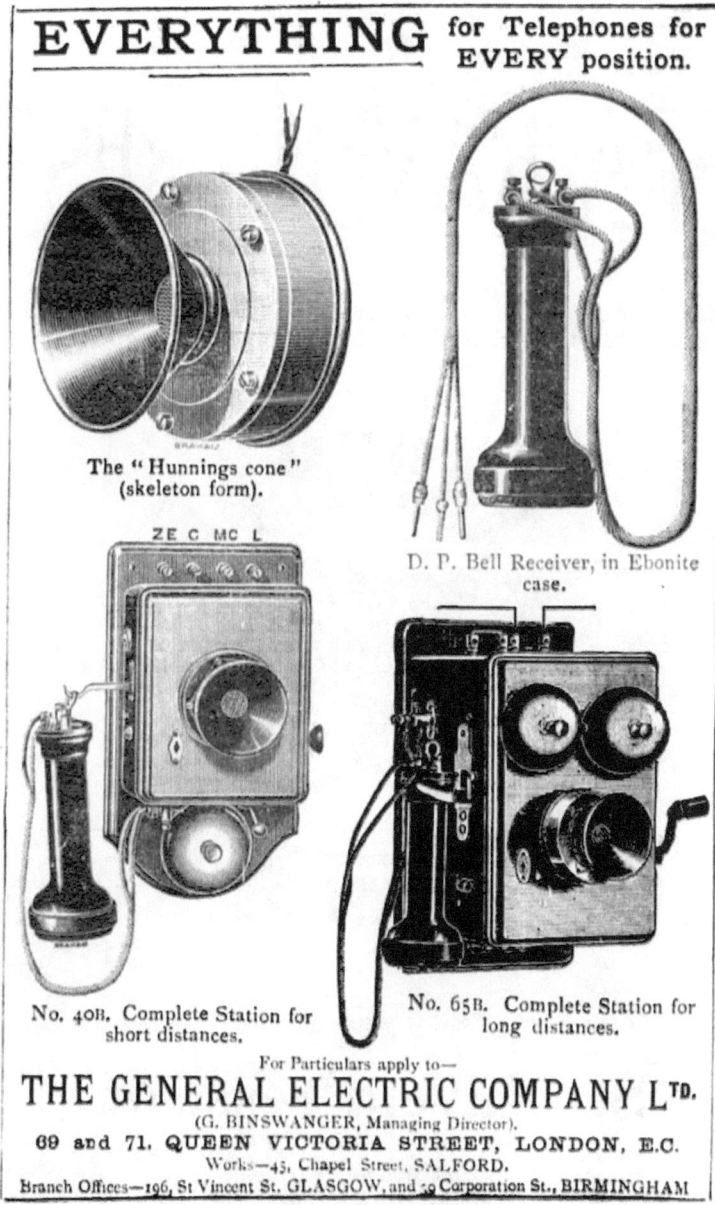

GE telephone and accessories magazine ad, 1895.

388 ∾ MANMADE

Nurture Your Mind, Body, and Spirit!
READ THE BOOKS OF

SEA RAVEN PRESS

Visit our Webstore for a wide selection of wholesome, family-friendly, evidence-based, educational books for all ages. You'll be glad you did!

Artisan-Crafted Books & Merch From the Rocky Mountains

SeaRavenPress.com

LochlainnSeabrook.com
TheBestCivilWarBookEver.com
YouTube.com/@SeabrookFilms
Rumble.com/user/SeaRavenPress
AmbianceGoneWild.com
Pond5.com/artist/LochlainnSeabrook

If you enjoyed this book you may be interested in some of Colonel Seabrook's other popular titles:

☞ SECRETS OF CELEBRITY SURNAMES: AN ONOMASTIC DICTIONARY OF FAMOUS PEOPLE
☞ EVERYTHING YOU WERE TAUGHT ABOUT THE CIVIL WAR IS WRONG, ASK A SOUTHERNER!
☞ THE GREATEST JESUS MYSTERY OF ALL TIME: WHERE WAS CHRIST BETWEEN THE AGES OF 12 AND 30?
☞ THE CONCISE BOOK OF OWLS: A GUIDE TO NATURE'S MOST MYSTERIOUS BIRDS
☞ VITAMIN D: THE MIRACLE TREATMENT FOR NEARLY EVERY DISEASE AND HEALTH ISSUE
☞ UFOs AND ALIENS: THE COMPLETE GUIDEBOOK

Available from Sea Raven Press and wherever fine books are sold

 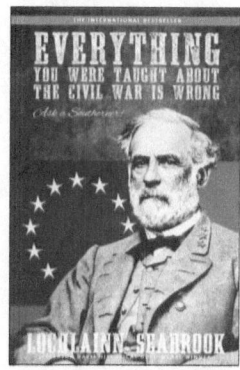 

 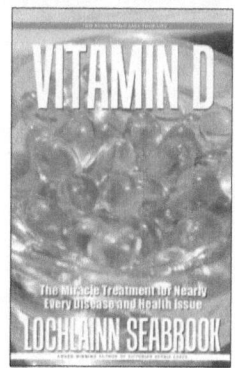 

ALL OF OUR BOOK COVERS ARE AVAILABLE AS 11" X 17" COLOR POSTERS, SUITABLE FOR FRAMING

SeaRavenPress.com

www.ingramcontent.com/pod-product-compliance
Lightning Source LLC
Chambersburg PA
CBHW020634230426
43665CB00008B/167